The Odyssey of Mary B

A true tale

John Durand

Puzzlebox Press

© 2005 by John Durand. Printed and bound in the United States of America. All rights reserved. No part of this book may be reproduced or transmitted in any form or by any means, electronic or mechanical, including photocopying, recording, or by an information storage and retrieval system, without permission in writing from the publisher (except by a reviewer, who may quote brief passages in a review to be printed in a magazine, newspaper, or on the Web). For information, contact Puzzlebox Press, PO Box 765, Elkhorn, WI 53121.

Cover design by John Durand. Cover © 2005 by John Durand. Cover image *English Ships in a Storm* by Johan van der Hagen © National Maritime Museum, London, used with permission.

Although the author and publisher have made every effort to ensure the accuracy and completeness of essential information contained in this book, we assume no responsibility for errors, inaccuracies, omissions, or inconsistencies therein. Any slights of people, places, or organizations are unintentional.

1st Edition

9 8 7 6 5 4 3 2

ISBN 0-9743783-1-3

LCControl Number: 2004094284

For further information contact

Puzzlebox Press

PO Box 765

Elkhorn, WI 53121

www.puzzleboxpress.com

This book is available at quantity discounts (20 copies or more) to educational and cultural institutions and organizations. Contact Puzzlebox Press at www.puzzleboxpress.com.

For my family, past and present,
and for Tom

Those who lived a hundred or a thousand years ago were then moderns as we are now. They had *their* ancients and those ancients had others, and we also shall be ancients in our turn.

Thomas Paine, *Rights of Man*, 1791

Author's Note

My interest in Mary B was roused by an account of her escape I read in Robert Hughes' popular history of early Australia, *The Fatal Shore*. Digging to learn more, I grew impatient with how history and fiction trivialized, romanticized, exaggerated, even caricatured this young woman. By any fair measure Mary B was remarkable, a true survivor in a hard world. I thought she deserved better, and set out to render a more complete and accurate account of her memorable experience.

The Odyssey of Mary B is replete with irony and coincidence. Were Mary B's story not a matter of record, elements of this narrative might be dismissed as improbable. But this is a true tale. As Dr. Samuel Johnson might say: "Depend on it."

Appreciation

I began writing *The Odyssey of Mary B* almost twenty years ago. To recall all who gave help and encouragement so I can thank them is beyond me. However, the support of several is indelible.

Agate Nesaule (*A Woman in Amber*) read every early draft of every chapter and provided much insightful and wise comment, and more. Thank you, Agate.

The reference librarians at University of Wisconsin Memorial Library tracked down many obscure works and helped me gain access to them.

Professor Robert Burrows of University of Wisconsin-Whitewater sent materials from Australia I was unable to access in this country.

Richard Parks of The Richard Parks Agency provided useful feedback and welcome support.

After *The Odyssey of Mary B* lay dormant for years, Jon Epstein, a friend from Peace Corps days, read it and was so enthusiastic he motivated me to do something with the story. Thank you, Jon.

To the many others who had a hand in bringing Mary B to life, whether acknowledged on this page or not, here she is at last, with my deep thanks to you all.

Also by the author
The Taos Massacres

Contents

Plymouth
Portsmouth
Santa Cruz
Rio
Cape Town
Botany Bay
Albion
Sydney Cove
Sydney
Kupang
Djakarta
London
Fowey

Afterword
Maps and illustrations
Chronology of Events
The Women on the Charlotte
Boswell's Farewell to Mary B
Notes

Plymouth
1786

*F*riends said Mary B's eyes were her best feature. And her hair. Her eyes were gray and frank and friendly and seemed to light up her face when she smiled. Her hair was the color of rough-burnished copper, gorgeous in sunlight. But as she stood in the docket on that fateful day in 1786, she watched the unsmiling faces around her through tear-filled eyes. In the chill, high-ceilinged chamber of Rougemont Castle she heard the drone of the clerk's nasal monotone: "...and of violently taking from her one silk bonnet valuing twelve pence and other goods valuing one pound, eleven shillings...."

Mary B wiped her tears with a sleeve and glanced at her friend Katie, who clutched the rail with white knuckles, her pout-face fixed vacantly on the scarred floorboards. Katie rocked forward and back, forward and back, bubbling spit out, sucking it in, bubbling it out, sucking it in. On the other side of Katie stood Seedy, slack-jawed, her expression vacant.

Mary B felt another wave of outrage. *I shouldn't be here!*

A dozen heads in the courtroom swiveled toward the docket. Peering over his half-glasses, Sir James Eyre raised a warning eyebrow. Mary B realized she'd uttered an involuntary cry of distress. But how could the soldier swear he'd seen her attack Agnes Lakeman? How could the constable say she'd tried to hide Agnes Lakeman's bonnet in her bodice? It wasn't true! Mary B felt another wave of outrage at Katie, the churlish soldier, the ogling turnkeys, the bailiffs, and stupid judges.

After reading the indictment, Sir James Eyre fixed the three women in the docket with a stern glare. Perched on a prominent nose, his spectacles did little to gentle his pinched demeanor, and although she knew his words were coming Mary B's vision dimmed and the sound of a rushing river filled her head when the judge addressed them each by name and said, "...You have pled guilty to felonious assault and felonious robbery on the King's Highway of one Agnes Lakeman...."

How long ago that seemed! For months the three women were held in a single unheated room in the Plymouth jail not ten feet square, surviving on two penny-worth of bread a day, sleeping on the floor atop dirty straw, having to beg for water.

During that time Mary B had seen only one family face, her brother, Little Bill.

"... and having pled guilty in full knowledge of the penalties for your felonious crimes and having placed your several persons at the mercy of this Court, without benefit of clergy, what have you to say why a judgment of death and execution should not be awarded against you?"

Death and execution! Mary B swayed and nearly fell. She grasped the rail and struggled to keep her composure. Expecting no comment from the three young women, Sir James Eyre scarcely paused before he continued the formula he'd recited a thousand times. "Therefore be it recorded: By the authority of His Most Gracious and Compassionate Majesty, King George III, this Court hereby sentences you Mary Broad, and you Katherine Prior, and you Cedelia Haydon, each and several, to be hanged by the neck until dead, dead, dead, at a date to be determined...."

Her eyelids fluttering wildly, Seedy voiced a little moan and crumpled to the floor like a marionette with its strings cut, her cheek hitting the rail with a sickening crunch.

At the pronouncement of her sentence Mary B felt a wave of panic. They're going to hang me! I'll be dead! She trembled. She could scarcely see. She broke into a clammy sweat.

The three women knew their sentences would be death by hanging. They knew the sentences would be commuted to Transportation, an expression meaning exile from England, usually for a period of seven years. But what if their sentences weren't commuted? For years there had been no Transportation because of the American War, and now the jails and hulks were filled to over-flowing with convicted felons awaiting exile. Maybe they'd have to hang her!

Prodded by the bailiffs, Mary B knelt with Katie to help Seedy to her feet and back to their bench. After the last of the prisoners were sentenced that morning the three women were returned to the jail. Mary B withdrew to the farthest corner of the large room that served as the women's prison. She hunkered with her fists clenched, her jaws tight, her eyes locked on the door through which men might come to take her to the gallows. She heard none of the usual clamor. When Seedy approached holding a hand to her swollen cheek and seemed about to speak, Mary B turned away. Katie stayed clear, knowing full well it was her doing that had made the three of them criminals.

As night came on, Mary B remained crouched in the corner like a frightened animal. The jail grew still. Time dragged, and once again she found herself reliving the strange turn of events that had plucked her from an ordinary life and plunged her into a living nightmare.

~~~~~~~~~

It all began with Katie.

Before dawn on that fateful July day a year earlier, a violent thunderstorm ripped the roof from the cobbled-up workshop in Dock where Katie and Seedy sewed Navy sails. Not being able to work, they spent the morning sipping gin in Seedy's room before wandering out to see Mary B, who ferried them for free across the wide expanse of Stonehouse Creek and then joined them on the spur of the moment to walk into Plymouth, a distance of only two miles.

In the old walled city they made their way to the commons behind the Guildhall, where a daily market of a dozen stalls sold the stuff of life. For a penny apiece they ate bread, drank small beer, and scratched between the eyes of a little black bull. Born with one minikin foreleg, the little bull's useless hoof dangled several inches above the ground. Katie and Seedy had heard of the little black bull - how its deformed leg was the Devil's work and would bring bad luck, how crops would fail, and storms drive in the fishing boats. But when the beribboned calf pranced in awkward play and then found Mary B's thumb and began to suck, the fun-loving young woman threw back her head and laughed. Such talk!

From the Guildhall they wandered down to the quay along Sutton Pool, where they drank rum with loafing sailors until Seedy began feeling queasy. Although Katie wanted to stay and drink with the sailors, sunshine and greenery beckoned Mary B, so they made their tipsy way to Hoegate Lane to sprawl on the grass by the cart track that meandered past the Citadel.

Katie's speech was punctuated with hiccups when she announced, "Look who's coming! The little bitch herself! Wants to make cow eyes at her soldier-boy, I suppose!" Mary B roused herself to look, but in the brightness saw only the green blur of an approaching figure with a parasol. She lay back again to Katie's prattle. "Never did a day's work in her life, that one! but

I wore my fingers to the bone. And then she sticks up her nose and calls me a little thief! Aye, little thief, it was! That bitch!" Mary B only half-listened to Katie's resentful tale, already a dozen times heard. "I'll 'little thief' her!"

As the young lady neared, Katie rose to her feet and limped to the puddled track. Katie's left foot turned inward and dragged a little. When she was tired she scuffed the toe of her shoe as she walked, so she usually wore a hole in the toe before the sole. "Good afternoon, m'lady," she said as the young woman neared. Katie's curtsy was awkward, her smile phony. Mary B sat up and giggled, her eyes sun-dazzled.

Agnes Lakeman halted a few feet from Katie, momentarily at a loss. Mary B thought the young woman of sixteen or so looked nice. Her green silk dress was trimmed with white lace, and a small gold brooch glittered on her bodice. A green silk bonnet covered her unpowdered brown hair. Although they were mud-spattered from the rain puddles, her green, silver-buckled shoes had been highly polished. Green stockings played peekaboo beneath her white, ornamental apron. She clutched her purse of matching green silk more tightly as a quavery voice betrayed her uneasiness. "Let me pass, you little thief."

Katie yowled with too-loud laughter and turned to Mary B with an exaggerated wink, "Hear that? You hear that? 'Let me pass, you little thief!' God Almighty, I can't believe it." Katie's voice was venomous when she replied, "Or what? You'll fart?" Mary B guffawed at Katie's outrageous rejoinder and looked to Seedy, whose translucent skin wore a moist sheen and whose eyes focused inward on some churning process. She looked about to vomit.

Agnes Lakeman backed away. "If you don't let me pass I'll call for help!"

"Oh, there's no reason to take offense now, Agnes," Katie said with an offhand motion, "We mean no harm, and for a penny apiece we'd let you pass, wouldn't we?" She flicked another wink at Mary B, who smiled and nodded.

Uncertain whether to retreat, Agnes took fresh hold of her purse and eyed the curious sentry on the wall of the Citadel. Seeing her glance, Katie's shoulders suddenly drooped. A contrite weariness that Mary B knew was false came into Katie's voice. "Oh, my, I'm saying it all wrong, Agnes, but sometimes the Devil must put words in my mouth." She sniffed. "But things have been so hard for us. We tramped all day looking for work because of the storm and we didn't eat and we're just beat down. I'm sorry, but I forgot my place." She made an awkward little curtsy and threw Mary B a sidelong glance. "Then some sailors down on the quay...well, we've had a bit too much to drink and I was wrong to vex you so." She bowed her head and made pretence of wiping her eyes with a corner of her dress.

The young woman nodded again, "Well, I do say, Katie Prior, you'd do well to mend your ways. Heavens! Such a show!" She sighed theatrically. "The Lord does forgive, though, and probably worse than you. Here," she said, undoing the ribbon of her purse, "I'll give you a penny each" - she paused to shake a finger - "But not for more drink, you hear? Drink is the Devil's brew! You silly girls! You gave me a fright! You're not to do that to your betters, you hear, Katie?"

"Oh, yes'm," Katie said, gathering her dress to curtsy again.

In Mary B's memory the next few minutes on the Hoe were dreamlike. For months she'd tried to put together the pieces of the puzzle in a way that made sense - Agnes Lakeman's lecturing finger, her purse like an open blossom in her other hand, Katie's good foot shooting up, perhaps deliberately, perhaps on impulse, kicking the young woman's hand and sending a jet of silver and copper coins shooting into the air, the sight of the coins suspended in glittery splendor for a long moment before they fell in a quick little shower onto the grass and muddied track. Mary B remembered seeing Katie's hand fly to her mouth as if shocked at what she'd done, then drop to her knees and scramble for the coins. Mary B could still hear Katie's cackle as she pawed the grass, and she'd never forget the dumbfounded look on Agnes Lakeman's face before she flung her parasol and pounced with a screech, carrying Katie into the mud where she clawed at Katie's hands to free the coins, all the while screaming, "Help! Thief! Help!"

Mary B also remembered collapsing onto her back, laughing so hard at the two women rolling in the mud that she held her aching stomach and even peed a little. That must have been when Seedy lunged upward with a groan to vomit a cheesy gush at the edge of the track, because Mary B remembered rousing herself to her elbows, looking from one scene to the other, and falling back again, helpless with laughter.

Katie was a biter, and her jaws snapped as the two women rolled back and forth in the mud, their arms and legs flailing.

When she found a piece of hand with her teeth and caused a piercing screech of pain and anger, Agnes grabbed Katie by the hair and began beating her head against the ground. Mary B stood, her laughter stifled by the seriousness of the battle. "All right now, Katie, that's enough! That's enough now!"

Good-sized, Agnes had Katie's arms pinned beneath her knees and was slapping her face with a rhythmic "Little! Thief! Little! Thief!" Crying with frustrated rage, squirming to free herself, Katie's teeth snapped at the hands that slapped her.

Mary B picked up the young woman's bonnet, come off in the fray, and stuffed it into her bodice for safekeeping, then stepped behind Agnes to pull her off. She was hauling back on the young woman when she heard pounding feet and a rough voice. Someone grabbed her shoulder, and from the corner of her eye she saw a swift dark shape and heard rather than felt a crunching blow on the side of her head. In a slow, reeling, dream-like movement Mary B saw herself going backwards, backwards, her watery legs useless, falling a long way until she sat heavily and lost her breath. She remembered still the pain in her stomach and panicky moments of suffocation, and how she'd accidentally put her hand in the warm goo of Seedy's vomit. Then, with an odd sense of detachment Mary B watched a soldier wade into the snarl of women wrestling in the mud. She remembered still hearing Agnes cry, "Help! Help!"

A kick from Katie's good foot propelled the soldier comically backward, his musket flying. He stumbled over Mary B's outstretched legs and tumbled into Seedy's lap, clutching his groin. Startled, Seedy locked his head between her knees in a scissors grip and began beating his chest. "Help! Help!" Seedy cried.

Suddenly the scene was peopled by a motley crowd, some laughing, some cheering on the combatants, others making clumsy efforts to pull one or the other away. When a constable and more soldiers finally separated Katie and Agnes, the women were marched off to the Guildhall, the two adversaries still hissing threats and curses. Only later did Seedy realize she'd taken off her shoes and forgotten them on the grassy Hoe.

The procession attracted several dozen followers as it wended back through Hoegate to the Guildhall, where only hours before Mary B had laughed at the antics of the little black bull. Idlers, urchins, and busybodies trailed behind. Others hung from windows and stood in doorways to point and laugh at the unsightly women and call "Shame!" At the

Guildhall a tall, itinerant musician strummed a rising scale on his little guitar to mock their ragged progress up the steps.

They were on their way to be locked up for safekeeping in the debtors' rooms when Mayor Nicolls heard the disturbance. He was not happy. Although the young woman in green stood muddy and bedraggled, the mayor took only a moment to recognize her as Jeremiah Lakeman's daughter. He stiffened with outrage when she began to weep about "that little thief...."

But Katie cut her enemy short with a stream of shrill, vituperative abuse so foul-mouthed the mayor ordered her gagged, which took two men. Even then Katie kicked at her captors. Disgusted, the mayor sent Katie and her cohorts down to the Clink and their victim home in a chair.

Although the Clink was reserved for felons, foreign sailors caught up in tavern brawls were often hauled off to its dismal confines. Located in the lower bowels of the antique Guildhall, the Clink lacked water, sleeping platforms, even straw. Its stone walls mold-covered, the only light and ventilation came from book-sized wickets in the iron-banded doors. In the adjacent cell the young women heard the muffled voices of foreign speech, by turns lamenting, swearing, and praying.

Mary B slumped to the floor, numb. Barefoot, Seedy sloshed back and forth through puddles that had seeped in from the morning's thunderstorm. A little taller than Mary B, Seedy had a bare two inches of headroom in the cell. She kept stopping to ask, "Why are they doing this to me, why?" At the prison door, Katie crouched like a predator in wait, staring out the wicket at the dim outlines of the mossy stairwell with fierce eyes.

After a time the light through the wicket failed completely and the metallic rattle of shackles in the other cell quieted. In the damp, fetid darkness Mary B felt sunk in a pool of utter despondency. Her head throbbing from a hangover and the blow from the soldier's musket, her ear ringing, her mouth and throat parched, her hand abhorrent with the sour stench of Seedy's vomit, Mary B wanted to believe she was having nightmare, that she would awaken in her own bed and go to work at the ferry service. She huddled in dumb confusion in the unfamiliar darkness, aware of Seedy's whining, of plaintive moans from the next cell, and the occasional scurry of unseen living things. Occasionally she was roused to awareness by a watchman's faint cry, "All's well, all's well."

Once she jerked fully awake when the man with the backward head sat up and grinned, a recurring dream rooted in a time when she was a young girl. She'd run to the scene just after a man was discovered at the foot of a seaside cliff near her home in Fowey. Crowding through the murmuring circle of curious on-lookers, she saw the victim lying on his stomach. One leg was stretched over his back at an impossible angle and his head was turned completely around. His battered face and lifeless eyes stared at a gloomy sky. Mary B shuddered at the visitation of the horrible sight and began to cry.

When at last the darkness began to melt and the tiny wicket in their prison door took shape again, Mary B woke to find Seedy's head in her lap and knew she hadn't dreamed her horror. She'd spent the night in a dungeon.

~~~~~~~~~~

The young women could not have run afoul of the law at a worse time. England was tough on the criminal class, and Plymouth especially so. Homeport of maritime heroes Sir Francis Drake and Sir John Hawkyns, place of embarkation for America's harried pilgrims, Plymouth had long tolerated the rough edges of a maritime economy. Smuggled French brandy, lace, and silk came ashore nightly in neighboring coves and inlets. But a nearby Navy yard was bringing too much change to the streets of Plymouth, and the town fathers felt put upon.

For decades this Navy yard existed as a small shipbuilding and repair facility known as "Dock," but in the aftermath of the American War "Dock" burgeoned into a major Navy station. Wet docks, dry docks, mast yards, sail yards, a gun wharf, rope works, warehouses, Navy offices, victualing offices, a huge Navy hospital - more growth and prosperity were being visited on the muddy streets of "Dock" than Plymouth had seen since the plundering heyday of Drake and his pirate cronies. Even now the first-rate *Royal Sovereign* rose on stays, a 102-gun behemoth of more than 2,000 tons, a ship far larger than Plymouth's out-of-date facilities could build. Shipbuilders, artisans, construction workers, merchants, victualers, opportunists, Navy men and Marines, and all that served them swelled its population day by day.

As the old walled city of Plymouth saw its maritime pre-eminence shrink beside the distending prosperity of "Dock" (soon to become known as Devonport), Plymouth was increasingly intolerant of ne'er-do-wells who caused trouble. In

8

this hostile climate the three young women incurred the wrath of Jeremiah Lakeman, who had an old bone to pick anyway.

Middle-aged, pigeon-breasted, florid-faced, mopping himself continuously, the Plymouth ropemaker announced to Mayor Nicolls (now sitting as magistrate) that he'd recently dismissed Katie from service as a cook's helper for suspected thievery. Now that she'd proved her criminal character he wanted the little thief and her cohorts punished, and the sooner the better! "Justice swift is justice fair," he opined, mopping his face.

Nicolls nodded. "But I can't," he said. "The charges against these women are capital offenses not a matter for this bench. If you want them punished now you have to reduce the charges."

"Absolutely not!" Lakeman sputtered.

The magistrate studied the aggrieved father. "And there's another problem, my good sir. This crime took place outside borough bounds, on Crown property. Maybe the charges should be assault and robbery on the King's Highway."

"Highway robbery? But that cowpath on the Hoe is no highway!" Lakeman sensed justice swift slipping away.

Nicolls raised his palm an inch above his desk in subtle reassurance. Whether it was actually a highway he couldn't say, but as the Citadel was Crown property the road around it could well be considered a King's highway. "Look," he said, trying to cinch his argument, "These women have no money for fines, and unless you reduce the charges this is a matter for the Assizes." Seeing Lakeman waver, the magistrate pressed on. Too bad the summer session was just ended, because now the women would have to remain in jail until the judges came around on the next Western Circuit in the spring, but meantime Lakeman could put his case together....

So vengeful was Lakeman he asked if the three could just be given a good flogging. He'd gladly pay the cost. After a much muttering, however, he finally assented to the magistrate's proposal, and with a few quick pronouncements the women were taken away.

The bill of indictment signed by Nicolls on July 27, 1785 charged the three women had participated in "feloniously assaulting Agnes Lakeman" and "feloniously putting her in corporal fear and danger of her life" and "feloniously and violently taking from her person and against her will one silk bonnet and other goods."

"That lying bitch," Katie hissed when she heard the terms of the indictment.

The young women were in a bleak situation. Poor, unmarried, of an age when they needed to be self-sufficient, Mary B and Seedy had followed Katie to seek their fortunes just weeks before and were almost without acquaintance in the area. Thus, several days passed before Fowey heard that three of its own had been jailed in Plymouth for highway robbery, although exactly who was unclear. But some weren't surprised when they later learned Katie Prior was involved. After suffering a crippling childhood disease, she'd been given to fighting and bullying, even with adversaries older and bigger. And Seedy Haydon...well, although a nice enough young woman she'd been a difficult birth and...well, sometimes the poor thing seemed to lack sense enough to come out of the rain.

But Mary B! How could she get mixed up in such a scandal? True, her father was not a Fowey luminary. A seaman in the coastal trade, a loquacious drinker, William Broad roused himself from his favorite chair in the Black Stallion only when feeling pressed. Like much of the maritime underclass, he sometimes smuggled, and was not above a little theft now and then to put meat in the family pot. But Mary B had always been so fun loving and was such a good worker! Why, the story was she'd out-rowed a swaggering youth to earn her job ferrying passengers across Stonehouse Creek!

Howsoever all that, as William Broad was at sea when Fowey learned the shocking news, Mary B's mother dispatched Little Bill to find out what was going on.

A husky youth of fifteen, Little Bill knew more about fishing than about officials. Trading his labor for passage on a coastal schooner, he arrived in Plymouth for the first time in his life almost three weeks after the incident on the Hoe. Once anchored and unloaded in Sutton Pool, the master pointed the youth towards the sky-bound steeple of St Andrew's church, and said the jail rooms were in the nearby Guildhall, which Little Bill could not miss for the clock in the tower. Little Bill found the Guildhall with no trouble, and the turnkey too, but whether the youth's sister was there the bull-necked fellow wouldn't say. His upper arms straining the sleeves of a greasy coat, he spat near the youth's feet and went back to paring a dirty thumbnail with a folding knife. "There's many that come and go," he said, "and I get no pay for answering questions."

Confused by the turnkey's churlish behavior, Little Bill wandered outside, wondering if he might glimpse his sister at a window. No sign. He thought of calling out her name, but decided that would be foolish.

In the marketplace he heard all about them. "Why, of course we know the three young women! Why, they beat Squire Lakeman's daughter to within an inch of her life! And oh! how the crippled one carried on, like the Devil possessed! No, we can't say for certain they're still there. Maybe they are. But have you seen the Devil's own little black bull?"

A legless man with an empty, leaking eye-socket hoisted himself by his hands near enough to pull at the youth's blouse. Transfixed by the man's empty socket, Little Bill scarcely heard what he said. "Pay the price!" the cripple told him, "You've got to pay the price!"

"What?"

The cripple craned back even farther, his head cocked like a bird, his good eye peering. "Are you such a stone, boy? You pay the turnkey a tuppence. You've got to pay!" He made a face and pretended he was going to bite Little Bill's leg. The youth jumped in fright, the cripple laughed, and bystanders took up the laughter. In moments the entire market was laughing as the bumpkin boy stumbled away.

Little Bill snapped a tuppence on the turnkey's table. "My sister, Mary Broad, I want to see her." He leaned forward, knuckling the table. The burly turnkey regarded the coin for a moment, then swept it up with a casual motion. He tilted back in his chair, crossed his arms, and sized up Little Bill, a smile playing at the corners of his mouth. "Broad?" he said finally, dragging out the word, "Big, good-looking woman?" Bill stared, such a description of his sister new to his ears. He nodded. The turnkey gave him a long fish-eye.

~~~~~~~~~~

Little Bill's visits were a tonic for Mary B. Each time he was let into her jail room Mary B felt her heart go out to him. She gave him long hugs and could not seem to get enough of holding his hand and tousling his hair, of just looking at him. At first he was embarrassed by her affection, but he came to realize how starved she was for the reassurance of his presence.

He brought her a few clothes, because her landlord in Dock had taken her clothes and scant possessions for unpaid rent.

11

Katie and Seedy suffered similar losses. The three now owned nothing but what they came by from jail-work or charity.

Little Bill tried to amuse Mary B by mocking their father in a falsetto voice, "Oh, the shame of it! Oh, the shame of it all!" Despite her low spirits, Mary B laughed at his mimicry. She saw the irony of their father playing the aggrieved parent to cronies whose histories of smuggling, poaching, pilfering, and sheep-stealing were common lore. "Oh, the shame of it all!" in his silly falsetto became Little Bill's motif.

"And what about Mama and Grandma? How are they taking it?" Mary B asked.

"We're all for you, you know that," Little Bill reassured her. But Mary B knew her mother would tip-toe around the family's two rented rooms, tight-lipped, beseeching Little Bill and her younger sister Dolly with worried eyes not to provoke their father lest there be a scene. And Grandma Spence would keep quiet as a churchmouse lest she provoke her son-in-law's ire.

Mary B didn't care if Little Bill made up stories to amuse her, if he exaggerated or concealed. News of her sisters and friends and Fowey provided a connection to a familiar, comforting world, even if only in her mind's eye. Each time Little Bill said good-bye Mary B felt such a welling up of love and such immense sadness and sense of loss she thought her heart would break. But the day he brought news of Grandma Spence's sudden death she felt she couldn't endure the heartache, that she'd disappointed her grandmother's last days.

That was the last time Mary B saw Little Bill. On a raw day in March, 1786, after Mary B had been confined in the Plymouth jail for eight months, Little Bill came to visit her again, and learned she'd been taken to the High Jail in Exeter's Rougemont Castle to stand trial.

~~~~~~~~~~

Katie shook Mary B awake. Sunlight streamed through the barred windows of the Exeter jail. Over Katie's shoulder Seedy grinned her wide, gap-toothed smile. They were excited. "We're going home!" Katie cried. Groggy from sleep, Mary B thought for an instant their sentencing the day before had been a mistake and that they'd be set free. "They're sending us to the hulk in Dock!" Katie continued, "We'll be with friends again! We're going home. Home!"

12

But the three women spent two more months in that sturdy old house known officially as Exeter's High Jail for Felons. During that time Mary B fell in love with James Martin and rode in a prison wagon with him and a dozen other convicts, shackled hand and foot, to the hulk in the Hamoaze that would be their new prison. There she was confined on a decommissioned warship named the *Dunkirk*, where only a few weeks earlier Marines from the nearby barracks had quelled an uprising of prisoners. Firing into the depths of the hulk, they shot and killed eight men, and wounded thirty-six.

~~~~~~~~~~

A year later, on a drizzly March day, the master of the *Charlotte* welcomed two Marine officers aboard for a tour of his ship. Fresh from the Deptford naval yard on the Thames, the *Charlotte* was a curiosity. Unlike disease-ridden hulks such as the *Dunkirk* that housed thousands of convicts in sloughs and backwaters, the *Charlotte* was a fairly new ship, modified to haul England's felons halfway around the world. At a time when the theft of a silk bonnet earned exile from England, the government of Sir William Pitt was spending upwards of £50,000 to rid the land of 750 convicted misfits, miscreants, and malefactors by sending them off to an unexplored wilderness called Botany Bay.

Among the beneficiaries of this enterprise was Captain Watkin Tench of His Majesty's Royal Marine Light Infantry, recently appointed to command a company of Marines that would protect the new convict colony. Technically a captain-lieutenant (a captain receiving a lieutenant's pay), Tench was one of many volunteers eager for a Botany Bay assignment, for the humdrum life of half-pay was no life for a soldier! Slim, fair-haired, thirty years old, with wide-set eyes and a mouth almost too small, he'd come to take a look at his new maritime home with one of his lieutenants, John Creswell. The lieutenant was sleek-faced, and going to fat.

"Pretty piece," Master Thomas Gilbert said, indicating a hatch cover of heavy oak planks reinforced with bolted iron bands. Tench and Creswell exchanged amused glances. Their guide was a man of few words. "Made special," Gilbert said. At Tench's quizzical expression the master added with a sigh of great effort, "In case they riot."

"Exercise," Gilbert said, making a circle with his hand. Creswell coughed behind his hand to hide a giggle and avoided

Tench's eyes. A few feet behind the mainmast, a three-foot-high barrier of planks ran the width of the bathtub-shaped ship. The barrier was topped with iron spikes and divided by a narrow gate. "Sentinel," Gilbert said, pointing to the gate. "No convicts outside the fence." Tench imagined shackled convicts shuffling in a circle around the hatch in an area further cramped by the longboat and jolly boat. How would his men fare for half a year without setting foot on land? God, where would they all fit? Marines, convicts, a few wives and children, the ship's company, the livestock and supplies - all crammed into this little Noah's Ark that looked barely a hundred feet long.

"Prisoners shackled in port."

"Wha..." Tench said, "Sorry."

"Prisoners shackled in port," Gilbert repeated.

"Women?" Tench asked, unconsciously falling into the man's laconic speech. Gilbert shrugged and started down the steep ladderway. Tench looked to Creswell, but his friend was avoiding his eyes to keep from laughing.

Below they found themselves in a cage of thick oak planks, banded and bolted with iron, the air redolent with creosote and quicklime and the ripe, fishy smell of the bilge. Although he'd spent many months aboard ship in the American War, Tench still reacted to the cramped headroom and over-powering smells of the below-decks. He looked at Creswell and saw in the pale gloom that Creswell's sleek face had blossomed with sweat. His expression was tense. In the first huge fleet England sent over to fight the American War, more than 2,000 seamen and soldiers had died of diseases brought aboard by recruits impressed from England's jails.

Thomas Gilbert chuckled, a low rumble that took Tench by surprise. Circling his hand around in the cage, Gilbert started to speak but Tench interrupted. "Ahah! You have to get into the cage down here before you can get out up there." Gilbert nodded with a pout, perhaps disappointed he'd lost his chance to explain the cage in fewer words.

Tench made a face like a smile and pulled open the gate that led from the cage. From experience he took care to doff his hat before stepping into the main cabin area, an expanse perhaps forty feet long. He bent his neck to view the scene from beneath his brows, already feeling queasy. Almost five feet ten, he had to stoop beneath the deckbeams.

Except for the planked deck, every surface was whitewashed. New-built compartments ran fore and aft along the sides of the ship. The compartments were divided into upper and lower bunks, perhaps five feet deep and six feet long. In the middle of the cabin stood a shorter row of compartments. "Men," Gilbert said, pointing with his chin towards the bow. He moved comfortably, his gray head clearing the deckbeams by less than an inch. Middle-aged, pudgy, Gilbert had a jowly face and small, appraising eyes. He wore a green frock coat, faded by the sun to the color of thin pea soup. Gilbert pointed to one of the compartments. "Three and three."

"You mean three men to a bunk?" His pudgy hands clasped behind, Gilbert nodded with the somewhat distracted patience of a teacher. Tench counted the compartments, did the arithmetic in his head, and heard Creswell give a low whistle. Creswell too had worked out the calculations.

"Wat, aren't we supposed to load more than 100 men?" Tench nodded. The chief surgeon for the new colony had come down from London the previous day with the list of convicts to be embarked. "But with just 15 compartments that means four in some bunks, not three!" Tench nodded again and caught Thomas Gilbert's eye, who simply shrugged. More than a hundred men in this cramped space, for half a year?

At either end of the convict quarters were stout bulkheads pocked with gunports. Tench nodded at the gunports and Creswell returned his nod. "In case they riot," Gilbert said, then pointed again with his chin, this time towards the stern bulkhead. "Women are back there."

"Like this?" Tench asked.

"Smaller."

Forty-eight people embarked aboard this so-called First Fleet would die before reaching their destination, but those who planned the enterprise actually expected more to perish. Some of the older convicts would simply give up, choosing death over the prolonged suffering in these cramped, stifling prisons. Others would die of maladies carried with them from the jails and hulks, or from accidents suffered en route. Still others would succumb to the hardships of heat, foul air, thirst, damp, cold, and weeks of an unrelieved salt-heavy diet. But not just convicts. The Marines and their wives and children and the sailors would live little better than the convicts. And some of them would also board these ships for the rest of their lives.

But those who masterminded the Botany Bay project would actually congratulate themselves on the successful voyage of the "First Fleet," not realizing that a second fleet, prepared under less careful watch, would kill more than a quarter of the convicts sent to sea.

Thomas Gilbert led them to the forward bulkhead and a solid oak door standing ajar. The door was strapped with iron, bolted top and bottom, and the barred, head-high look-in hole was fitted with an iron plate that swung over the hole and bolted fast. Gilbert pushed the door open and stepped through.

"Here?" Tench asked, "My men are here?" Gilbert nodded.

They stood near the ladderway of the forward hatch, the configuration of compartments the same as for the convicts. "Forward," Creswell said with disgust. Tench nodded. The bow of a sailing ship was the least desirable place to live. Wind filling the sails and pushing the ship also carried the ripe accretion of odors from all the leaks, dribbles, splashes, spills and discharges of the ship's crowded humans and livestock. Gilbert said nothing, but pointed to a bunk. "Just two." Tench counted. Two to a bunk, there was room for 48 men.

"And wives?" Creswell asked.

Gilbert shrugged again.

Creswell blubbered his lips in exasperation and sought Tench's eyes, "Wat?" he cried, his voice expressing pleading and disappointment.

Feeling he needed to explain Creswell's disappointment, Tench said to Gilbert, "Everything's been changed so many times! First we were told our entire battalion was going on the *Sirius*. Then they decided to mix men and women convicts on the transports, so they needed to assign our Marines as guards. Then they said some of our men could take their wives along." He smiled ruefully at Creswell. "Long is the way, and hard, that out of Hell leads up to light."

Creswell smiled his own rueful smile in return, and slowly shook his head. "You and your Milton."

As they continued their tour, Tench sank into a funk. Suddenly the enterprise seemed overwhelming. Despite fresh whitewash, the below-decks was dreary, oppressive, and malodorous. Even the banter and bursts of song and the whistling of seamen going about their chores in the chilly, light rain couldn't relieve his mood. Finally, with brief thanks to

Gilbert and some quick arrangements to load their company's baggage next day, the two officers were rowed back to Dock. There they sought a table by the fire in Mrs. Brandystone's public house.

"Shit!" Creswell said as soon as they were settled and each had drunk a good draught from a bottle of claret.

Tench smiled through his distraction. "Could be worse."

"How?" Creswell asked, a small smile showing his appreciation of their laconic conversation.

"Shooting," Tench said, and both men laughed.

Tench had seen little of Creswell since the War, so the two comrades in arms talked about the fortunes of common acquaintances until their bottle was empty and the afternoon grew late. From the pub they strolled together as far as the barracks on Stonehouse Lane, where Tench made a familiar excuse and they parted.

Tench started down the road to Plymouth, but before reaching the city he turned off on a cart track and headed towards the Citadel. It was nearly dusk when he turned off again to gain the rise called the Hoe. There he stood on a cliffside path overlooking the broad reach of Plymouth Sound.

To the east he watched a rising moon break through scudding clouds, bathing the battlements of the Citadel in pale light. The reflection of the huge orb on the water pointed a shimmery finger at Tench. To the south the distant, yellow glow of the Eddystone lighthouse pointed a fainter finger. To the west, some distance up the Hamoaze, a bonfire flared at water's edge on Mount Edgecomb. Tench could make out tiny silhouettes moving back and forth across the firelight, perhaps dancers. Did he imagine the faint music of a pipe? The bonfire too sent a narrow finger across the water. Tench amused himself with the notion of three illuminations all pointing their shimmery fingers at him. Perhaps there was a poem in that, something melancholy like "Last Night on the Hoe."

He turned to stroll the cliffside back to the barracks, pausing again to watch the figures in the distant firelight, speculating whether their fire might be a smuggler's signal. Ahead beckoned the yellow lights of Dock. He decided to go see the widow after all. He owed her a final farewell.

On a whim he tied back his coat, unsheathed his sword, and began to slash the night air in imaginary combat. He liked the

17

swish and whir as he advanced, retreated, pirouetted suddenly to fend off a treacherous foe behind. He sprang to thrust, and thus dispatched several dusky enemies before stopping to fill his heaving chest with the chilly night air. Sheathing his weapon with a self-satisfied grin, he set off for Dock with long, purposeful strides, ready to say good-bye to his latest love.

~~~~~~~~~

His company's baggage stowed aboard the *Charlotte*, Tench had given his Marines leave to buy last-minute "necessaries." Rich with a year's advance pay, their spirits were high. Now, at the end of the blustery day three privates made their unsteady way back to barracks along Dock's puddled Fore Street, quite drunk. On a whim Private Quint had stopped at a stationer's to buy a quarto-size, leather-bound, blank book and writing materials. With only the rudiments of reading and writing from church school, he was nevertheless loquacious about his plans.

"I think I'll call it, 'The True Journal of Private Peter Quint,' and then something about Botany Bay."

One of his companions asked, "And what about us, Quint, you're going to put us in your book, aren't you?"

Quint glanced over at Porter's flattened features, which along with his short body and long arms put Quint in mind of a monkey. "Well, let's see..." and as he embellished his response with imaginary strokes of a giant pencil, Quint said, "'The True Journal of Private Peter Quint...', ah, now, should I put the Marines in?"

"Aye, the Marines for certain," his other companion nodded, his rum-dulled eyes fixed on the muddy road. He was tall and hatchet-faced, a gangly fellow whose uniform never seemed quite to fit no matter how he took in tucks or let out seams. His Adam's apple jumped when he spoke, and he walked with a peculiar, head thrusting motion that had long ago earned him the name of Stork.

"Right," Quint said as if to himself, not paying much attention. "All right then, listen to this: 'The True Journal of Private Peter Quint, of His Majesty's Royal Marines, ah, being the true story of his voyage to Botany Bay, ah, with his mates.'" He emphasized the last phrase. "Aye, I like that," he said.

"What about our names?" Porter asked. "You're going to put our names in too, aren't you?"

"Sure, you'll both be in it, and Corporal Baker and Sergeant Motherwell...well, blimey - everybody!"

"But not officers, right, Quint? What do you think, Stork, no officers, right?"

Collecting himself, Stork concentrated with a furrowed brow. "No, no officers. They got their own books."

"And this is going to be our book, right, Quint?"

"Right," Quint said, clapping Porter on the back, "Right."

They progressed in silence awhile, brawny Quint swaggering a bit as he contemplated his project. When he spoke again, he was ruminative. "Now if I was to get two shillings on a book, and if I was to sell ten books a day, why! That's a pound a day!" He whistled softly. "And if I was to sell...." He halted and put out his arms to stop his companions. "Listen, mates, how many people in London?" Conspiratorial, he lowered his voice, "And what would you say if I was to sell a book to just one of every fifty, say, or even one of a hundred? I think I'd get a hundred pounds! Easy! A hundred pounds! Me, Peter Quint, with a hundred pounds! Blimey! Think on it, a hundred pounds right here in these two lovely hands." His cohorts stared in boozy wonder at the thin, flat parcel he held up to the heavens like an offering.

Porter said, "See here now, Quint. If Stork and me are in the book we should get something too, right?" He reached over to yank at Stork, whose weaving frame and frowning face betrayed an interest tending to wander off in an alcoholic fog. "Listen, we should get something too, right, Stork?"

"Right," Stork said, taking a wobbly step for balance.

The glitter of a promising future in his eye, Quint looked fondly from one comrade to the other. His face grew a wide grin, his missing foretooth a black gap in his sturdy face. "All right then, mates, we'll do it! We'll share the booty! Mates stick together, right? And we're the best mates that ever was."

"Right!" his friends answered in ragged chorus. And with that, Porter, Stork, and Quint continued arm in arm in lurching step. Smiling knowingly from one to the other, Quint broke into song with a husky, vibrant voice,

> *One Friday morn when we set sail*
> *Not very far from land*
> *We there did spy a pretty fair maid*

> *With a comb and a glass in her hand*

Porter and Stork joined in,

> *In her hand, in her hand*
> *With a comb and a glass in her hand*

Quint pointed to the red door of the Running Fox, and as the three privates veered into the Marine-rich public house they swung into the chorus again.

~~~~~~~~~~

The night had again brought violent wind and pounding rain, but the morning sun was bright, the air chill, and the wind brisk. Mary B emerged on the rapidly drying deck of the hulk and looked anxiously for Little Bill. She wore her only remaining dress, a shapeless, shabby rag, and she clutched the ends of an ancient wool blanket over her head and shoulders. It was Friday, March 9, 1787, and she was about to board the *Charlotte* for the long voyage to Botany Bay. Just shy of twenty-two years old, she was three months pregnant.

Under the proprietary eyes of the turnkeys, a few women had already found husbands, relatives, or friends who came to say goodbye, but most had no one to see them off. Poor and illiterate, from poor and illiterate families, who would write their letters? And what would they say? "I'm bound for Botany Bay. Please forgive me for being a disappointment?" For most of the women, getting entangled with the law meant being cut off from their families and friends.

Mary B scanned the crowd again for signs of Little Bill, certain he must have known this was her last day on the shores of England. No one. In a corner out of the wind, Seedy nuzzled a man from her past. Shaken by a sudden fit of shivering, Mary B swallowed her disappointment and tried to concentrate on the ship alongside.

Much smaller than the hulk, but fairly new and freshly fitted, the *Charlotte* seemed to promise better things. But Mary B had learned to be cautious. Almost twenty years of life in Fowey hadn't caused such suffering as Mary B experienced in the twenty months since that fateful day on the Hoe. The worst was the hulk, for among the 350 prisoners crammed into its stinking recesses, illness and death were never distant, and one or two or more died each week. How often had she seen a simple chill or fever or festering sore begin the slippery slide to a fatal end? Even so, being wrenched from the familiar dirt and smells and sometimes harsh discipline of the hulk and prodded towards an unknown future aboard the *Charlotte* was a frightening prospect.

Mary B had learned other lessons too, sometimes painfully. Living amid greed and bullying, her trust frequently betrayed, she learned that convicts who eat, sleep, work, talk, and laugh together are nevertheless engaged in individual struggles to survive. And she was still learning the subtleties of a painful lesson - sex was her most valuable resource, but also the most troublesome - and in some ways the most worthless.

"We always pay," Katie growled, "Pay, pay, pay. He gets a bellyache and we pay. He can't get a hard-on and we pay!"

Mary B regarded Katie's lowered eyes and thin, muttering lips. Her friend's pouty face was white with tension and cold. Katie was speaking of Hugo Blackpool, the pot-bellied, sour-faced superintendent of the hulk, who paced the afterdeck of the *Charlotte* with a short, pudgy man in a badly faded, green frock coat. They were arguing.

Mary B shifted her feet and look away. She'd never liked the bitter, quarrelsome, petty side of Katie, much preferring their whispered, giggly confidences. How different it was now with Katie! Before the incident on the Hoe, Mary B had looked up to Katie. Somewhat older, Katie had been her tutor in Dock, a scary place for a naïve young woman like Mary B. But since that fateful afternoon on the rampart of the Citadel, Mary B sometimes dreamed violence against her friend so vivid she'd jerk awake, panting and sweaty. But she didn't burden Katie with all the blame for their terrible situation, because Mary B knew she herself wasn't blameless. After all, lying half-drunk in the sun, laughing and nodding at Katie's confrontation of Agnes Lakeman, she'd done nothing to hold things back. Her mindless passivity had encouraged Katie. No, Katie had started the trouble, but something Mary B recognized but did not yet understand about herself made her an accomplice. Some part of Mary B's character as real as the air she breathed was also to blame.

Not responding to Katie, Mary B studied the remnants of her shoes, bound together with twine. She wondered if the rumor was true that they would all get new clothes.

On the deck of the *Charlotte*, lounging Marines gaped up at the waiting women with leering curiosity. In their new, bright-red uniform coats, their polished brass gorgets and gleaming black leather hats with white plummets, their bright-white cartridge belts and leggings and powdered hair and beribboned pigtails, the Marines looked ready for parade. But the sight of their muskets with fixed bayonets jarred Mary B. The cruel weapons brought back stories she'd heard of Marines storming the *Dunkirk* to regain control of the floating prison. Bloodstains remained unscrubbed for months as reminders to the convicts.

A few of the soldiers visited with their families, who clung to them tearfully. Some had drunk themselves into a befuddled leave-taking, a confusion of bravado and backslapping and wet kisses. Nonetheless, knowing they wouldn't see their sons or brothers or husbands or fathers again for three years or more, knowing the stories of Botany Bay's savage cannibals and man-eating beasts, even these drunken farewells had meaning.

A girl of thirteen, dressed as a domestic servant with mobcap, apron, and kerchief, approached the women. Tall for her age, fresh-faced, with the disappearing, slightly rounded features of adolescence, she wore a new-looking, gray wool cloak and carried a small, unpainted wooden box by its securing cords. She was tentative. Putting down the box, she called out, thinly at first, then more boldly, "Mary B. Mary B!"

Mary B turned at the sound of her name. She took a moment to recognize her sister. "Dolly!" she exclaimed with surprised pleasure, "Dolly, *you* came!" She took the girl into her arms and held her tight, murmuring over and over, "You came, you came." Finally standing back, she looked anxiously into Dolly's tearful eyes.

"I'm glad to see you, Mary B," Dolly said shyly. Her eyes misty, Mary B smiled and embraced her sister again.

After they separated, Dolly turned and pointed to the box. "Mama said I should bring this to you. Little Bill made it, but he said it isn't very good." She forced a little laugh.

Mary looked at her closely, then towards the waterfront, then again at Dolly, her expression questioning. Dolly dropped her eyes. "Dad wouldn't come, and he wouldn't let Mama come."

Nodding, Mary searched her sister's face for hints of the unspoken. Sighing heavily several times, swallowing rapidly, Dolly's voice was choked when she finally blurted, "Little Bill's gone!" She collapsed into Mary B's arms.

Mary B held her, patting her back, kissing her head, shushing her softly. "They had a terrible fight," Dolly said when she recovered, "and Little Bill ran away. I thought he'd come back, but he didn't. He's been gone three weeks."

Mary's blanket slipped unheeded to the deck as her shoulders sagged under the weight of utter sadness. Little Bill, Little Bill, she thought, what have you done? Now she knew why he'd acted so silly on his last visit, clowning around with nervous energy as if afraid of a moment's quiet. He was putting on so I wouldn't see, she thought, but why didn't he come to see me when he ran away? Where did he go?

"But I came, Mary B, even though Dad said I couldn't." Dolly paused and took a big breath. "After Little Bill ran away and we heard about Botany Bay, Mama got some things together. But Dad wouldn't let her come." A long pause. "He said it was the shame." Dolly broke into sobs again.

Mary B held her. "I know, I know." She shushed Dolly until she quieted. Then Mary B asked the question she had to ask, already dreading the answer. "What did they fight about, Dad and Little Bill?"

Without answering, Dolly sobbed harder, and Mary B knew. She heard herself repeating, "I know. I know. I know."

Suddenly Mary B pushed Dolly back and forced a gay voice. "How good you look, Dolly! How grown up! How pretty! Why the fellows must be after you all the time!"

Dolly managed a little laugh through her tears.

"And when I get back," Mary B added, "Well, I suppose you'll be married too. My, what a lucky fellow he'll be!"

Dolly blushed and laughed again, more naturally, and the sisters began to joke and tease and giggle and caress until there came a silence that grew long. They looked into each other's eyes, and then Dolly looked down, her shoulders once more shaking with sobs.

Mary B spoke urgently, "Listen, Dolly, I'm going to be all right. I'll get through this, I know I will. I always have, you know. But I've learned that you must be strong, Dolly, and use your head, understand? You hear me, Dolly? You must be strong and use your head!" As she said this, Mary B shook her sister gently, as if trying to coax the significance of her words into Dolly's understanding.

Head bowed, still sobbing, Dolly nodded.

"Tell Mama I'm sorry for the shame, Dolly, and thank her for the things." She looked over her sister's head into the distance for a long moment. "Now give me a squeeze and off you go. We'll be loading soon."

Unwilling to raise her eyes, Dolly embraced Mary B with a steely grip until the older sister at last forced Dolly's arms down and away and turned her around. "Go now, Dolly," Mary B said, kissing the back of her sister's head, "Go!" She gave Dolly a gentle push. "Go, Dolly. Goodbye, my little sister!"

Dolly paused a few steps away, head still bowed, her shoulders jerking, then suddenly stripped off her cloak and turned with averted eyes to thrust the garment at Mary B. Then she bolted for the side. Mary B watched her disappear down the ladder. A few minutes later she saw Dolly huddled in a waterman's boat. They waved several times until Dolly mounted the stairs of the rickety wharf, took the arm of a young man in a blue jacket, turned to look at her for a long moment, then waved again and disappeared into the crowd.

Her throat tight, Mary B donned the cape, bent for the box, and returned to her place near Katie, who was silent except for her heavy sighs. She became conscious of Katie's eyes. "Sometimes things are hard," Katie said finally.

Mary B looked at Katie with a faint, sad smile, glad she still had Katie for a friend. "Aye, Katie, sometimes things are hard."

~~~~~~~~~~

Shackled hand and foot, James Martin moved onto the deck of the *Dunkirk* with the other men, their chains a doleful chorus of metallic scraping and rattling. Some carried bundles or boxes with clothes and personal articles, perhaps books and food. Most wore rags, but some were in finery. Most were silent, brooding, and frightened; others showed their uneasiness by ceaseless chatter, filling the air with the reassurance of their own voices. Usually talkative, James Martin was silent. Of above-average height and slender build but with broad shoulders, James had friendly blue eyes and a ready smile that had earned him the boyhood name of "Sunny Jim."

He looked about nervously, hoping he wouldn't be surprised by the appearance of his wife. Relieved at not seeing her, James looked for the auburn glint of Mary B in the clutch of women crossing over to the transport. When he failed to see her his heart sank. Wasn't she going to Botany Bay after all?

24

Marines on either side of the gangway helped the women board. They joked and carried on. "Whoops, my hand slipped." "Why, here's a pretty one." "Hello, my lovely. Now there's a leg." "Hello, sweetie, Quint's my name." "Hello, love, mind your manners now."

Then James saw a vaguely familiar figure in a gray cloak mount the gangway. When she paused to look back, her pale face framed in the hood of her cloak, his heart leapt. But he raised his hand too late in recognition as Mary B turned and disappeared down the gangway. So she *was* going!

For the hundredth time James wondered if they'd be able to get together on the voyage. Lucky even to be on the same ship, he knew life aboard the *Charlotte* would be very different. Bribery on the *Dunkirk* was a way of life, and a man and a woman could find ways to get together, even if only for a few minutes. But aboard the transport taking them to Botany Bay, Marines would guard them.

"There she is in all her beauty," Will Bryant said. James was momentarily taken aback, thinking Will meant Mary B. Then he realized Will meant the *Charlotte*.

"Aye," James said, running his eyes over the ship, no different than hundreds like her that creaked along the Atlantic and European trade routes. She was a three-masted, medium-size brig. Lacking a figurehead and fancy galley, plain to the point of severity, the *Charlotte* was a utilitarian tub. "A Quaker beauty," James said.

Will snorted a half-laugh, and James swelled slightly at the success of his joke. A little older than James, with a solid frame and a level gaze, Will Bryant was the kind of man James liked having as a friend.

After rattling down to the *Dunkirk* from the Exeter jail in the prison wagon with Mary B, James had soon learned about Will's reputation. Claiming to be a shirttail relative of a prominent family, and something of a local hero, Will Bryant could be a charmer. Now James regarded Will's face. The man possessed a somewhat fleshy nose, firm lips, and a mop of curly black hair going to gray. His brows were heavy, and whorls of gray marked his whiskered cheeks and chin. But Will's eyes were what first brought James up short. When Will's smoky, hooded eyes challenged his own in some early disagreement, James felt his bowels loosen. Since then James had seen a devious side to Will and occasional outbursts of anger that made most convicts give Will Bryant plenty of room.

Feeling James's eyes, Will returned his look. James smiled, "Better than the hulk though," he said, and Will nodded.

And for a time it was. Before mounting the gangway to the *Charlotte*, each convict slid first one foot then the other into a special anvil to have the pins of their permanent shackles struck off with expert ease. Unbuckling the heavy leather belts through which his leg chains ran, he tossed the entire assembly onto a mounting pile of sweat-stained, smelly leather and wear-burnished metal. Removing a man's shackles took perhaps a minute. But after a few unencumbered steps down the gangway to the deck of the *Charlotte* he was almost as quickly fastened into iron bonds again, this time with beltless shackles locked by key. The shackles on the *Dunkirk* belonged to the contractor operating the hulk; those on the *Charlotte* belonged to the contractor transporting him to Botany Bay.

As the convicts prepared to descend into their new prison, each stopped at a small table behind which Lieutenant Creswell sat with pen, ink, and carefully ruled paper to check their names from the list sent down from London. If suspicious of a convict's bundle or box, he motioned to Sergeant Motherwell to make the prisoner lay out his belongings for closer inspection. Standing close by, clasped hands behind his back, Captain Tench gave a sign of "yes" or "no" as Motherwell held up a questionable object, thus deciding with a moment's thought whether a convict's book or keepsake or item of food would remain with its owner or be locked away or cast aside. Anyone wearing an old wool garment or carrying an old wool blanket was made to toss the item on a growing heap; old wool harbored vermin and was not allowed below-decks.

Convicts who had nothing to wear but old wool were hustled below and made to hand up their clothes. "You'll get new clothes and blankets," they were told. But that would be in Portsmouth. Meanwhile, they sat naked until others might share extra clothes out of kindness.

Mary B seemed flustered when Creswell asked about her newly acquired box. "I don't know what it is, sir," she blurted.

"You don't know what's in your own box?" Motherwell's question was a mild, teasing taunt.

Taller, Mary B looked down at the sergeant's close set eyes. "No, sir, I only just got it." She motioned with her head towards the hulk.

"Well, then, let's take a look, shall we?"

Mary B knelt and began struggling with the tight little knots. After a few moments of fruitless effort she leaned back as Sergeant Motherwell impatiently drew his sword and with a few deft flicks left the cord in pieces. Tench stepped closer. In the box Mary B discovered a tin of tea, two cones of coarse, brown sugar wrapped in paper, a hard cheese, a length of kersey and another of cotton, two needles, thread, a length of ribbon, a cloth purse with a half-dozen shillings, and her grandmother's *Book of Common Prayer*. She picked up the prayer book and caressed its cover.

Tench watched her self-absorbed concentration. He found the young woman's face and demeanor strangely moving. "Very good," Tench said finally, "You'll have something useful to read. Go below now." Her expression puzzled, Mary B looked up in a manner Tench found oddly disconcerting.

"Well, what is it, what's the matter?"

She searched his eyes, her mouth working to form words. "Well?" he said, his irritation growing at her silence.

Finally, her eyes misty, she blurted, "But I cannot read!"

Tench looked away in embarrassment.

Mary B started for the women's hatch when Motherwell made a grab at Dolly's cloak. Their eyes locked. As Mary B pulled with her free hand, Motherwell turned to Tench with a questioning look. Tench pursed his small mouth, then shrugged. Motherwell let go. Mary B felt eyes follow her below.

By early afternoon the convicts were loaded - 20 women, two infants, and 107 men. The women brightened when they saw their quarters, so much cleaner and roomier than on the hulk, with more bed space than they needed. After the *Charlotte* was towed clear, another ship called the *Friendship* was warped alongside the hulk and the transfer of convicts repeated. By late afternoon, after both prison ships were loaded and ready for departure, the chilly wind had died to a whisper. Longboats towed the transports past the bustling shipyard of Dock, then past the wide expanse of Stonehouse Creek where Mary had ferried her passengers to the Navy hospital, then past the fortifications at Devil's Point. There they were treated to the

Division band and a cheering crowd of Marines, wives, and well-wishers lining the works to see them off.

Below-decks, convicts cried out with fright when the vessels heeled sharply with the rush of people to portside to wave and whistle and call back. Ever the clown, Quint mimed diving off to join the shoreside crowd, which drew a round of good-natured horseplay among his mates. But as the figures ashore grew smaller, the sounds fainter, the light poorer, an eerie silence fell over the ships, and the Marines and their handful of wives set about finding mindless, private things to do.

After the longboats cast off their tows, the two convict transports rocked gently in the deepening darkness, waiting for wind. Vaguely luminescent through a thin cover of clouds, the rising moon was just bright enough to bathe the ships in soft romantic moontones.

~~~~~~~~~~

As the 217 occupants of the *Charlotte* rocked gently in Plymouth Sound, Arthur Phillip sat with a cloak draped over his nightshirt, writing in the warm glow of an oil lamp. He was creating a new world. In his solitude he liked the comforting scratch of his quill across the creamy paper. With an occasional sip of dark sherry to nourish his thoughts, words came easily.

Professionally he was at a fine point in his life. Designated governor of the new Botany Bay colony, he possessed both great opportunities and considerable power. Little he asked in his new assignment was not granted, but the daily drudgery of dealing with dullwits and bureaucrats to get the First Fleet underway had worn him down. For a colony expected to raise its own food he'd been issued only a few poor-quality gardening implements. For a voyage halfway around the world, he'd received no anti-scorbutics to forestall scurvy.

Such stupidities, multiplied a hundred times and percolating down to the smallest details, had occupied Arthur Phillip for weeks. The evening was rare when he could be alone in his rooms near Covent Garden and arrange his thoughts. Once at sea he'd have more time to fill in the still-vague outlines of his little colony, but then he'd be cut off from the Home Secretary. And to be cut off was to be vulnerable to London's incessant sniping, carping, and second-guessing. No, before he said best set forth in some detail his thoughts and proposals and pass them on for review, whether his superiors read them or not, understood them, or even cared.

With his ships soon assembling at Portsmouth, Arthur Phillip's thoughts were on the convicts. He was convinced that at some point he should divide his fleet and send an advance force to prepare a settlement site. "*By arriving at the settlement two or three months before the transports,*" he wrote, "*many and very great advantages would be gained. Huts would be ready to receive the convicts who are sick, and they would find vegetables, of which it may naturally be supposed they will stand in great need, as the scurvy must make a great ravage amongst people naturally indolent and not cleanly.*"

He laid his pen aside, sipped sherry, and imagined the transports sailing into Botany Bay. There would be the Union flag, the Marines drawn up in parade, the band playing, the boom of a salute from the *Sirius*, an answer from his battery. He pictured rows of neat, thatch-roof cottages with tidy garden plots, happy natives in canoes paddling out to greet his ships with songs and garlands. And he, Captain-General and Governor of the Botany Bay colony, would assemble the convicts to read them the official proclamations and launch them on their new lives of fruitful enterprise. Arthur Phillip liked to daydream such scenes.

He picked up his little hourglass, a gift from his wife. Of rosewood and brass, the device was inscribed on one end to "My Beloved Arthur, on the Occasion of His 40th Birthday" and on the other *Omnia tempus revelat*. The Latin inscription was a cruel irony. Just two months later they'd learned of her wasting illness. Now, when the world lay dark and dead and Arthur Phillip felt gloomy, he sometimes sat alone and held the hourglass to his ear, listening to the faint hiss of the falling sand. He found the sound companionable and comforting.

He sucked another sip of sherry back and forth a couple of times in the gap of his missing buttertooth and resumed his writing: "*Huts would be ready for the women. The stores would be properly lodg'd and defended against the convicts in such manner as to prevent their making any attempt on them.*" Aye, solid storehouses would be necessary to protect the liquor, food supply, and tools. And the women too would need protection. At least some. Among such thieves and lowlifes, nothing would be safe. But he would build a colony that was ordered. Yes, ordered and disciplined.

His thoughts moved back to the women. "*The women in general I should suppose possess neither virtue nor honesty. But there may be some convicted for thefts who still retain some degree of virtue, and these should be permitted to keep together, and strict orders to the master of the transport should be given that they are not abused and insulted by the ship's company...."*

~~~~~~~~~~

As the *Charlotte* and *Friendship* lay quietly in placid Plymouth Sound, lucky Quint drew the first watch of the women's prison. Beneath his feet huddled the convict women, mostly young. Small wonder that fantasies ran amok among the Marines and sailors. True, there were orders and rules. Sergeant Motherwell had read the rules from Tench's order book: "No intercourse will be permitted with the female prisoners except in the line of duty. No insults, profanities, or imprecations will be directed towards the prisoners."

To general laughter flat-nosed Porter asked if an "imprickation" was what he thought it was. But among themselves the Marines concluded, if the women connive in bending the rules a bit, well, what's the harm? Besides, the sailors all talked of taking shipboard "wives." So lantern-lit Quint, his missing tooth a black gap, shot many an inviting grin through the hatch grate as he paced his few deliberate steps back and forth above the women's prison. Not that anyone expected the women to break out. Quint's job was to keep the men from breaking in.

Guarding the male convicts was another matter. Two Marines stood above their hatch with loaded muskets and fixed bayonets, for twice within memory convicts being sent to exile had seized their ships. One had put ashore just a few miles from Plymouth. For weeks the area was in uproar over dozens of escaped felons running loose in the countryside. But they were rounded up by ones and twos as, perhaps inevitably, they stole to survive and were caught again.

"Remember the *Swift*, remember the *Mercury*," Arthur Phillip cautioned in his orders, referencing the two seized ships. As a result, convict men would live in shackles whenever the transports were in sight of land.

Quint's amorous forays were suddenly jolted by a blast of wind that heeled the *Charlotte*. From below he heard the women cry out in alarm

He called down, "It's all right, m'dears, we're just getting under way." He knelt to peer down. Inches away, Seedy's frightened face stared back. Quint jerked back. "Blimey, you gave me a fright there, miss! Are you all right?"

Seedy nodded yes, then no.

"Are you afraid, then?" Quint asked.

Seedy nodded. "Don't let me drown!" she begged, "Please, don't let me drown!"

As sailors scurried aloft in the stiff, cold wind, a half dozen convicts were brought up to work the groaning capstan to weigh the anchor. The bow sank slightly, and ponderously the *Charlotte* began to move.

For the next two hours of his watch, as the *Charlotte* plowed toward the Channel, Quint heard moans and whimpers, cries and prayers, retching and curses as his beauteous captives were initiated into the joys of sailing. By the wee hours of the morning the convict transports were in the Channel, plowing eastward towards rendezvous in Portsmouth.

~~~~~~~~~~

Tench lay abed in the six by seven-foot stern compartment that would be his home for the next ten months (four months more than he ever dreamed). Hands behind his head, he luxuriated in his sense of satisfaction. On the move at last! Without a war or squabbling of some kind it was hard for an officer to advance, and Tench had chafed since going to half-pay after the American War. Now was his chance! And not just to advance. He was also going to write about this bold enterprise, a book that would make people sit up and take notice.

He let his mind drift over the long day's events. Blackpool! What an ugly pig! The sour-faced superintendent of the *Dunkirk* was lucky Tench hadn't run him through! A replay of their confrontation had certainly enlivened the officer's mess that evening. Tench yawned, remembering with pleasure the cadenced oars of the Navy longboats towing the ships and the farewell crowd at Devil's Point. Then the wine-soaked mess of flush-faced officers drinking to each other, the King, Botany Bay, "the lags, gentlemen, because without them we wouldn't be here," Arthur Phillip (the Commodore, the Governor), the King (again), the wind ("may she blow soon"), the Royal Marines, Major Ross, success, "my Meg," "my Poll," "our sweethearts...and our wives...."

Tench had excused himself. Aship or ashore he usually ended up alone. Like Milton, one of his favorite poets, he preferred small company. He knew some saw him as aloof and moody, and sometimes chided his withdrawals. Tench sighed. His solitary nature might be a detriment to his career, but the need to be alone was at times more powerful than his ambition.

Balance was the thing. If you're too brave, you're dead. If too cautious, you're a coward. Too forward, a climber. Too easy-going, a victim. The surgeon who'd come aboard that day seemed like a balanced fellow. At least he read books, judging by the two heavy boxes he had lugged aboard.

As Tench drifted on the edge of sleep he slipped into a familiar vision, a swirl of nameless faces. That day he'd seen so many tense, frightened, convict faces. The woman with a prayer book she couldn't read. Eyes that evoked a dim memory. Oh, Tench thought, suddenly wide-awake, where had he put the letter for Martin? In his coat pocket? Imagine, a letter announcing you have a new son just as you sail for Botany Bay! What are the chances he'll ever see him? And how many others are leaving women they love, or hate, crimes committed but undiscovered, unresolved conflicts, unfulfilled passions, unkept promises? He thought, thus is landlife brought to an end by beginning an ocean voyage. Yawning, he decided he'd start his journal next day.

He remembered his recitation, a daily chore. Memorizing a poem or passage each month helped keep him easy in table talk. For March his project was a little piece by Goldsmith. He had it by heart now, and tried it quietly to himself for pleasure:

When lovely woman stoops to folly,
And finds too late that men betray,
What charm can sooth her melancholy,
What art can wash her guilt away?
The only art her guilt to cover,
To hide her shame from every eye,
To give repentance to her lover,
And wring his bosom - is to die.

Good poem. Perhaps he could use it in Portsmouth.

Portsmouth
1787

*O*h, Lord, Katie crowed, "Remember that monkey leering down the hatch? I wanted to throw a pisspot at him!" Her words got the women on the *Charlotte* laughing again. Clustered near Mary B's compartment, they'd worked themselves into such silliness that almost anything would set them off. Weak from laughing, a couple women lay on their backs, holding their sides with their mouths working in silent, aching laughter like fish out of water struggling to breathe. "Stop! I'll piss myself!" Liz Bason gasped. She squeezed her hands between her legs and rocked to and fro with a comical, cross-eyed expression.

Someone shrieked, "Then we'll throw your pissy dress!" A remark that set them all off again.

Their laughter brought relief. The previous night they'd anchored near the Isle of Wight at a place called the Mother Banks. Now, with their fears of ocean sailing quieted, their insides settling, they relived those first harrowing hours at sea. "Well, I liked him," Seedy said when the laughter quieted, "He made me not so scared."

"Oh, Seedy, you liked that monkey?" Katie said. "Why, did he show you his tail?" As she cackled at her own joke Seedy laughed too, but with some puzzlement.

She was the only Seedy among the *Charlotte*'s complement of women convicts, but with four Marys, four Anns, and two each of Margarets, Elizabeths, Hannahs, and Janes, they looked for ways to keep themselves straight. The solution for Mary B was easy. She'd been called "Mary B" since childhood, and remained so on the *Charlotte*.

Mary B touched Seedy's arm. "Never mind, Seedy. Katie just wishes she had a good-looking Marine interested in her."

"Speak for yourself, lady!" Katie said with sudden archness. "Any Marine monkey comes near me with his tail out is going to lose it," and she made a scissors motion with her fingers. A ripple of reaction ran through the group at Katie's waspy edge.

The women were quiet for a few moments before Liz Bason asked Mary B how she was feeling. "I think I'm over my morning sickness," she answered, "But for the last three days it would be hard to tell!" To demonstrate she grasped the edge of her compartment and made her own cross-eyed, puff-cheeked, queasy face. Ready to revive their good cheer, the women laughed, and then began taking turns acting out their seasickness stories. Finally the woman they'd already named Sick Liz (to distinguish her from Liz Bason) said seriously, "Oh, God, I didn't think I could be so sick in all my life. I thought I was puking up my guts!"

"Smelled like it too!" one of the Bristol women called from the other aisle.

"Phooey on you and your crotch-rot!" Katie yelled back.

"Hush, Katie," Mary B said, "We don't need that kind of trouble. We've all got trouble enough."

"God's truth," Liz Bason suddenly cried as she burst into tears. "Jacob's got the flux so bad...." Jacob was her infant son, born on the *Dunkirk* five months earlier. Liz was one of a handful of married women whose name on the Botany Bay list meant leaving their husbands behind, although in fact Liz's husband had already abandoned her. Rumors that pregnant women and women with babies would not be sent to Botany Bay kept Liz happy, and for a time she was Lucky Liz. Then the rumors proved untrue. In fact, officialdom thought sending infants at the breast or babies in the womb to Botany Bay made much sense. For no extra expense the little mouths would serve as the first building blocks of the new settlement. Native materials, so to speak.

Mary B reached out. "I've got some sugar, Liz. Maybe some sugar water will settle his little stomach."

Liz was touched by Mary B's offer. "But I don't have anything to pay you back with."

"Never mind," Mary B reassured, "When my baby comes you can tell me what to do."

Katie started to speak, but the group quieted at the noise of the hatch being lifted, the clank of scabbards, the sight of black boots. Resplendent in red, gold-frogged coats and tight white breeches, bearing short-plumed hats under their arms, Captain Tench and Lieutenant Creswell surveyed the women.

"Ahem," Tench said, peering through the gloom, "I want you to listen carefully."

The women shifted to hear better. Several were barely covered by what remained of their clothes from the hulk. "I'm Captain Tench of His Majesty's Royal Marines, in command of the guards aboard this ship." He proceeded to tell them the surgeon was sending down water and soap for bathing and then would examine them. He told them they were fortunate because they'd get new clothes and their rules were less strict than the men's. But he expected good behavior. If they behaved well they'd be treated well. "Otherwise..." he said, letting the word hang. He said the men were under strict orders not to harass or intimidate or abuse them, so consorting with Marines or sailors would be not be tolerated.

The longer Tench spoke, the more Latinate he became, and the women begin to wobble under the burden of his language. Smiling benevolently, he concluded, "I am going to assume that these expressions of generosity will be reciprocated by your virtuous behavior, so that, ah, the felicities of your good conduct will also be enjoyed by we who are charged with the responsibilities of your well-being."

Clumsy silence. The women look expectant. Perhaps he thought they'd applaud. At last he gestured to Creswell.

Referring to a small, black notebook, Creswell provided the women with ten minutes of details about their daily routines. Each group of six women would elect a mess matron and provide a daily cook's helper. The water ration at sea would be three quarts a day. When they were in port they'd receive four. If they had tea they could make tea if the cook-shack was open and not busy. Their slop buckets were to be emptied morning and evening, and cleaned and rinsed and sprinkled with lime. Their bedclothes would be aired every day....

As Creswell droned on, the women exchanged looks, at first skeptical, then with slow smiles and nods. Compared to their lives on the *Dunkirk* they'd be living in luxury! When Creswell finally finished he was surprised to receive enthusiastic applause. His sleek cheeks blossoming, he looked to Tench, uncertain whether he was being mocked.

~~~~~~~~~~

Katie was sick. Mary B heard her retching at the slop buckets next morning. In a few minutes she was back, coughing softly and groaning as she pulled herself into bed. She drew her blanket up to her chin and coughed again. Mary B recoiled from the smell of vomit and the foul breath of Katie's troublesome teeth. She whispered, "What is it, Katie, are you hot?" She put a hand to Katie's brow. Katie felt damp and clammy.

"No," Katie said, her voice hoarse from vomit.

Mary B stroked her brow. "What? Are you achy?"

"No, I'm all right." Her speech had a catch.

"What? Are you wormy?"

Katie hesitated, then burst forth, "Oh, Mary B!"

"Katie, tell me, what's the matter?"

"I was such a fool!"

"What do you mean, Katie? Why?"

"I never should have believed him."

"Believed who?"

"Blackpool! I never should have believed him."

"Oh, Katie! You too?" Mary B sank back with a huge sigh. They were silent for a long time. They heard the sounds of stirring, whispers, and the thump and scrape of feet overhead.

"He promised, Mary B! He promised I wouldn't go!"

Mary B found Katie's hand under her blanket and squeezed. "My God, he must have tried with all of us."

"But I believed him!" Katie began sobbing quietly. "I wanted to believe him, Mary B. I didn't want to go!"

As if Katie were a child, Mary B pulled her close and patted her back. "It's all right, Katie, none of us wanted to go."

"But your baby's different," Katie said. "You've got James and he can take care of you!" Mary B didn't respond. Her thoughts were back on a day in late November when Blackpool summoned her to his over-heated little office in the ramshackle superstructure of the hulk. In a tiny anteroom a sallow-faced clerk sat working at accounts. With scarcely a look up, he tossed his head to motion Mary B through.

Blackpool's room was gaseous from a glowing brazier. The superintendent wore his usual sour face, as if he'd bitten into a lemon. Middle-aged, jowls sagging like a hound's, his eyes were reddish and rheumy as he surveyed Mary B from behind a small table. He twirled an iron key by a loop of cord. Without preliminaries he told Mary B she was going to Botany Bay.

Her heart sank. So it was true! With almost childish optimism she'd carried a secret hope her sentence would never actually be carried out and she'd serve out her time in England, right here in Dock, where Little Bill would keep her connected. She'd never taken seriously the rumor that convicts were being sent to a place called Botany Bay.

Mary B sagged with the burden of Blackpool's message. Why her? With thousands of convicts in the jails and hulks, why her? "What?" she asked, still dazed.

"It's a long way," Blackpool said, "A long, long way. You'll not be coming back to England, I think."

"How far?"

"The end of the world."

She could hardly believe her ears. Where was the end of the world? Past America? Past the places where they got tea? As Mary B's thoughts swirled, Blackpool talked of how hard it would be to leave home and family and never come back. How would she ever get money for passage back?

He paused and began twirling the key again, studying her, then snapped up the key in his hand and leaned toward her. "There might be a way," he said, his voice confiding, "I hear a woman with child won't be sent. It's too far and too hard."

Mary B pictured Liz Bason and others who struggled with infants on the hulk. "But I don't want a baby," she cried.

Blackpool cleared his throat, "Or they might let me keep a few special ones back. A few."

Not meeting his look, Mary B nodded.

"Listen," Blackpool soothed, "You're a fine looking lass, and a man might be able to find a way to help you...unless you want to go. Do you want to leave England forever?"

Eyes tearing, Mary B looked at her grimy, bare feet. She sniffed and wiped her eyes with the heels of her hands, roughened and cracked from unraveling old rope to make oakum. "Picking oakum" was their prison labor. Her fingernails were rimmed black. She worried a damaged nail as she shook her head. She knew what would happen next. Blackpool would come around the table or indicate she should sit on his lap.

Blackpool came around the table and put a hand on her shoulder. He raised her chin with a bent forefinger. His eyes moved over her face. "Listen, love, you can save yourself from

Botany Bay." He wore a coaxing smile as he waggled her chin between thumb and forefinger.

Mary B forced her head down against the pressure of Blackpool's hand as if it were important that she watch herself rub one hand over the top of the other to scrub off some of the grime. Finally she blurted, "It's all too fast! I need time!"

Blackpool leaned close to kiss her on the forehead and squeeze her shoulders. "There's not much time. Too long might be too late." His hand slid down to her buttock. She felt suffocated by his nearness and the foul air. "'Tis a terrible fate, Botany Bay." She could feel the swell of his prick.

"I need time!" she suddenly cried as she twisted free. His arms still half-raised from his broken embrace, Blackpool regarded her for a long moment, then shrugged and returned to his seat. "Very well, but don't take long. As soon as they send down the lists I'll have to tell them who I don't want to go. It's up to me, you know." He picked up the key and began twirling it again. Mary B found herself staring at illusion of the shadowy circle. He'd dismissed her. She never spoke to him again.

Katie was quieted. From the compartment below Mary B heard Seedy whispering to Liz Bason about the Marine again. "Don't you think that's a nice name, Liz? Quint? It's got a nice sound, like when it's real quiet and a drop of water falls in a water barrel, quint! like that! Can't you just hear it, quint! quint! Little drops of water." They giggled.

Katie whispered, "Mary B? Are you awake?" Mary B squeezed Katie's shoulder in answer. "Mary B, listen, there's something I want to tell you, something important. I've wanted to tell you for a long time. Are you listening?"

Mary B squeezed again. Katie's breath was unpleasant. "Mary B, I know you're mad at me. I wouldn't blame you if you hated me. It's all right if you hate me. If I was you I'd hate me. But you never say anything - anything! I don't even know if you hate me. But even if you do, I want you to know I'm really sorry I got us in trouble. If I could take it back by going to Hell I'd burn a hundred years, Mary B, I would. But I can't. I don't even know if there is a Hell. But listen - I'm scared. I've got this baby inside that's making me sick and it's going to get bigger and bigger and then it's going to want to come out and it's going to hurt so much!"

The last words were strained as she began to cry again, but after a moment she collected herself and continued, "I don't want it, Mary B, but I think it's a punishment for the bad I did - like a punishment I've got to carry!" Katie raised up so she could look down at Mary B in the half-light of the new day. Her eyes were swollen. Her whisper was unnaturally loud. "Listen, Mary B, I owe you and you don't owe me, and you can hate me if you want and I'll understand, but please, I need you to help me! I'm so scared! Please, Mary B, will you help me?"

Mary B pulled Katie down to the nest between her arm and her side. She rubbed Katie's shoulder and kissed the top of her head. "Katie, Katie, Katie," she said, "Katie, Katie, Katie, Katie."

~~~~~~~~~~

After their usual morning meal of a grainy porridge called "burgoo," the women had an hour on deck. Their bedding aired, some washed clothes, but most found a corner in the spring sun to sit and chat and sew or simply stroll the enclosure on deck that confined the convicts, an area they called "the pen."

The pen was a midships window on a busy maritime world. In the distance rose the now-familiar steeples of Portsmouth, a sight as common as the constant ship traffic, everything from small craft and merchant ships like the *Charlotte* to dramatic warships from the busy Navy port and the giant Indiamen gliding in from the disease-ridden ports of the exotic Far East.

On this overcast and blustery day the women had come topside for a few minutes, then retreated to the warmth of their prison. Wrapped in Dolly's warm woolen cloak, however, Mary B stayed above to pace the pen alone. Now four months pregnant, burdened by enervation, listlessness, and depression, on some days she could barely drag herself out of bed to do her duty as mess matron. She hated the changes taking place in her body - her swollen ankles and feet and puffy face. But lately she'd felt better, and this morning was restless. She wanted to be active, free of the women's prison, so stifling with the malodorous air of crowded living and unwashed bodies. She'd already tried to store a little fresh air for below by giving her blanket its daily shaking, and now, hands pressing the small of her back, she paused in her pacing to twist from side to side and bend back until she felt her stomach muscles pull. She seemed unable to find relief from a feeling of strangulation in her limbs and belly.

She was looking over the bulwark at the clutter of Portsmouth when a sentinel approached and put a hand on her shoulder. "Captain wants to see you," he said, pointing to a lone figure on the quarterdeck. Mary B glanced around. She was the only woman in sight. A clutch of Marines sat out of the wind rubbing pipeclay into their gear to make it bright white. They lapsed into silence as she passed. Climbing the steep way to the quarterdeck, conscious of eyes following her progress, Mary B was careful to hold her dress lest she trip. Her heart hammered. What had she done?

Captain Tench stood near the bow, a knit shawl draped over his uniform. His foot on a storage locker, his elbow on his knee and his chin in his hand, he seemed utterly self-absorbed. Mary B approached and stood quietly. A full minute passed. At last Mary B broke the silence, her voice nervous as she curtsied and said, "You wanted to see me, Sir?" Her Cornwall accent made the question come out as "You wanted to zee me, Zir?"

Tench straightened. When he turned, his widespread eyes took a moment to focus. He looked her up and down. Then his small precise mouth smiled. He fidgeted with his shawl. When at last he spoke again his tone was formal, his words somewhat hurried, "The surgeon reports that you're with child."

Mary B stood expectant, searching the captain's eyes in that disconcerting way he remembered from the first day. She nodded with a bob of her head. Tench shifted his gaze so he looked behind her to the massy Isle of Wight, "And it seems you're not married." Silent for a long moment, Mary B at last bobbed another nod. "Well, then, perhaps you'll want to earn some money." As he spoke his eyes flicked repeatedly from her face to some place above her head. "I mean washing and mending and that sort of thing. Do you know how?"

Mary B hesitated, then nodded.

"I have no servant, you see, and I'll be fair - say six-pence a week. That's fair for my needs. I'll provide what you need, of course, I mean the soap and so forth."

Mary B was bothered that Tench wouldn't meet her gaze. She wanted him to look at her without his eyes sliding away. What did he really want? Other men had already reached understandings with some of the convict women, for there was no rule against the women doing laundry or sewing for men aboard the *Charlotte*. But some arrangements included more than domestic work, particularly when the officers were absent, which was often. As wiggy Marines made their protracted farewell rounds in Portsmouth, many furtive, thrusting liaisons had already occurred under the winking guise of a customer delivering dirty laundry or clothes that needed mending. But Mary B was pregnant!

At the long silence Tench finally returned his eyes to her face, seeming to focus somewhere just below her eyes. "Well?"

She studied the deck, then fixed Tench with that frank gaze he found so unusual. "Is that all the Captain will be wanting?"

Tench was startled. "What?" he exclaimed, thinking Mary B was spurning his offer, "You mean you won't? But what about your child? You'll need things for your child!"

"Aye, sir," Mary B said, "That I know, but what else might the captain want? I mean besides laundry and such?"

Tench looked stupid. Mary B helped him out. "Sir, you must know that some would do more. But not me, sir! I will not!"

"Good Heavens!" Tench said, stepping back to collect himself. He started to fidget with his shawl again, looked down with surprise at his busy hands, and cast the wrap aside. "What do you take me for, woman? I was only trying to help! Do you think - do you think I'm no different than some backstreet.... Good God, I don't know what to say!"

"I beg Captain's pardon, sir." Mary B curtsied and put out a hand of reassurance, then drew back when she realized what she was doing, "Truly, sir, I meant no offense. I'd be grateful for Captain's favor, truly I would. I just need to be sure. I'm in no place to be proud, sir, but I'll only do what's right."

Hands behind, head down, Tench paced a little circle, "I only want to help! How you came to your condition is no concern to me. I'll not judge that. And I'm not going to pry into why you're here. That's all said and done. But you must make the best of your situation, and if ...Look! I'm simply giving you a chance to make things a little better." He continued pacing, finally stepping up to look into her eyes. "But of course if...."

"Oh, no, sir! I'll be happy to do your service and do my best to please you, sir, and I thank the Captain for thinking of me."

Tench seemed relieved. He squeezed out a little smile, "Well, very good then. I think this arrangement will be of advantage to both of us. I never was one for sewing buttons, though Heaven knows I'm sure I could if I had to."

41

Mary B smiled. Seeing her smile, Tench's own smile widened. She laughed then, her eyes crinkling in a way few men had seen lately.

As he held his smile, Tench realized that despite her prison pallor, a warmth showed through Mary B's face that seemed to emanate from a basic good nature. His eyes moved over her face with new interest and he smiled even more broadly. Mary B suddenly dropped her gaze. An awkward silence followed. They'd almost broken an invisible barrier.

Tench spoke first. "Ah, yes, well, then, all settled, is it? I'm sure things will work out. I'll send a man down."

Mary B looked at Tench with a more relaxed expression and bobbed still another nod.

"Good enough, then" Tench said, "You may go back to whatever you were doing."

Mary B made a little curtsey. She looked directly into Tench's eyes when she said, "Thank you, Captain, I believe you are an honest man, sir!" and with that she turned and headed back to the pen, the open-mouthed stare of a startled Captain Tench following her progress.

While Mary B was speaking on the quarterdeck, James Martin and Will Bryant came up with their own mess groups for exercise. At first James wasn't certain it was Mary B he saw, but when he glimpsed her hair and recognized her laugh he suddenly felt jealous. How he missed the sound of her laugh and the delicious smoothness of her soft skin! He hunched against the chill wind, shivering.

"Isn't that your Mary B up there?" Will Bryant asked, his arms akimbo against the chill.

James watched Mary B's return. He sighed, "Ah, she's not really my woman, Will. That was something...well, it wouldn't be right now, would it, especially now...." With sudden vehemence he cried, "Jesus! Will, how was I to know my wife was going to have a baby? She never said anything! Why didn't she say anything? How was I to know?" He chewed on a knuckle. "Damn! I'm sure in the suds now! Mary B's having my baby and I don't know when I'll ever see my wife again...I just don't know!" Will Bryant studied James, his half-grin appreciating the wry humor of James Martin's dilemma.

"Back off, gents," the sentinel said, "Let the prisoner through." The convicts made a murmuring aisle for Mary B.

42

"Hello, Mary B," James said at her approach.

Mary B stopped, momentarily confused. "Oh, hello, James, I almost didn't know you." She put her hand to the side of her head to indicate his haircut. The male convicts on the *Charlotte* had received such severe haircuts their bony skulls were stubblefields.

James scratched his head, grinning. "Well, ain't I a handsome one for all that?"

The sentinel interrupted Mary B's polite laugh. "Move along now, prisoner!" Mary B smiled at James. She hesitated, wanting to ask about a rumor that his wife had a new baby.

"I said, 'Move along now!'"

Mary B gave James another little smile and had started towards the women's hatch when she felt a hand on her elbow halting her progress. She turned to find a pair of penetrating smoky eyes looking into her own. Momentarily startled, she didn't immediately recognize Will Bryant.

"Will you be my Mary B?" he asked. She pulled, but he held her just firmly enough to keep her back. "Please?" His smile was a smirk. His eyes mocked. She pulled with firm, steady pressure against his grip. Mary B glanced back at James. Shoulders hunched against the wind, hands stuffed in the waist of his trousers, James watched with an awkward half-smile. Mary B suddenly twisted free and hurried on. Near the hatch she looked back. Will Bryant had turned away and some of the convicts were smiling at something he must have said.

~~~~~~~~~~

The fleet's departure rumored to be imminent, Malvina Thornapple lumbered aboard in late April. But no sooner did she set foot on the *Charlotte* than she misstepped on a coiled line and blundered into Tench. Struggling for balance, the two went careening across the deck until Tench righted himself, steadied Malvina, and politely inquired after her well-being. Blushing deeply, Malvina gathered her skirts and curtsied so low she had to be helped upright by her husband.

Malvina had kept ashore as long as she could to lay in a goodly supply of hams, smoked sausage and dried fruits and Dutch cheeses and flour, potatoes, lard, salt, sugar, and various other condiments, thanks to the largess of relatives who also appreciated a good feed. She arrived on the *Charlotte* with boxes, chests, demi-barrels, hampers, children, and complaint, and as she eased her embarrassed bulk down the ladderway, still red-faced from the mishap on deck, she wheezed a familiar stream of faultfinding. "Heavens, why do they make stairs like this?" she puffed, "Do they think we're monkeys? Dobber, don't hold my arm like that! It doesn't do any good! Get below so if I fall you'll catch me. Heaven's Sake, I have to tell you everything!" She stopped to catch her breath partway down and looked around, a fat person stalling for time. "Heavenly Days, I can't see a thing! What's a body to do if I can't see? Dobber, are you there? Why don't you say something? Where's Ned? I never know where that boy is! I'll go to my grave chasing after him. Ned, where are you! And Nell, where have you gone to? Dobber, why can't you tend those kids? Heavens, I feel faint! I'll fall and kill myself sure! Dobber! don't just stand there like a lout! Get up here where you can do some good! Take my arm! Why do I have to think for you! Ned! you foolish boy, there you are! Now wait before you come tumbling down these stairs!"

Malvina finally made it safely to the bottom and waddled on Private Dobley Thornapple's arm to the special quarters for a married man, a blanket draped across a compartment. "Good Heavens, Dobber, this can't be it? What have you done? I'll die in here for air! Ned, fetch my other basket! If I don't eat something, I'll faint! Oh, Heavens! This is a chore! Ned, go find Nell. She shouldn't be running around. Dobber, who's that looking out that peephole! Why are those men leering? Nell! Ned! come get behind me, quick! Dobber, are those the convicts! Help! We'll be murdered in our sleep! Good Heavens! I'm going to faint! Oh, Dobber, how could you do this to me?"

~~~~~~~~~~

In the pre-dawn darkness of Sunday, May 13, 1787, James Martin, Will Bryant, and a half-dozen other convicts were rousted to man the capstans. Winding in the heavy anchor hawsers had become familiar work for the capstan crew; the *Charlotte* was moved several times to take on stores from this dock or that ship. For their labor the convicts received a cup of grog. Against the black mass of England's southern lands the lamps of Portsmouth were tiny yellow specks. A chill wind blew from the east. It was the hour of commitment for the First Fleet, the time of no turning back. After the previous day's departure came to a bumbling stop because of last-minute problems and miscommunication, everyone knew that when anchors next were weighed the fleet would be off to Botany Bay. The groaning capstans signified a government enterprise that, like a ship under sail, would not easily be brought to a halt.

But as government enterprises went, Botany Bay was small. A large enterprise had been England's protracted campaign against the American rebels across the Atlantic, an enterprise that employed hundreds of ships, thousands of men, and cost millions of pounds. Nonetheless, Botany Bay had a certain scale, and the previous three weeks had been filled with the bustle of last-minute preparations.

What occurred on the *Charlotte* was typical. To relieve overcrowding in its prison for men, twenty had been transferred to the *Scarborough*. Then for nearly two days convicts winched up the future colony's iron - iron rods and bars, blacksmith forges, anvils, hammers, bellows, tongs, chisels. They loaded casks of water, casks of salt meat, coals for the cook stoves and coppers, private stocks of food and spirits for the officers, crates filled with chickens, ducks, geese, pigs, bleating sheep and goats, great trusses of hay, Lieutenant Creswell's Jersey cow, bricks for ballast, and bales of old sailcloth from the Portsmouth Navy yard.

The bounty of England poured into the First Fleet. Axes, wire, shovels, nails, adzes, pick-axes, planes, hoes, rakes, boxes of beads, mirrors and hatchets for the savages, coopering tools, sawyering tools, more than 4,000 books, pamphlets and tracts from the Society for Promoting Christian Knowledge, great iron cooking pots, a pianoforte for the surgeon on the *Sirius*, twenty tons of candle tallow, barrels of tobacco, oats for the horses, and Arthur Phillip's four greyhounds. The disgorging of England's warehouses went on and on in measurements long disappeared - puncheons of flour, tierces of salt beef and salt pork, firkins of peppercorns and ginger, of lard, of butter, demi-johns of spirits, barrels of molasses, of dried peas, of rice and sugar and salt fish, cheese wheels and chests of apothecary supplies, quires of paper and powdered ink for reports and records. On board the ships came a great telescope for the observatory, bronze cannons, powder, and iron cannonballs for the defense of Botany Bay, tents for the Marines (complete with wooden stakes and ropes), a three-room canvas house for Arthur Phillip, bolts of cloth, bundles of extra clothing, and crates of cooking utensils. As a maritime nation often at war, England knew much about preparing for foreign adventures. The Naval Agent and clerks responsible for coordinating the expedition's

45

supplies had worked themselves to a frazzle preparing for this sliver-mooned morning of departure.

Nevertheless, as the convicts at the capstans strained at their bars to raise the two-ton anchors and the *Charlotte* dipped her bow and got slowly under way, loose ends dangled.

Item. Somehow the Marine commandant, Major Robert Ross, forgot his battalion's powder and ball. His Marines possessed no more ammunition than what they carried in their individual pouches, twenty-eight paper cartridges and musket balls per man.

Item. En route to Botany Bay, women on the *Charlotte* were to sew bales of old sailcloth into tents for temporary shelter in the new colony, while women on other ships were to sew shirts and dresses. But someone forgot to send needles and thread.

Item. Did Arthur Phillip know this agricultural enterprise was without a single plow?

Item. Did he know the Provost Marshall for the new colony had missed his ship?

Item. Did he know the records of the convicts remained in London? Lacking their records, Arthur Phillip would know nothing more about the convicts than what the convicts chose to tell. Who could have foreseen that this last oversight would link the lives and fates of so many aboard the *Charlotte*?

The convicts at the capstan wore new wool caps as protection against the raw wind, for a week earlier the surgeon had ordered the men's hair cut close a second time as a precaution against vermin. Their stubble-heads raised smirks and laughter among Marines and sailors alike; the convicts responded with sheepish grins. No one yet knew the role lice played in spreading jail fever, an often-fatal form of typhus so common in England's jails and hulks. Convicts sent down from the Woolwich hulk and Newgate Prison in London must have carried lice onto the *Alexander* in their hair and clothing, for at least ten had suffered from crushing headaches, fever, rash, and dysentery before dying painful deaths.

Most disease was thought to originate in bad air, and the standard countermeasure was ventilation and fumigation. Thus, the surgeon ordered the *Alexander* off-loaded, fumigated, disinfected, and whitewashed again. The jail fever disappeared. Science had triumphed! But it was a matter of *post hoc ergo proptor hoc*, for lice remained the unindicted culprit.

The men on the capstan also wore new clothes. Those embarked from Plymouth had been the poorest clad of all the convicts, and Arthur Phillip took it upon himself to issue new clothing at the Mother Banks. In addition to their new wool caps, each man received a nightcap, two coarse linen shirts, a blue kersey waistcoat and jacket, and duck cloth trousers. All received shoes, knit stockings, and slippers. The women were issued dresses and overlaying petticoats, but some were so poorly made and ill-fitting they immediately started to fall apart. Nevertheless, most convicts were in decent dress for the first time in months, and in some cases, years. And better fed.

At Arthur Phillip's insistence the prisoners began receiving fresh beef, more than two pounds per week, about two-thirds of a Marine's rations. The meat was the poorest cuts - bullock heads, necks, shanks, and tails - but eating so much meat was a new experience for most convicts. In the jails and hulks the daily fare was a pound of bread and occasional small beer. Sometimes the bread was turned into brewis, a bread-soup made by boiling the bread to the consistency of gruel. Any other food had come from a convict's own pockets or from charity. Arthur Phillip had also provided the convicts with what greens he could - onions, beets, turnips, potatoes. Accustomed to a simple fare of bread and little else, the new rations raised havoc with the convicts' digestive systems. And with relations between Arthur Phillip and Major Ross.

The conflict arose when Arthur Phillip sent word that he wanted Marines to oversee feeding of the convicts; he was concerned that the contractors hired to transport the convicts to Botany Bay would skimp on food to increase their profits. Thinner rations meant sicker convicts. After fighting with London bureaucrats to get better food for the convicts, Arthur Phillip wanted to insure his prisoners received it.

Major Ross bristled at Arthur Phillip's order. The job of his Marines was to guard the ships and keep the prisoners under control, not be nursemaids. Ross wrote a letter of complaint, tore it up, and hurried to London. Arthur Phillip listened patiently to the Major's complaint, then told Ross that sick and malnourished convicts would only make everyone's job harder. Ross was obstinate. At last an exasperated Arthur Phillip bristled. "While you're on my ships you'll do what I say!"

Normally pink-faced, Major Ross purpled. He bit his tongue and rode angrily back to Portsmouth. "We're not going to play nursemaid!" he told his assembled captains. "No Navy blueback is going to make us play nursemaid to a bunch of criminals! Let the lazy dogs handle their own rations. If they cheat each other, that's their worry." The captains looked at each other and said nothing. They knew better than to get caught in the middle.

James Martin and a dullard boy named Hartop were among those picked to begin weighing out food and distributing water on the *Charlotte*. "Keep an eye on them," Tench told Lieutenant Creswell. But as the convicts went about their daily chores, Creswell could find no fault. Like shopkeepers out to build a trade, they were conscientious, attentive, energetic, and good-natured, and glad of something to do. James put on a show by using Hartop (whom he called "Tartop" for his stubbled head of thick, black hair). "Why, look here, gents," James might say, "Meat again! I didn't know there was so much meat in the world! Meat, meat, meat! Did you ever eat so well, Tartop?"

His adolescent helper gloried in the older convict's attention. Most of the time he had no idea James that James sometimes made fun of him. Tartop stood at hand with a bucket or basket, drooling through a fixed, half-smile, happier than he'd ever been. Tartop came to love James.

Anchors raised and secured, the *Charlotte* joined the convoy struggling westward through a narrow channel called the Needles. Far ahead, stern lanterns rose and sank in the white-capped Channel. In a couple hours they'd see sunlight wink on the windows of south England. Topside, a strange calm pervaded. For a time no sentry ordered the capstan crew below. No mate chided his watch for being lax. Seamen went about their familiar tasks with quiet-spoken words. How different this leave-taking from Plymouth! No music, no well-wishers!

In the glimmer of dawn the convicts seemed at one with the sailors and Marines and those who'd come up to see the slow disappearance of dying watchfires on the shouldering coasts. The poignancy of the moment seemed to blur distinctions. All were joined in a common bond of leaving behind all they knew.

Huddled on the lee of the "women's caboose" (where the ship's cooking was done in fair weather), the capstan-crew convicts nursed their grog and half-listened to James Martin. "Reminds me of the first time I was ever on a ship," he said, "Dublin to Bristol when I was fourteen."

The hardest thing James ever did was leave his family's six acres in County Antrim. Endangered by rackrents, their land

48

was incapable of feeding yet another grown mouth. Thereafter, "Sunny Jim" had lived almost half the years of his life as an itinerant laborer, not easy when towns and parishes jealously watchdogged their local economies. James lived rootless. With a boyish, natural charm, he became adept at convincing prospective employers he could do any job, whether he'd ever done it before or not. He also became adept at rolling up his bundle, hoisting up his pants, and setting off on a new adventure - his solution to any number of awkward situations.

"For ten years I never wrote home," James recounted, "Really, what would I say?" He stared at the deck. After a silence he suddenly cried, "Goddamn it! How was I to know my wife was going to have a baby?" He took a gulp of grog and swished it around in his mouth. He took off his wool cap and scratched his stubble hair. Leg irons rattling, he moved to the rail and stared down at the slide of dark water. When he came back he said, "Well, Hell! Maybe I'm glad to be getting away from all that! Aren't you glad, Will?" He peered at Will Bryant and his other companions. "You bet!" James said, "We all are!"

"Yeah, we're getting away alright!" Cox said. Normally a beautiful, clean tenor, his voice was tense and bitter. A rat-faced ship's carpenter caught trying to smuggle fine lace and a few pair of silk stockings, Cox had been a convict for five years and was serving a life sentence. Life sentences were often remitted for good behavior after eight years, but Cox was one of the convicts who'd taken over the *Mercury* and escaped; he'd not see remission of his sentence for many years. "Getting away from what? We'll be a year chained up in this pig-shit tub!"

"Easy, Cox," Will Bryant said.

"Well, we will! Listen," he whispered urgently, "We're only a half mile from land. They won't stop if just a couple of us jump. Come on, Will, you with me?" Will didn't move. "Anybody?" No one moved. "Ah, shit," Cox said, "You guys got no balls!"

"Just go, Cox." Will's voice had a sinister edge. "What're you waiting for? Here, I'll hold your cup." Will held out a hand. Cox coughed and wiped his eyes with a sleeve. He coughed again, then laughed a nervous little laugh.

"Well?" Bryant said.

Cox coughed again and buried his nose in his cup. A Marine approached to shoo them down to their prison.

~~~~~~~~~

49

From their vantage on the afterdeck, Tench and the chief surgeon of the Botany Bay project watched the capstan crew file down the ladderway. "So you think we're really ready?" John White asked. He was remembering a balmy evening when he and Tench shared a bottle of port. Since becoming acquainted in Plymouth the two had become friends, and their discussions about the Botany Bay project had filled many an otherwise lonely hour. On that particular evening White speculated that the Botany Bay project would not prove disastrous as some predicted. After all, the dying on the *Alexander* had stopped, and with a better diet convict health was improving. Tench agree. Indeed, some convicts seemed more than ready to be off. As the *Charlotte*'s mail censor Tench had a fairly good idea what convicts were thinking, and although some were whiny and begging, most were simply tired of having nothing to do. Few received visitors, and while a handful were chosen for work, the rest did nothing. They sat below-decks in irons all day in semi-darkness, idle except for the brief times they were allowed up to air their bedding and rattle around in a shuffling circle in the convict pen.

Perhaps Tench was right. For most convicts, life at the Mother Banks was barren of meaning. Some had been on ship for four months. Yesterday they'd waited for the day to end because it would bring today, and today they'd wait because it brought tomorrow, and tomorrow they'd wait because it brought the next day, and the next, and the next. At least the prospect of change promised relief from their boredom.

Rumors of change had thrived for weeks. "Did you hear, the expense of Botany Bay has got so great it's all being called off?" "Did you hear, we'll all be pardoned and set free?" "Did you hear, we're not going to Botany Bay, we're going to Halifax?" "Did you hear, Arthur Phillip is down from London and we'll sail any day?" "Did you hear, the women are getting off because they don't have clothes?" "Did you hear, we're going to wait and join a second fleet?" "Did you hear...?" "Did you hear...?"

But to those who observed the greater world, only one rumor seemed believable. Departure was imminent. And when Arthur Phillip came down from London in early May and his hurried inspection of the six convict transports found everything in order, realists knew they'd soon sail.

"Well, I don't know about the convicts," Tench answered with a grin, "But I'm ready."

Santa Cruz

1787

*T*hus, in that luciferous dawn, weeks behind schedule, six convict transports, three supply ships, and two ships of the Royal Navy comprising the First Fleet set out for Botany Bay. And Tench was right. Most were ready. The endless days of idleness and rumor had grown wearisome.

But what a beginning! Unlike the convicts from Plymouth who'd sailed to Portsmouth on the *Charlotte* and *Friendship*, most others arrived at the Portsmouth docks in prison wagons. When the ships passed the Needles lighthouse and breasted the first swells of the English Channel, these men and women experienced their first big-water sailing. As the transports pitched and rolled in the billows, the convicts huddled in claustrophobic semi-darkness, moaning and groaning and vomiting and turning their prisons into foul-smelling sickbays.

The fleet was barely in the Channel when Arthur Phillip sent a temporary Navy escort back to tow the *Charlotte*, which fell ever farther behind. The *Charlotte* appeared to be a bad sailor. But every ship setting forth needed cargo shifted and sails adjusted until the vessel was "trim" and sailed efficiently. While under tow, convicts labored to shift the *Charlotte*'s cargo of iron, but to Arthur Phillip's consternation this ship and one or two others would lag for the next eight months.

Such laggards brought trouble. At sea the only communication was by line-of-sight. If ships were close enough, messages could be shouted back and forth, but mostly a fleet used signal flags. Fluttering from the aftermast of the 600-ton *Sirius*, Arthur Phillip's flagship, these signals provided visual instructions to his fleet. Some were general: "Draw closer," "Spread out," "Sail in a line," "Anchor here." Special flags also enabled Arthur Phillip to signal a specific craft.

Signal flags worked fine when seas were friendly and the light was good. But on a dark night of ocean swells and spumy waves the fleet spread out to avoid collisions, while lanterns blazed in the topmasts to help keep everyone together.

Storms could be big trouble. Scattered by a storm, a fleet might not regroup until everyone straggled into a rendezvous port, sometimes weeks later. But what if a ship never showed up? Its fate might remain conjecture forever. Given these uncertainties, a convict transport might find itself sailing all alone, vulnerable to seizure by its cargo of convicts. That's why Marines sailed as guards. Who were these Marines - this half-strength battalion of four companies, just under 200 men?

When the call went out for volunteers for the Botany Bay project, Porter, Stork, and Quint had just completed their first two months of a three-year enlistment. Men often enlisted as a way of escaping something worse - jail, debt, too many children, a nagging wife. Porter joined because a magistrate gave him a timeworn choice of the Army or jail. Quint and young Stork joined for the three-guinea bonus, a substantial sum. Other recruits were simply gullible. Receiving a bounty for each enlistee, recruiting officers were notorious for devious ways.

But even recruiting officers and bonuses failed to keep up with desertions. The life of a private soldier (for so they were called) was difficult at best, and privates often deserted because they could not afford to remain soldiers. From their nominal pay of six pence a day, a corrupt system deducted widows' funds, hospital funds, paymaster funds, regimental funds, and other ambiguous funds that sucked up a goodly portion of a month's pay, which was usually paid late. Porter, Stork, and Quint were lucky to net a pound or two a year. From this remainder they were expected to buy their shirts and stockings and gloves, their shoe blacking and pipeclay and whiting, their hair powder and powderbag and puff. They were also forced to supplement their food, because the same corruption that shaved their pay also shaved their rations.

After assignment to Captain Tench's company, Porter, Stork and Quint marched and wheeled and wheeled and marched to practice maneuvers that would bring them into line of battle as a close-knit unit of a dozen muskets. They drilled for four to six hours a day with their 14-pound weapons, a shorter Marine version of a standard-issue weapon called the "Brown Bess," corrupted from 'brawen buss," an ancient blunderbuss. Under the supervision of a sergeant and corporal, a squad practiced loading, priming, aiming, and firing until their dozen weapons delivered successive volleys of three-quarter-inch balls as one - no easy feat. Loading and firing was a nine-step process that took a novice two or three minutes to complete. By the time he sailed for Botany Bay, Quint could load and fire twice a minute. Marines also learned the cutlass for fighting aboard ship and endured endless bayonet practice - parry, thrust, parry, thrust. Traditional infantry warfare still saw disciplined lines of troops advance to the beat of drums on a battlefield raked by cannon fire, then charging in a do-or-die assault with bayonets.

So long as they trained or served ashore, Marines were considered ordinary infantry, but once assigned to a ship they fell under Navy regulations. Seasoned Marines considered shipboard life better than life ashore. Although their quarters were crowded, their duties were light, and they saw the sea. Their rations had remained unchanged for 100 years - a salt-heavy weekly allotment of 7 pounds of full-grained bread, 4 pounds of salt beef, 2 pounds of salt pork, plus peas, oatmeal, butter, vinegar and cheese. Plus a daily pint of watered rum called grog. Items to prevent scurvy, usually sauerkraut and lemon juice, were not yet parts of official rations.

Like Sergeant Motherwell, Quint might have found a home in the Marines. A bright, engaging fellow, he might have been promoted. As a sergeant he would have netted six or seven pounds a year, enough to keep a family. But Quint was only twenty, and still thrilling at the excitement of life's discoveries. And what was a more exciting discovery than love?

Since that first night when her canted, skeletal face and fear-haunted eyes beseeched him from the women's prison, Seedy had fascinated Quint. "Don't let me drown! Please don't let me drown!" she pleaded. When Quint knelt to put a finger through the grate, she seized it with such desperate strength he winced, but he smiled and tried to crook his finger in response. "There, there now, love, have no fear, Quint's here. He'll take care of you." While she continued holding his finger a willing captive, Quint reassured her, "That's all right, love, that's all right." As the motion of the *Charlotte* grew more regular, Seedy's panic slowly subsided, but the two young people continued touching through the grate until footsteps announced Sergeant Motherwell. Thus was Quint smitten.

He did everything he could to be near Seedy. He delivered dirty laundry, he drew and emptied wash water, he invented lame excuses. Not only Seedy's ready smile fascinated Quint. Her canted face intrigued him, a face that curved from top to bottom like a quarter moon, so the right corner of her mouth and her right eye were noticeably closer than their opposites.

Her complexion possessed a translucent quality and seemed to glow. He thought her wide mouth and strong teeth (with good gaps between) were lovely, along with her delicate ears and white neck that throbbed with her pulse. Her disjointed body fascinated him, for when she stood or turned her arms and legs and hands and feet seemed to flap.

But most of all Quint liked Seedy's eyes. They were deep-set, and even in light of day they moved in shadow. They were dark brown, simian, soft and moist and vulnerable, with a hint of helplessness that Quint sensed rather than saw. Oh, when Quint first looked straight into the depth of Seedy's eyes he felt such a surge of compassion and such desire to enfold and protect that his heart swelled until he thought he might explode. Smitten, indeed!

"There, there now, General, don't be so hard!" Seedy once said when he puffed out his cheeks and chest in frustration at her imprisonment, at her awful inaccessibility. She placed a gentle hand on his chest and blew softly on his face, "Look now, General, you're all hot! You can just be my Quint, can't you? My big, strong Quint?" He melted, smiled back, and with a quick look around, reached out to squeeze her hip.

"Aye," he said, "Quint's here, have no fear!" and they both laughed quietly. They knew they would become lovers. Later, when Quint was being flogged in Rio, he bore each whistling slash of the cat with a martyr's resignation, Seedy's canted, concerned face a spiritual haven for his soul.

~~~~~~~~~~

A Marine in Quint's squad provided the first real laugh on the *Charlotte*. Just two days out of the Mother Banks, Corporal Baker dropped a musket he was taking out of an arms chest, and the loaded weapon discharged with a belch of black smoke. Its errant ball shattered Baker's left foot, smashed through a crate of innocent geese (killing two) and traveled on to puncture a waterbutt, which released a vigorous spout onto Lieutenant Creswell's sleeping cat. With the musket's BANG! Baker's OW! and the cat's MEOW! the fowls and livestock set up a wild barnyard racket of cackling and honking and mooing that lasted a full minute. As word later spread of how Corporal Baker had shot himself and two geese with one ball, convicts and keepers alike shook with laughter.

After two weeks of sailing south the days grew warmer. Somewhere over the horizon the coasts of France, Portugal, and

Spain slipped by. With fresh air and their shackles removed and seasickness for most a thing of the past, convict spirits rose. Conversation grew more animated, laughter more frequent. Whistling and song were common. The stubble-heads clapped and stomped and swung their partners to the twangy music of Jews harps and crude stringed instruments.

More sedentary, the women sang. At first they sang only with those they knew. Mary B knew her group from the Exeter jail. Others had come to the *Dunkirk* from the jails of Taunton, Southhampton, and New Sarum. As the days passed, the women began finding common ground in their songs and stories. Soon they were talking of towns and markets and festivals, of families and growing up, and finally, after much hesitation, how they came to be aboard the *Charlotte*.

Of course they embellished and exaggerated. Of course they omitted and lied. Not one deserved her punishment. Their crimes were someone else's fault, a terribly misunderstood series of events, a miscarriage of justice. Mary B was speechless when Katie told how Agnes Lakeman had framed her. In Mary B's mind, Katie's guilt was clear. Katie had sometimes bragged of her petty thefts while working for Jeremiah Lakeman. Deep down Mary B thought that in the way of things Katie's punishment simply caught up with her.

But what of Mary B? What had she done? What was catching up with her? When dispirited she sometimes felt she deserved to be punished. And why not? What had she ever done for the good of the world? Nothing! Living with scarcely a thought about anything serious, she'd laughed and gossiped and trusted that tomorrow would bring as good as today. And that was the end of it. Now she realized she should have taken life more in hand. If only she'd.... But despite these doubts, she usually came back to the same place. Her punishment was undeserved. It was wrong for her to be sent to Botany Bay.

Whether they told the truth about their pasts or not, the women grew closer, except for a handful from the Bristol jail - foul-mouthed, mean-spirited, mocking women who bullied the meek and taunted the rest. Whenever possible the Bristol gang was given wide berth.

Strangely enough, for some women on the *Charlotte* life had never been so good. For the first time in their lives they had adequate food and new clothes and a warm place to sleep and lived as safe as anyone can on the high seas. And with almost ten men for every woman, they felt special. Those like Seedy who'd sometimes sold their sexual favors began to realize they didn't have to give up their bodies simply because it was expected. Of course a few savored the excitement and power of being a valued sexual object among so many men. Sailors and Marines vied for their attention. Picking out their favorites to keep an eye on for their futures in Botany Bay, some women spent hours discussing the relative merits of their choice - how this one was tall, that one bold, how this one had a nice voice, a pleasant smile, how that one's sly touch sent shivers down her spine. Of course they never disparaged the apple of another's eye. Although John might have rotten teeth, bad breath, a pock-ravaged face, and only two fingers on his right hand, if he was someone's favorite the women were sure to agree John was a fine prospect indeed.

At times the women toyed with the men. On deck for their daily allotment of sunshine and fresh air, they sat with their sewing and teased the Marine guards by singing:

How can I keep my maidenhead
My maidenhead, my maidenhead
How can I keep my maidenhead
Among so many men? Oh!

A private bid a shilling for't
A shilling for't, a shilling for't
A private bid a shilling for't
The captain he bid ten! Oh!

But I'll do as my sister did
My sister did, my sister did
But I'll do as my sister did
An' soldiers I'll have none! Oh!

I'll give it to a handsome lag
A handsome lag, a handsome lag
I'll give it to a handsome lag
For just as good again! Oh!

~~~~~~~~~~

Not everyone thrived on the sunshine, fresh air, and salt meat. A sallow young sheep thief named Ishmael Coleman had kept to himself at the Mother Banks and given away most of his food. At sea he stopped eating altogether. Soon dead, he was sewn into his blanket, weighted with a ballast stone, and his wasted body sent with an insignificant splash into his watery grave. End of sentence.

Others ailed. Liz Bason's little Jacob continued barely to hold his own. Poor Malvina Thornapple! She suffered from the moment she lumbered aboard. A snack of little sausages, salty cheese, and bread on that first day was her last nourishment. Vomiting her snack, she'd suffered since. The very thought of food was enough to make her gag. Worse, the smell of her daughter Nell would set her stomach and bowels churning. She was unable even to clutch her favorite child for comfort! Malvina lay alone in her blanket-draped compartment, groaning and whimpering, too weak to complain. Her stomach and bowels empty, her body began drawing its sustenance from her rolls of jolly fat. Day by day she wasted away.

But while Malvina waned, her friend Hester Thistlethwaite waxed. Also a Marine wife, Hester was reborn on the salty sea. She discovered the deep gave her appetites she never knew she had. Tall and angular, skinny and childless, a woman with a sharp face who usually dabbled at her plate, Hester took over management of Malvina's children and used the opportunity to eat often and with gusto. She'd provide Ned and Nell with a nibble of cheese or biscuit from Malvina's abundant stores, and take the opportunity to carve herself a slab of ham or hunk of sausage or to stuff a potato or two in her pocket. Mealtimes saw her guiding her charges with guttural noises and hand signals, too intent on keeping her own mouth full and her jaws working to bother with speech. She seemed never to get enough to eat.

Neither could she get enough of the foredeck. She loved to stand in the foremost bend, braced against the rise and plunge of the ship, her sharp face pointed like a bird-dog's, her nose a-quiver with excitement and anticipation, retreating from her vantage point only when sailors and Marines came to empty their bowels in the bow chains. She discovered excitement in the motion of the *Charlotte* ever plowing through the water and the sea ever spreading in ceaseless rise and fall. The air and motion and sea-spray seemed to release atavistic juices and urges. One afternoon, unable to restrain herself a moment longer, she sought out her puzzled husband for an urgent coupling in their compartment, an occasion followed by others so frequent that Willard Thistlethwaite became the envy of his mates.

~~~~~~~~~

Mary B needed to talk with James. She needed to straighten things out with him and find out for once and for all whether the rumor was true.

She felt better now. With the bulging of her belly and the regular kicking of life, her lethargy and depression vanished. Her swollen ankles and hands returned to normal and she felt like taking charge. Although frightened by the unknown forces controlling her body, the thought of carrying a new life excited her. Good Lord, a new human being! Now she loved to lie abed with her hands resting lightly on her womb, conjuring up the little person within. She'd feel serene, and take on a faraway look, as if she could see something invisible to others. Would he be a boy? Would he have her auburn hair or be blond like James? Would his eyes be gray like hers? Or blue? She hoped her baby would have James's intense blue eyes.

She had to find a way to talk with James.

~~~~~~~~~

Late on a hazy afternoon, twenty-two days out of the Mother Banks, the First Fleet anchored off the little town of Santa Cruz on Tenerife, one of the Canary Islands, the Fortunate Isles of ancient times. Arthur Phillip felt fortunate. He'd brought his eleven ships almost 2,000 miles without mishap.

Ocean travel still left much to luck. Not only did ships depend on wind to blow them where they wanted to go, navigation was still an arcane science. The biggest problem was longitude. For generations mariners had determined how far north or south they were by noting the position of the sun, or the moon and stars. But how far east or west on this spinning globe was something else! For decades the British government offered riches to anyone who could figure a way to determine longitude out of sight of land. And for decades the riches went begging. But by the middle of the 18th century the development of fairly accurate clocks had advanced the art of determining longitude. The trick was Greenwich, a naval observatory near London. If a sailor knew what time the sun rose at Greenwich and kept his clock on Greenwich time, he could determine where he was at sea by comparing the time of his sunrise with the time of the Greenwich sunrise. If his sunrise came twelve hours later than sunrise in Greenwich, his ship would be exactly halfway around the globe from Greenwich.

But the best clocks still ran fast or slow, and a skilled navigator using good equipment could at best determine latitude and longitude within a few miles. And there were other problems. Charts drawn even a few years earlier using cruder measurements might show an island or reef miles from its true location. With the hazards of such uncertainties, sailing at night in the vicinity of land or shoal water was asking for trouble. When ships approached unknown hazards at nightfall, prudent sailors shortened sail and lay back until daylight. And Arthur Phillip was prudent.

Not only had his ships arrived without mishap, but the hundreds aboard had fared well also. In fact, the voyage saw the health of most convicts steadily improve. But not all. Six more convicts had joined Ishmael Coleman in watery graves, including five from the *Alexander*, a ship that continued to suffer from careless hygiene.

At Santa Cruz the convicts received fresh beef, onions, potatoes, even beer! With the holiday fare and abundant water once again, a festive mood permeated the transports, and laughter and song wafting across the water from the transports oftimes puzzled the Spanish sentries on the pier. Convicts? Was this how England treats her convicts?

The day the *Charlotte* was brought near the pier to refill her waterbutts, Mary B had a chance to talk with James. With the women taking advantage of abundant fresh water to do laundry, James and other convicts carried buckets to serve the Marine wives first, then the convict women. The Marine wives were clustered in a chattery group on the foredeck, where Malvina Thornapple sat shakily among them. She'd risen from her sickbed soon after the anchors splashed down, some twenty pounds lighter. Trying a little porridge, she'd immediately vomited, and was afraid to try any other nourishment. In short, her disposition was not improved. "Lord, they smell!" she said when the convicts delivered the first buckets of water. She gazed longingly at the brown hills of Tenerife. Word was that a community of English lived in the little town. If it weren't for being bound to Dobber on this hateful ship...!

"Be quick," Quint growled as James approached Mary B, a wooden bucket sloshing in each hand. Seedy had begged Quint:

would he please let Mary B and James talk? Quint turned to study the Marine wives on the foredeck.

"Pour it in this big tub, James," Mary B said.

As he slowly poured the first bucket, James smiled, "You're looking good, Mary B, you've got a glow."

Sorting clothes, Mary B returned his smile and rubbed her belly. "I can feel the baby."

"Oh, that's good!" James continued to pour slowly, his eyes on the water.

"I can feel it kicking, it must be a boy."

"Oh, that would be nice."

"If it's a boy, maybe he'll look like you."

"Well, that would be something."

"Would you want me to call him 'James'?"

James finished pouring his bucket and shook a few last few drops into the tub. "Well, I've never thought about that," he said, studying the inside of the bucket.

Mary B's face clouded. "I could, James. After all, he's your baby. That would be fine with me."

James remained silent, studying his empty bucket.

"You know he's your baby, don't you, James? There wasn't anybody else."

"Well, I guess I don't," he said finally, looking at her directly, "I don't really know if there was anybody else."

"Well, there wasn't, James! How can you say that? You were the only one, and you said...."

"I know what I said! But it's different now." He said this last quietly.

"Different? What's different now?"

"Well, Hell, we aren't together, for one thing! We can't even see each other! It isn't like we're together. We're not! And besides, what have I got? Nothing!"

"But I'm not asking you for anything!"

"Yes, you are! I can tell. You're really asking if we can get married. Well, I can't. I've already got a wife, and you know it."

"And I'm telling you I've got your baby! Our baby! Our baby is alive in me!"

60

"Well?"

"Well, it's your baby!"

His protracted inspection of the bucket from every possible angle completed, James picked up the other bucket and began to pour slowly. "Well, I appreciate it's my baby, but I'm no good for you. I've got a wife already, and you knew that. Besides, I've got nothing - and I'm not regular, especially with a woman. You know that, and I never said I was."

Mary B stiffened. "I'm not asking you to marry me, James, I never asked that."

James stopped pouring, leveled the bucket, and looked at her, "Well, what do you want?"

Mary B turned to face the sea. Quint tilted his head, not wanting to miss a word of their spat. James began pouring again. When Mary B spoke it was to the sea. "Well, why shouldn't I want to get married? I can get by now - I've got friends, and we help each other." She turned. "But after we get to Botany Bay it's going to be different. I'm afraid, James! What am I going to do after we get to Botany Bay? At least if we got married we could be together there!"

James dumped the rest of the water into the tub. "Jesus! I can't get married again. I've already got a wife. I'd go to Hell." He took a deep breath, "Besides, do you know how far we still have to go? Christ! We could all get drowned! And you talk about getting married!"

Mary B glared. "I didn't say we had to! I said, maybe we could! You left your wife! And this is your baby! And we'll probably never get back to England. It's the end of the world, James. So who would know?"

"Mary B!" James said her name as if she'd uttered a blasphemy in church.

"Damn you, James, you courted me! You courted me with your smiles and your jokes and your laughs! You...you...."

"Well, yes, but.... Well, yes, I admit I liked you the first time I saw you. But that doesn't mean anything now. It's all different now. Besides, you said you didn't want to go to Botany Bay and if you had a baby you wouldn't go. Well, you've got a baby and you're still going. Just like the rest of us!"

"Damn you, James, I didn't ask you for a baby! I didn't ask you for a damned thing, and I'm not damn asking you to damn marry me!" And with that she swooped to upend the tub and send a cascade of water over James's shackled ankles. Quint did a little dance to get out of the way.

"Jesus, Mary B!"

"Get! Me! Some! Water!"

Quint took in the scene. He rolled his eyes and slowly shook his head. "Get her some water," he said resignedly.

In a few minutes James returned with two more sloshing buckets. "Well, do you want these in the tub, or should I just pour them over my head and save you the trouble?"

Mary B looked at him steadily for a long moment, then her face relaxed and she gave him a small smile of apology. "Oh, James."

He grinned sheepishly. "I'm real happy for the baby, Mary B. I mean, I'm proud you have a baby in you that's mine. I'll help you all I can and I swear I'll help to see after your little one, too. I'll help to see after you both, Mary B, but...."

"I just wanted you to know, James. I just wanted you to know that I've got your baby in me. Maybe he'll look like you. Do you want to call him 'James?'"

James shook his head. "Nah," he said, scratching his jaw, "When he gets bigger he might ask about me, and what would you know to tell him? Nothing. You only know me for a convict. You don't really know anything about me. I had a life before, you know. And I don't know anything about you. All we really know is we were in jail together and on the hulk together and now we're on this boat together, sneaking talk over a laundry tub, and you still don't know anything about me." James looked pensively at the horizon. "We can't sit and talk over a table, or in bed, or on a bench outside after supper like my folks used to do. If it was just you and me like that...."

After an awkward silence, Mary B said softly, "James?" James didn't respond. "James, what is it I don't know about you? You said I don't know anything about you. I knew you were married. I knew you had to leave your land. What else don't I know?"

Water buckets forgotten, James stood silent, his head down, his shoulders slumping as if his hands bore burdens of immense weight.

"James? What?"

Slowly, James began to shake his head, his gaze fixed on some spot in the bottom of the laundry tub.

"James?"

At last, as if he could scarcely lift the weight of his head, he raised his eyes to look at Mary B with an expression of profound sadness. "I'm Catholic," he croaked.

"I don't care!"

"And I've got a wife."

"I don't care, that was back there and those were just words! Popish words! And you left her!"

"And she's got my baby, a baby boy, and he's baptized - he's James."

Mary B looked as if someone had punched her in the stomach. So it was true!

"Quint!" Sergeant Motherwell roared from the afterdeck. "What the hell's going on down there!"

"Whaa?" Quint jerked around. "You there, Martin! Shut your mouth and get to work! How long you going to take with that damned water?"

~~~~~~~~~~

Tench stumbled aboard the *Charlotte* with his coat flung carelessly over his right shoulder. His buckler loosened, his hat and short wig a bit askew, he waved away a question from the pimply junior officer and threw an off-hand salute to the sentry. His saluting hand held a small bouquet of red carnations and a bloodstained handkerchief, which he returned to his left upper arm. A string bag containing a loaf of fresh-baked bread and a bottle of sack dangled from the hand of his wounded arm. He lurched rather tiredly aft, humming one of the little airs the guitar and violin ensemble had played through the long dinner at the Governor's Palace. He raised his bloody handkerchief long enough to sniff the bouquet presented by the beautiful Señorita Esperanza de la Marchicorde, an impulsive gift as she was whisked away after the dessert ices.

Spying light through the louvers of John White's door, Tench knocked, then entered clumsily at White's, "Yes?"

His voice hoarse, dreamy, Tench said, "John, I'm in love."

Wrapped in a blue robe atop his narrow bed, White closed his book over a finger and smiled, "I say, it does look like Cupid got you with his arrow. Is your love coming with us to Botany Bay?"

Tench dropped the string bag on White's tiny table and eased himself into the little straight-backed chair alongside. He regarded White with bleary eyes. "Ah, John, perhaps I'll resign my commission and offer my invaluable services as an English officer and gentlemen to the Spanish Guard. And if I can't sell my services..." - he paused to sniff the flowers again, then exhaled with a blissful sigh - "...I'd be poor, and live on the perfume of my lovely Esperanza's bosom." He thought a moment. "I'd spend my only fortune, John, my poor little life, putting into words the meaning of Esperanza's sweet, coy smiles. And with that brief fortune too soon spent, just to see her limpid Spanish eyes regard me o'er the brim of her small fan once more - aye, but once! - on my deathbed I do swear I'd sell my very soul."

White laughed aloud, "Bravo, Wat, bravo! You're well struck for sure! But I suspect the wound is not so deep as you may feel. What happened?"

Tench growled and surveyed White's confining cell. "Anything to drink?" White motioned to an obvious bottle on the table. Tench set his bloody handkerchief and the bouquet aside and poured himself a generous half-tumbler of port. "You know, John," he said, "It may have been just the time at sea, but I was beginning to see the finer points among some of the women on board. But John, Señorita Esperanza is *non pareil*."

White chuckled. "What else, besides your falling in love?"

Tench groaned again, this time more genuinely. He took a mouthful of port, picked up his bloody handkerchief, and held it against his seeping wound. "Oh, it was hard duty, John, hard duty. To be smothered by the perfume of a beautiful young woman, to see her breasts nested like sleeping little white doves...oh, it was hard duty, John, hard duty!"

White chuckled again. "All right, Wat, tell me! What happened? Are we at war over the women?"

Tench drank another good draught and stared at his outstretched boots. "You were lucky not to be there, John. We drank to royalty and heroes and everlasting friendship till we were lucky we found our chairs again!"

White chuckled politely.

Tench sighed and poured himself more port. "Really, John, it was ridiculous. But the best of the night was when our idiot Major Ross asked if there were still many canaries in the Islands!"

White looked non-plused. When Tench saw his expression he explained, "John, the Canary Islands aren't called the Canary Islands because of canaries. 'Canary' is a corruption of the Latin, meaning dogs. The Romans found the place over-run with wild dogs."

White roused himself and put aside his book. "Really! Now that's interesting! Wild dogs! *Canaria*, of course!" He slid his legs over the side of his bed and gently began to unbutton Tench's shirt. He unbuttoned enough so he could pull the shirt from Tench's shoulder, then raised his lamp to examine the wound. "Wat," he said, gently testing the edges of the wound with his fingers, "Don't you imagine most people think these islands are named after canaries?" He moved the lamp closer on the table and began taking things from his medical box.

"Yes...yes, I suppose they do," Tench said.

"Well, then, if most people think so, doesn't that make it rather true? I mean, isn't it reasonable that these tropic islands would be named after canaries? Who'd ever think of dogs?"

"And then, of course," Tench continued, "as time goes on, fewer and fewer people will ever know the islands were actually named for wild dogs, and pretty soon what sounds true becomes true, even though it isn't. And so books are written saying the Canary Islands were named after a large population of canaries that were hunted to extinction or perished in some series of natural disasters or...."

Laughing, White raised a hand, "Hold, sir, you take me too far!"

The two laughed together, realizing they were bantering as they had so many times since making acquaintance in Plymouth just three months earlier. Both avid readers, they shared books and discussed them with great energy, and were a regular sight strolling the afterdeck in deep discussion of a hundred subjects.

Watching White's progress as he swabbed the wound, Tench inhaled sharply through his teeth. "What would you do without your oil of tar, use a hot iron?"

White chuckled. "It's great stuff, Wat. We ought to use it more, but I suppose heroic Marines prefer hot irons." He finished swabbing and retrieved a bandage. "So what did you talk about?"

"Triangulation."

"What?"

"We talked about mapping and surveying, which is all triangulation." White again looked non-plused. "Surveying is all triangles, John. You see, if you know two sides, you can figure out the third, like we learned in school. Triangulation is important to a lot of things like geometry and map work. It's like your oil of tar."

"Ah, well, I never did understand that stuff. Not my angle, I suppose."

Tench snorted. "I thought your studies would have encompassed more."

White snorted in turn. "Not as a rule." He finished tying Tench's bandage, lifted his shirt gently into place, and slid back on his bed.

Tench picked up the bouquet and sniffed it again. "You know, I must be getting old, John. Here I am, in love again with an impossible object. I'm always doing that, always finding someone who's just been spoken for, or another man's wife, or someone who's going away, or finding someone just when I'm going away." He sighed. "There's something missing in my life, John. I think it's real connections. And sometimes I get really tired of it." He rose with some effort, his mood somber. "And tonight I'd find much more comfort between the covers of Señorita Esperanza's bed than between the covers of any book."

White laughed gently. "Oh, Wat! I haven't known you long, but I'd venture that if I took a book, any book at all, and said, 'Tench, this is the last copy of this book in the world, and you must choose between your lovely Señorita Esperanza for the night, and the preservation of this book for posterity...' - well, Wat, there's absolutely no doubt in my mind that you'd choose... the book."

Tench smiled ruefully and inhaled the bouquet again, the memory of Esperanza's perfume mingling with the scent of the flowers. He drained his glass, groaned, and put the glass upside down over the neck of the bottle. "Ah, John, her hair smelled like springtime! And when she said goodnight and looked at me over her fan I thought my heart would melt right into my boots. I felt so delicious in my loins I wondered if I'd started to drool."

White laughed gently again. They both were silent, immersed in private thoughts. Finally White stirred. "All right now, Wat, for my log, what happened?"

Tench rolled his eyes. "No medal here," he said, wincing as he shrugged his wounded shoulder. He gathered his things and stood, momentarily weaving, "One of the sentries on the pier was helping us into the longboat and accidentally poked me with his bayonet." He shook his head. "The officer of the guard was so upset he booted the poor fellow right into the water. That upset our leader, of course. He wants no incidents. But the officer was so embarrassed.... Anyway," he continued, fumbling behind for the latch, "Getting back to important things, I believe you're right. It's a sad choice, and I wish you hadn't burdened me with it, but you're right. I'll sail away and forever leave the lovely Señorita Esperanza behind. Books must be my comfort." As he said this he wore a mock doleful expression, which changed to a grin when he said, "I hate it!" He paused. "I hate it, and I love it!" He laughed. "But how did this matter become a choice between just two things? I want more! I want more choices! I want books and Esperanzas and more!"

"And you shall have them." White smiled back and yawned again. "When you come back from Botany Bay you'll have books and Esperanzas and more, for England's honors will be yours. You brave Marines will be the new heroes of the realm!" Tench returned his yawn and his own little smile. "I suppose," he said, "And what about you? Won't you receive honors and preferments too, gold and precious jewels?"

White chuckled as he shook his head. "I'm afraid not, Wat. I don't wear a pretty red uniform."

Tench stuck out his tongue, turned, and began to close the door behind him. But he backed in again and nodded towards the string bag with bread and wine he'd left on the table. "Oh, that's for you. A gift from the Governor of these fortunate isles. *Para Juan White, el medico.*"

White looked puzzled.

"What can I say?" Tench asked with a shrug, easing out the door once more, "They seemed to know a lot about us." But a moment later his head and arm reappeared, and he tossed a flower from Señorita Esperanza's bouquet to White's bed, where it landed between his legs. "*De la mano de mi amor*," he said, closing the door behind him once more.

White picked up the flower, the stem warm from Tench's fingers. He sniffed the blossom tentatively. "*De la main de mon ami*," he said softly, and placed it between the pages of his book as a place marker. He closed the book and pressed it tight, his eyes focused inward. He blew out his lamp.

~~~~~~~~~~

Governor Miguel de la Grue Talamanca Branciforte loved coffee. He stood at the window in a long, heavily brocaded dressing robe, enjoying his second cup made from a rich, dark, finely ground powder, then laced with fresh milk and sugar. He inhaled the heavy, sweet aroma as he watched the English ships make their slow way through the headlands to the open sea. To the end of the world, he mused. To the end of the world with a cargo of convicts. What were the English up to? Was there gold there? Perhaps a primitive civilization like the Incas, rich with treasure? He shook his head. There had to be something. Control of the mysterious Pacific? A challenge to his own country for the Philippines? The Dutch for the Indies? The French or the Russians, even the Americans? And why convicts? As slaves? Why not just enslave the natives?

After all the fuss of entertaining the English, just what had he learned? They'd arrived as forewarned, ships loaded with convicts. But did they really think they could build a self-sufficient colony in two years with criminals! Arthur Phillip was a foolish optimist!

But not self-sufficient in wine. His Excellency chuckled to himself. The English had lugged aboard enough of his wine to last five years! And drunk enough at his table to have headaches for a week!

The last English vessel was disappearing around the headland, a plump little transport heading west. For what it cost, his report would be thin. He picked up a little bell from his desk and rang for his secretary. He'd best get his report on the convict fleet off to Madrid. With luck they'd receive it in two weeks, giving them ample time to make mischief for the English at Cape Town.

Rio
1787

*A*s the *Charlotte* fell away from Santa Cruz on that crystal-clear Sunday, the sun-drenched mass of 12,000-foot El Pico, hitherto hidden from Mary B's view by clouds and crowding hills, loomed higher and higher above the receding land. My first mountain, Mary B thought, smiling at the sunlit peak so much more immense than anything she'd imagined possible. My first mountain and my first palm trees... she felt a thump ...and my baby! What am I going to do about my baby?

Clearly James could not be depended on. Damn James! Did he expect her to be a washerwoman the rest of her life? So like a man, to have his pleasure and skip out scot-free! She wondered now, did he really have a new son back home? Or did he just say that? Really, he didn't have to go back to England. For all he knew his wife could be dead by then. And if he never knew whether his wife in England was alive or dead what was wrong with being married to someone else in Botany Bay? And so what if he was a Catholic? On the hulk he'd been ready enough to forget his wife and what the Pope said! "I need to be married," she said aloud, and pictured someone being good to her, someone who was kind and close and a helper. Someone steady. Not like James!

Seedy had her Quint, of course, but how dependable would a Marine or even a sailor be in Botany Bay? Their enlistment might end, or their ship sail - and whoosh! He'd be gone. "Fare thee well, Mary B." Why, some of the women prisoners could hardly wait for the chance to take up again with their men from the hulk. At least those men weren't going anywhere. Oh, damn James!

"You only know me as a convict," he'd said. Aye, and what would she know about any other convict? Maybe she'd discover another wife! And when that man's time was up would it be "Fare thee well, Mary B - but have no fear, there's plenty left to pick from." Aye, a whole boatload of scoundrels, right under her feet! Oh, damn James!

During this rumination the face of Will Bryant hovered at the edge of Mary B's mind, as if some inner voice reminded her of his unlikely words at the Mother Banks, "Will you be my Mary B?" At an impasse in her thinking, she let his face come.

Although she was unsure why, something about Will Bryant put her off. Perhaps it was his bluster or his smirking superiority. She remembered the grins of the other men after he stopped her that day in the Mother Banks. Maybe it was all a game with him. But what if he didn't mind her having James's baby? So what if Will Bryant was less than perfect? At least he was somebody.

She had so many things to think about! Maybe she should just try to go it alone, stand on her own two feet and make her own way, men be damned! And then maybe if that special someone came along....

She straightened. El Pico looked so solid and strong. Whatever happened, she wanted her baby to see this splendid sight, and she imagined holding him up to see. "But he won't be a baby then, will he?" she suddenly realized. If she served out her sentence and could make her way back to England he'd be a boy of five or six, even older. Tears came to her eyes once again with the reminder that her mother and Dolly and Little Bill would never see her baby as an infant.

~~~~~~~~~~

El Pico remained in sight for two days, the sunlight on its rocky mass creating the illusion of a summit bright with snow. The mountain became the Canary Islands, and the people on the *Charlotte* seemed never to tire of looking back to watch its slow diminishment. And when at last the skies turned cloudy and the sinking pyramid on the northern horizon was lost in haze, they felt let down.

Days grew steadily warmer. Some were almost idyllic, and the *Lady Penrhyn*'s crew celebrated crossing the Tropic of Cancer by dunking the convicts in a barrel of saltwater, a ritual made more fun because all the convicts were women. The sailors and Marines became so engrossed in the lusty rite their neglected ship nearly collided with the *Charlotte*.

Sea life grew increasingly bountiful. The voyagers saw flying fish, dolphins, porpoises, occasional whales, frigate birds, and other birds of all descriptions. Some of the men fashioned fishing tackle baited with offal and hooked an occasional fish that relieved their monotonous diet. Little Ned Thornapple seemed never to tire of testing his luck with a baited hook, and when he caught his first fish, a nice bonito, he peed his pants in excitement.

Under pleasant skies the fleet made steady, uneventful progress, sailing more than a thousand miles in a week's time. Then the *Supply*, Arthur Phillip's little, general-purpose Navy ship, sighted the Cape Verde Isles, a spattering of islands off the westernmost bulge of Africa's Slave Coast. For two days the ships picked their way through the islands to reach Port Praya. There they lined up to follow the *Sirius* to anchorage. But after a few tentative moves to bring the ship in, Arthur Phillip abruptly hoisted the signal to put to sea again. Puzzlement and consternation. Port Praya offered the last chance to take on fresh fruits and vegetables before heading down to Rio, a leg of 2,500 miles across the Equator. Not to top off water and fresh stores was inviting trouble.

And it came.

Skies soon grayed. Winds became listless and variable. The air grew heavy. They had entered the Doldrums. As the heat became oppressive, progress slowed to 40 or 50 miles a day. On some days the wind died completely and the fleet drifted back toward Africa on a strong easterly current. Day after day the skies remained lead-colored, the air heavy with moisture. Rain fell in unexpected torrents, with fierce thunder and lightning. Although ducks and geese fluttered and preened with nervous gusto during the rainstorms, the livestock seemed terrified by the wild, incessant flashing and crashing and the sheer volume of water that plunged from the skies. Minutes after a sudden downpour ceased the heavy tropic heat returned. The ships became sodden; a film of moisture coated every surface. Unprotected wood grew waterlogged. Sails and rigging dripped continuously. Animals swayed with lassitude, heads bowed and noses to the deck.

Temperatures in the below-decks rose steadily - 75, 80, 85, 90, 95 degrees. Mary B and the women sat nearly naked, glazed with sweat, their eyes stuporous, avoiding even minimal movement. Convicts began to faint, and were incredulous when John White forbade their daily time on deck. He worried that a sudden rain shower might cause them to sicken.

Day after day the First Fleet creaked towards the Equator, averaging less than two knots. As rain seeped the warm liquor of animal, vegetable, and human residue through the decks, bilges began exuding a reeking miasma. The stench grew overpowering. John White ordered frequent, almost continuous pumping of the bilges by rotating crews of convicts. Torsos glistening, some convicts poured buckets of seawater into the bilge-well to flush the slime, while others pumped it out. The tedious work continued hour after hour, day after day. White also ordered gunpowder soaked in vinegar burned in small dishes to sweeten the air in the below-decks, and small coal fires to help dry it. He gave out oil of tar for swabbing bodies and bedding and sleeping spaces to kill the proliferating vermin. He ordered a hole cut in the grate of the women's hatch to funnel in whatever air moved by way of a special canvas wind-catcher. When Quint saw the ship's carpenter sawing through the grate he began thinking of the possibilities.

As days dragged and the fleet struggled to make mileage, Arthur Phillip ordered water rations cut in half, from three quarts a day to three pints a day, and one more for cooking. The convicts grew quarrelsome and surly. Naked men sprawled in their stifling prison, moving only to relieve themselves or wipe their sweat or take their turn at the pumps. Incongruously, three convicts huddled over one of John White's coal fires to melt scraps of pewter and a few brass buttons.

"What did I do to deserve this?" Will Bryant ranted, "They're killing us for stealing handkerchiefs. They could as well hang us in England as kill us like this!" The mutter grew. Tench doubled the Marine guard.

One afternoon, seeking a diversion among his enervated companions, James Martin asked, "Who's here for what?"

No response.

"You know why I'm here?" he asked rhetorically, "Old iron! I'm sweating like an Indian for rusty, old iron! And I don't even know why I took it. It was just there and I thought it might be worth something." He smiled ruefully. "Eleven iron screwbolts! Eleven iron screwbolts worth two shillings six..." he held up a finger, "...and some old lead worth two shillings." His voice dropped, introspective. "For four and six I'm on a boat bound for Botany Bay for seven years."

"Screwbolts," the doltish Tartop giggled. "Screwbolts" he said again, and then again. He liked the word.

The other men were silent for a long time before a slight, gray-haired, gray-faced, sad-looking man said almost to himself, "I can't go cheese no more."

"What's that?" James asked.

Nate Mitchell looked up from where he sat slumped against a post. "I said I can't go cheese no more. And I used to love cheese. I used to eat cheese the same as the next man. But I can't go cheese no more. I'm here for cheese."

Nate sighed loudly, and slowly raised his eyes. "I can still see that cheese sitting on a bench outside a dairy, a little wheel covered with a white cloth. I could smell it from the road! God! my stomach growled like a wild dog. I remember how I stood on the road studying that cheese. I could almost feel my knife cut through it. I could almost taste that...that clean, salty taste. I could almost feel it squeak on my tee...."

"All right!" James interrupted, "Enough about cheese!" The men laughed, a rare occurrence in the Doldrums.

"Well," Nate continued, unfazed by the interruption, "I don't know how to say it, but next thing I knew I was watching myself take that cheese. It was like I was watching someone who wasn't me. Then I saw myself go back on the road till I came to a shade tree. All this time it was like I was watching myself from outside. I was so calm and easy. Well, when I cut that cheese my mouth watered so bad I couldn't swallow fast enough and I had to spit so I could eat." He was silent, his eyes now fixed on the deck near his sprawling legs. Finally he stirred and looked around. "Well, I ate about two pounds before the dairy woman came with a constable. She'd been watching me the whole time through a window. She could've yelled. I would've moved on. But I think she wanted me to take that cheese. I think she wanted to sic the law on me just to see it happen. I don't know.... Anyway, I got put in jail for stealing eight pounds of cheese."

James asked, "Eight pounds?" Nate nodded at his outstretched feet. "And how much was it worth, that cheese?"

Tears welled in his eyes as Nate looked up, "Three pence a pound! And I didn't have money to buy a penny's worth! No, but I ate two pounds, and for that six pence worth of cheese I'm going to Botany Bay."

James slowly shook his head. He spoke with mild amazement, "Well, Nate, that sure beats me. Yes, sir, that sure beats me. Here," he said, reaching into his compartment, "You know what you get, Nate?"

Nate looked disbelieving.

James smiled. He held forth a soft, white glob of cheese. The men watched Nate regard the little piece of food. James proffered the piece of cheese as if encouraging a dog to beg. James was smiling, "Come on, Nate, come on! You can do it!"

The men began to encourage Nate. "Go on, Nate, take it!" "You won fair and square, Nate." "Go on, Nate!"

Finally Nate stirred. He slowly got his legs under him and stood. He looked from the cheese to James. At last, ever so slowly, he reached out a hand.

"Eat it, Nate!" the men encouraged, "Go ahead, eat it!"

Nate's hand trembled as he slowly brought the cheese to his lips. He hesitated, swallowed several times, then took a tentative, little nibble of one corner of the morsel. He chewed slowly at the front of his mouth, the prisoners riveted on his expression. He tried to swallow, looked as if he might choke, then tried again, and succeeded. He regarded the cheese a long moment, then took another little nibble. At his second bite, the prison filled with applause, yelling and whistling, and gray-faced Nate Mitchell, smiling shyly, nodding appreciation, bit again into the cheese.

~~~~~~~~~~

And still they creaked on through the oppressive Doldrums. Some days they made less than 40 miles, some less than 30. Arthur Phillip ordered water rations cut to two pints a day. Sailors stretched canvas rain-catchers to refill the waterbutts. During unpredictable downpours they worked frantically to capture the precious bounty, sometimes garnering hundreds of gallons for their efforts. Draped through the hole cut in the women's hatch, the wind-catcher brought rainwater down in brief abundance. The women drank their fill and rinsed off greasy sweat. The rain was a cool, sweet contrast to what they drank from the moldy waterbutts. They stored the extra water in cooking pots and wash buckets.

The night Quint slid down the rain-catcher to be with Seedy the decks were crowded with those seeking sleep under the sky. Sailors, Marines, and wives lay everywhere. They sprawled among the beasts and fowls, beneath the water-catchers, beneath the deck boats, wherever room might be found to escape the insufferable conditions below. On rainless nights they slept the darkness through, muttering meaningless dream words to an indifferent sky. Sometimes, startled by fierce claps of thunder, a few fat, splattering drops might send them below in a wild scramble out of all proportion to the threat, as if pursued by nightmare demons come to life.

On this particular night a pale half-moon rose through the haze. On the western horizon cloud-draped lightning fluttered. Once again the convicts had asked to sleep on deck, and once again John White had denied their request. Now, their dinners of boiled salt beef and rice moved well along the digestive process, those on deck lounged in small groups. They smoked pipes and sipped tea or their daily grog as they visited in quiet conversation. Perched cross-legged atop a capstan, a sailor strummed "Greensleeves" on a guitar as the Marine wives hummed in soft accompaniment. Malvina Thornapple stroked the hair of her daughter Nell, whose head rested in light sleep in her mother's lap. Thumb in mouth, Nell twitched from time to time at some febrile fright. Never well, Malvina felt somewhat feverish herself.

The slow, almost imperceptible rise and fall of the *Charlotte*, the muted voices and music, the resigned sigh of Lieutenant Creswell's cow lying down for the night made a tranquil scene. In the below-decks, exhausted by heat and thirst, convicts sprawled in shallow-breathing torpor. The women congregated around their wind-catcher for an occasional faint puff of air. Whether to crowd beneath the wind-catcher or to lie alone was a matter of endless discussion. Heat from another's body could be as irritating as the discomfort of a damp deck or prickly confinement. The women shifted from place to place, trading one benign torture for another.

On the afterdeck, their backs against the taffrail, Tench and White were at ease on a blanket as they finished a near-empty bottle of canary. White's servant, a slight, young man named Broughton, snored nearby. The lucky catch of a dozen bonitos had provided a special mess for the gentlemen aboard the *Charlotte* - baked bonito garnished with lemon, boiled rice, sauerkraut, and onions soaked in vinegar - a meal eaten with civility and good humor that were rare in the Doldrums.

Sipping canary from delicate stemmed glasses, Tench and White had talked quietly of White's inspection of the transports. On the *Alexander* he discovered that mold from unpumped bilges had crept up the interior of the ship to invade the officer's compartments and turn their brass buttons black. When he raised the grates to go below to the convict quarters, the billowing stench from the prisons sent him reeling. Arthur Phillip had blistered the *Alexander*'s master.

"The Commodore can't stand things not going to plan," White said, referring to Arthur Phillip, "He's beside himself that we've made only 900 miles in two weeks."

White's use of the term "Commodore" was ironic. The Admiralty Office had turned down Arthur Phillip's request to be designated a commodore for the voyage to Botany Bay. Those in the Admiralty Office, ever sticklers for protocol, reasoned that two ships escorting a bevy of convict transports and supply ships hardly constituted a Navy squadron.

White and Tench then talked of books and a production of *The Tempest* Tench had seen in Portsmouth before sailing. "The Commodore would love to be Prospero now and command the wind," White joked. Then sleepiness began to overcome them. The silence between their exchanges grew long. So replete was their repose they might have been lovers falling asleep in bed. His eyes closed, Tench's words trailed off. White was silent, perhaps dozing. Bathed by pale moonlight, the scene was utterly peaceful - faint voices, the languid strumming of the guitar, Lieutenant Creswell's cow blowing chaff from her hay.

His thoughts now random and sluggish, Tench saw the deck as a stage filled with people come up to play their little parts. Every character had a story. Every life was unique. But what play had so many players? From the mass of characters only a few could be chosen - the tallest or the most attractive, the most evil, the loudest, the ugliest, the wittiest. The rest must be background - a crowd of convicts, a squad of soldiers, a watch of sailors, a gaggle of wives, all blurred figures in the background, taking on life only if the playwright gave them life.

Tench snorted awake and looked over at White. With his chin against his chest, his face hidden in moon-shadow, White's breathing was soft and regular. Palm up, fallen away from the stem of his near-empty glass, White's hand rested on the blanket between them. His fingers were half-curled, vulnerable, and inviting. As Tench reached down to remove White's glass his hand brushed the surgeon's sensitive fingertips. A few nights earlier Tench had seen White stare down from the stern for almost three hours, transfixed by the phosphorescent waters drifting by. What had he seen? What did he hope to see in those eerie clouds of mysterious light?

White sighed softly in his sleep. Tench let his eyes close again, and soon saw the familiar soft, rose-colored swirls that came to him with sleep, like blood gently stirred in water. In the swirls he began to see the indistinct faces of soldiers, sailors, convicts, women, children, silently swept around, carried ever more swiftly into a deepening vortex, a never-ending rush of silent, beseeching faces. Faces from the *Charlotte*, from the War, from the past, drawn to his unconscious by forces as vast and mysterious as the ocean currents. Tench slept.

~~~~~~~~~

The strong, cold wind from the southwest that rushed through the darkness to shock the sleepers hit the *Charlotte* broadside, heeling her sharply. The dozing helmsman jerked awake and spun the wheel as rude shouting sent groggy sailors racing aloft. Wind hammered the canvas and thrummed the rigging and a cold hail-laced rain peppered the decks. As befuddled sleepers scurried below, the ship's canvas whipped and cracked. A weakened spar snapped. Its broken end plummeted over the side and the fouled rigging entangled poor Lily, Lieutenant Creswell's Jersey cow, who tried to rise, and fell heavily to the deck again.

With the first rush of wind, Tench started awake and snatched at his things, both he and White pulling at the blanket in brief tug-o-war. Rain plastering their hair, Tench's grin was broad as he yelled, "Prospero got his wind!" and they laughed in their dash for shelter as White's lithe servant stooped comically again and again to retrieve one of the wine-glasses rolling wildly on the heaving deck. Dodging around and over the more bumbling sleepers, Tench saw the women's wind-catcher pull loose, flap wildly, and glimpsed a bare torso engulf the canvas and sink to the deck. Safe in the companionway, their soaked shirts sticking, giggling at the pleasure of their adventure, Tench raised the near-empty bottle in salute, "To Prospero!"

~~~~~~~~~

The *Charlotte*'s heeling and the wailing wind awakened Mary B from fitful sleep. She felt a downpour of cool, refreshing air. "Ahhh!" the women exclaimed. Now inured to the extreme motions of the ship, they bathed in vitalizing air that poured down the wind-catcher. Almost immediately their sweat-glazed bodies began to cool, and when the cold rain shook loose a

77

million droplets from the fluttering canvas, they showered with shivery delight. They stood beneath their hole to the heavens - a happy, close-packed mass of goose-pimpled, wet flesh.

Suddenly the wind-catcher disappeared. "Oh, no!" they cried. A moment later they were dumbfounded by a pair of bare feet poking through the hole and then a human figure sliding down the canvas into their midst. The figure landed with a thud, almost bowling over a couple women. "Ow!" someone cried. A brawny, bare-torsoed man stood before them.

"Good evening, ladies, I've come to call on Seedy." In a flash of lightning Quint's missing-tooth grin gleamed. "Well, well," he said, eyeing the naked glistening flesh in the instant's light, "I wonder where my Seedy might be. Do I have to feel my way?"

Although used to the traffic of men in their prison, the women stood transfixed. A man had never actually sneaked among them! In another flash of lightning, Quint saw Seedy directly before him, her arms crossed over her bare breasts, her deep-set eyes shining with happiness. The gap between her two front teeth was a visible black line in her grin. "Ohhh," Quint breathed. Reluctantly the women drifted back to their compartments, giggling or tsk-tsking in good humor, but not before a couple of them stroked Quint's slippery back with light fingers and teasing comments. Quint enfolded Seedy in his arms. "Mmmm," he murmured when he smelled her damp hair. He felt her body tremble against his. He tasted the slightly salty moisture on her neck.

"Oh, Quint," Seedy breathed. She ran her hands down his broad, muscular back and smelled the rainy, sweaty scent of his chest and felt his prick swelling through his breeches. She raised her face. They kissed softly, then with open mouths. Seedy's delicate tongue licked his lips, stroked his own tongue.

"Ohhh," Quint groaned, leaning back and raising Seedy so her vulva pressed against his bulging groin. Seedy kissed him quickly - once, twice, thrice - then slid out of his arms. "Come," she said, "Let's let them have the air."

She led him by the hand to the rear of the prison where stacked bales of old sailcloth were stored for making tents. A mound of canvas lay on deck from which the women were cutting patterns. Seedy turned and stood on her toes to kiss him again while fumbling with the string that tied his breeches. He tore the string away and his swollen penis throbbed erect, hard as a musket barrel. "Ohhh," Seedy sighed.

She sank to her knees, running her hands lightly down his chest, his stomach, along his hips, his muscular thighs. She held Quint's penis gently between praying hands, smelling sweat and urine and his natural unguents. He smelled all right. She kissed the hot knob lightly and licked an oozing drop of semen. "Aye," Quint whispered, fondling her tiny ears, running his fingers through her damp hair, "General Quint of the Royal Marines reporting, and my private stands at ready."

"Oh, my!" Seedy said, "Why, this little soldier has a fever! He's so hot! He won't die, will he?"

"I hope he dies a dozen times, but if you pray for him like that, he'll rise to fight again!"

"But then I'll have to bury him a dozen times, and that's hard work!" Quint laughed softly, lifted Seedy, and fumbled at the knot of sleeves around her waist. She stood patiently while he freed the knot with trembling hands, and when her dress finally slid from her hips to fall in a soggy garland, Quint's hands gently explored her back, under her arms, felt the soft swell at the sides of her small breasts, the smooth, jutting ledges of her hips. He sank to his knees. He kissed each breast gently, then her navel. He kissed where her pubic hair met the smooth skin of her womb. With the tip of his nose he traced the border of hair along the top of her pudenda, then down one side on the delicate soft skin of her inner thigh and up the other. He smelled the sweet pungency of her sexual juices and the clean, sharp smell of urine. She smelled all right.

"Your holy ground," he said, settling back on his haunches and looking up at her. "Three-sided, like the Trinity. My private wants to be buried there." Seedy tugged gently at Quint's ears and he stood before her again.

"But he's so big! He might go hard into the grave."

"Have no fear. He's a sinner, and he'll go into that holy ground like a sinner to his pew. Then he'll go to Heaven."

"You mean he'll lie at rest?"

"Oh, no! He's like a spirit, see, and wants to move about. Ohhh," Quint groaned, "We've got to bury him before he shoots his musket off!"

"Oh," Seedy said with an odd rising inflection as Quint laid her carefully back on the pile of canvas. She guided him into her shuddering body. "Oh!" she said, "Oh! Oh, Quint! Oh! Oh! General!"

"Aye, and you're my sweet, sweet Seedy. The sweetest one I ever knew! Sweet, blessed Seedy. Ohhh!"

~~~~~~~~~~

Too soon, a night of rapturous love-making passed. Too soon the time was gone for caresses and tender words and soft kisses, for Seedy's small head to rest on Quint's chest, for his restless hand to stroke her arm, her smooth back, her hips, her buttocks. Too late Quint realized he'd out-loved the dawn. The decks above were alive with sailors bustling about and with officers giving orders and assessing the havoc the storm had wreaked on their livestock.

Quint later tried to explain his presence in the women's prison by saying he was reefing the wind-catcher when he slipped and fell through the hole, knocking himself senseless and not fully recovering until near morning. Tench thought his story plausible. It must have been Quint he saw struggling with the wind-catcher. But when Motherwell revealed the wind-catcher stays had been cut, Quint was in trouble. Where were the bruises from his fall? Why didn't the women call for help? What about the scratches on his back? They looked like fingernail marks! Arrested, Quint was confined in shackles in the dark recess of the anchor-cable hold.

Porter sneaked down to bring Quint his blank book, Quint's only request. "I'll take my punishment like a man," Quint said. "I did what I did, and maybe I did wrong, but I'm not sorry for what I did."

"Was it good?" Porter asked, his simian eyes busy at Quint's face. "Was it worth a flogging?"

Quint looked away and didn't answer.

"Well?" Porter persisted.

Quint stared at Porter, a warning in his eye.

"Ah, well...." Porter said.

Quint stretched out as best he could. After a while he held up his book at arm's length as if it were a baby, then laid it on his chest, and caressed the smooth leather cover. He thought of what he'd write and pictured himself showing his book to his mother and Seedy. The two women stood side by side, their arms around each other's waist. Soon the sounds of life above came to Quint as through layers of muffling blankets. Hands folded over his blank book, Quint's chest rose and fell in peaceful sleep.

The matter of Quint's punishment wasn't settled right away. Arthur Phillip ordered his fleet to pile on sail and take advantage of the favorable winds. Several days passed before Quint learned he'd stand trial at next landfall. Until then he'd remain shackled in the anchor-cable hold.

~~~~~~~~~~

On New Year's Day in 1502, a Portuguese explorer threaded his little ship between abrupt granite peaks in wide-eyed wonder, believing he had entered the sea-outlet of a great river on the mysterious South American continent. Reluctant to risk his vessel among the reefs and small islands, he christened his discovery Rio de Janiero (January River), and continued his voyage down the coast of Brazil. It was into this vast, hidden bay that the First Fleet threaded its way 285 years later, on August 6, 1787.

The Portuguese inhabitants still called their city St. Sebastian, but the maritime world knew it as Rio de Janeiro, or simply Rio. With 50,000 whites (and countless black and Indian slaves) Rio was the seat of Brazil's colonial government, a major military base, and its most important port. And special to Arthur Phillip. Earlier in his naval career (with the blessing of the British government) he'd sold his professional services to the Portuguese Navy, and from Rio had commanded a small Portuguese warship in the war with Spain, showing boldness, skill, and valor. The Vice King of Brazil was effusive and generous in his thanks for Arthur Phillip's service, and for a time the Englishman lived as something of a local hero. He'd also come to love Rio's granite buildings and its kissing closeness to the water.

But that was then and this was now, and as Arthur Phillip led his ships to anchor he fretted. The drudges in the Home Office had not finished his credentialing letters before he sailed. Now he was limping into port with a battered, leaky fleet asking for special favors, and without credentials. Who knew what might have happened while he drifted for weeks in the Doldrums? England and Portugal might be at war!

~~~~~~~~~~

An almost vertical sun beamed through the grate to paint soft-edged shadows on the women gathered below. Eyes closed, Mary B watched tiny white specks and shapes drift across a sea of orange. With a bustle of activity and the sudden rush of an anchor let go they'd finally come to rest. For almost two months

81

they'd suffered from heat and bad air far worse than the worst of England's jails and hulks.

The last few days had improved, however. After Quint and Seedy's concupiscent rite on the night of the big squall, providential winds pushed the fleet sometimes more than 100 miles a day. With full water rations restored, the below-decks cooled off and dried out, and convicts were again allowed up for fresh air and exercise. Life became more tolerable. Nonetheless, everyone was ready for an end to the ceaseless motion, the endless boredom, and the relentless diet of porridge, rice, salt meat and salt fish and hard biscuit gone weevily.

"Look out below!" a voice laughed from above. It was James! Mary B broke from her reverie and squinted at the bright hole to the sky. She saw the shadowy outline of a head haloed by the sun. "Look out below!" James laughed again, a rich, full-bodied, happy laugh. A torrent of round, sun-spangled fruit poured down, bumping down the ladder, drumming the deck, bouncing and rolling, some bursting on impact, big as human heads. "Eeee," the women cried.

The torrent began again as James laughed, "Oranges! Oranges like you never saw!"

"Oranges!" one of the women cried. "I know oranges!" She snatched one that had ruptured and tore it apart and buried her face in the fruit, greedily sucked, chewed, swallowed. "Good!" she panted, her eyes fixed on the fruit, "Sweet! Good!"

Others began tearing the fruit open. Some who'd never eaten an orange tasted tentatively, then they too began devouring their bounty. They ripped off mouthfuls of rind and pulp and worked themselves into a frenzy of sucking and chewing and swallowing. Juice ran down their chins, wrists, and forearms. Juice soaked their dresses. Stopping only to cast the rinds and wipe their hands, they ate three and four of the giant oranges apiece with animal sounds of feeding pleasure.

Mary B ate until she felt her stomach couldn't hold another bite. A half-eaten orange lying in her lap, she rested her head against a post and sighed with pleasure. The air was rich with sweet orange and the sharp, acidic aroma of rinds. Over her bulging womb she looked at her bare, outstretched feet nearly buried in thick, meaty orange scraps, the sun bright on remnants of fruit. She lifted a leg and saw the tangle of white and orange shift and slide, the colors clean and fresh. Her front was soaked with juice, her arms and chest were sticky, her feet and ankles glistened. Bits of pulp and fiber stuck to her skin and her dress. Some had even gotten in her hair. She belched, smiled, and looked at Katie, who still gnawed with a determined stare at an orange in her lap, willing it to stay until she could devour it. Mary B laughed. Katie looked up, joined her eyes, and laughed too, and in a few moments the entire prison was filled with the women's pealing laughter.

"More oranges!" James called, "More oranges, compliments of Captain Tench!" Yet another cascade of fruit poured down the sky-bright hole, and then another and still another, until fruit began to heap amid the women, who now lay weak and helpless from over-eating and laughter, their eyes teary with effort and thanksgiving.

Perhaps thanksgiving for life. Not long before, an old convict no one really knew had sighed quietly in his sleep when he saw a shadowy figure approach. The old man had straightened his bony frame and pressed his hands as if to rise, then sighed with resignation when the figure slowly raised a hand over his face, willing his eyes to close and his body to sink into deep, deep rest. The old man sighed again, a prolonged, expiratory sigh of relief, and next day from the deck of the *Charlotte* he joined Ishmael Coleman and all the others gone to watery graves.

Liz Bason's Little Jacob teetered in the Doldrums. One day he'd rally enough to reward his mother with weak little smiles and soft coo's; the next day he'd turn feverish and fretful and cry until some of the women yelled to shut him up. The strain on Liz Bason showed. With each reversal of her baby's health she grew more haggard and dispirited. John White came down to look into Little Jacob's pyrexic eyes. He listened to his tiny pattering heart, and gave Liz some oil of tar. "Try rubbing it on his chest," he advised.

Some of the women sacrificed a bit of their own precious water to bathe Little Jacob's hot body or to wet a sucking rag. "Maybe I should just let him go," Liz ventured, "What kind of a life is this? Maybe we'd both be better dead!"

"You'd at least give us peace!" one of her old Bristol jail mates yelled. Like the sound of shattering glass in a tavern, the remark brought sudden silence, then an immediate taking of sides. Within minutes the women's prison had erupted into an uproar of screaming argument, than threats and pushing, and finally a melee of slapping, kicking, pulling, wrestling, and biting. Mary B retreated to her compartment, then saw Katie squirming in a pile and jumped down to pull her off. "Katie! Remember your baby!" Wild-eyed, Katie stared at Mary B as if she didn't know her, then dove into the wild pile again. The guards at the hatchway stood rooted in disbelief and stared at each other. At last Sergeant Motherwell rushed down to pull some of the flailing women apart. "Jesus Christ!" he shouted up at the stupefied Marines, "Get down here and help!"

Until tempers cooled on the night of the big squall, that incident was the first of many quarrels and small fights that erupted over trifles among the women. While most of them tried to maintain a peaceful equilibrium, one especially vicious clique seemed to feed on whispered spite, mean laughter, and pointed jibes. An older woman from Bristol named Hannah took Liz Bason under her protection and preached at every opportunity of God's Grace and Infinite Mercy. From her well-worn Bible she told of ancient struggles between despair and hope, greed and charity, temptation and virtue...struggles like their own. Hannah preached that anything was possible through faith. Her favorite maxim was "Faith overcometh Adversity, Faith overcometh All."

Earlier, Katie had feuded with Hannah and told her to shut her big mouth. "Go preach to the slop buckets, you old bag!"

But Hannah wouldn't be deterred. "It's my Christian duty," she'd say.

The only woman among them who could read well enough to make her way through the Bible, Hannah said she had a Christian duty to enlighten the ignorant.

"And was it your Christian duty to steal cloth?" Katie taunted.

"Yes, I admit I'm a sinner," Hannah said, "But I've seen the Light and now I know my Christian duty." And Hannah was among the first to put down her Bible and lend Liz Bason a helping hand.

So Little Jacob, supported by many mothers when his own mother faltered, lived perilously on in the dim, quarrelsome confines of the *Charlotte*, and finally, blessedly, began to recover in Rio.

And not just Little Jacob. Everyone thrived. As the weather remained sunny and pleasant, convicts and Marines enjoyed a cornucopia of fresh food. Those with money bought oranges, limes, bananas, and even more exotic fruits from vendors who paddled small boats out to the transports and tossed up samples. An English penny bought oranges enough to gorge a half-dozen. Fresh beef appeared in quantities that boggled the convict mind, more than a pound a day, and such vegetables as they hadn't seen in months. With each day's dawning, memories of their suffering in the Doldrums receded farther.

Given the freedom of the deck during daylight, the women were endlessly entertained by Rio's Catholic beauty - gleaming, ivy-draped monasteries, convents, and churches crowning the steep hills, a spread of reddish, imbricated roofs and colorful buildings of white and blue and pink. So unlike England! They loved the color and movement on the large square at water's edge. It seethed with horsemen, strollers, squads of soldiers dressed in blue, sedan chairs, carts, and an occasional carriage. They were awed by the sight of so many black bodies in this city of slaves. Slaves did everything. Jugs and casks atop their heads, a continuous procession fetched water from an ornate, pyramidal fountain at the waterfront. Other slaves met fishing boats coming to anchor and plunged through the small surf to get the best fish for their masters. Because Rio lacked a pier, slaves waded between lighters and shore carrying cargo on their heads. On the square, hundreds more struggled at a shuffling trot under sacks, boxes, barrels, and baskets loaded with sugar, coal for cooking, coffee, rice, rum, and the other stuff of commerce. Slaves pulled the carts and sledges, moving to rhythmic chants or the shaking of an overseer's rattle. Some herded cattle and sheep across the square to market or to slaughter, while others were themselves herded to the slave market, chained together by iron collars. Even the women wore iron collars with long, protruding spikes.

In the distance the women heard the constant clanging of church bells signal devotions, births, deaths, and celebrations. Sometimes a religious procession made its way across the square to the sound of flutes, cymbals, drums, and chanting. With abundant fresh food, visual entertainment aplenty, and the security of lying at anchor, life seemed good for Mary B.

~~~~~~~~~~-

Quint was flogged on the third day in Rio. With his shackles removed, he was allowed to bathe, shave, and don his parade uniform. Porter and Stork helped whiten his cross-belts and his gaiters, blacken his boots, and powder and tie back his hair. "You want to look your best," Motherwell said, licking his thumb to wipe a smudge of blacking from Quint's cheek, "You're going up before Captain Phillip himself."

Quint looked puzzled. "You're Navy, son, and the first Marine to go among the women." He backed off and looked Quint up and down, nodding approval. "You'll do me proud now, won't you, son? You won't let me down under the lash, will you?" He reached to pick a speck of lint from Quint's shoulder.

Overwhelmed by the attention and support from Porter and Stork, from his squad and his mates, Quint squared his shoulders and shook his head, his throat so tight he was unsure he could speak.

In shackles once again, standing tall before Arthur Phillip on the *Sirius*, Quint answered in a voice that sounded almost natural. Had he heard the orders read concerning conduct with the women?

"Aye, Sir."

Had he understood the orders?

"Aye, Sir."

Had he gone among the women knowing he was violating the orders?

"Aye, Sir."

Did he have anything to say in his own defense?

A flood of half-thoughts boiled up in Quint - that it seemed right, didn't seem right, that he was sorry, wasn't sorry, that he was ashamed, proud, satisfied, afraid, loved his mates, loved Seedy, wanted to help, wanted to be a good Marine.... "No, Sir."

Arthur Phillip regarded the brawny young soldier standing at attention, the trembling hem of Quint's coat the only sign of his nervousness, the Marine's determined gaze fixed on some faraway point. He took in the spotless uniform and the carefully tended accoutrements and the shining, healthy, youthful face now making a gulping sound as Quint swallowed. My God! Arthur Phillip thought, I love this boy! He's what a young English soldier should be!

Arthur Phillip conferred in undertones with Major Ross and John Hunter, the captain of the *Sirius*. Their three wigged heads rolled back and forth like three white yarnballs as they leaned this way and that. They separated. With a nod to his secretary to make the record, Arthur Phillip said, "Private Peter Quint, you have admitted to violating expressed orders regarding conduct with the female prisoners with full knowledge of your actions. The prisoners are Crown property, and our mission is to deliver them safely to New South Wales as if they were the private possessions of the King himself."

Arthur Phillip flicked an eye to Major Ross. "Private, your conduct has endangered our mission. You have broken the King's trust! We cannot permit, and I will not permit, anyone under my command - soldier or sailor - to break the King's trust." He paused and looked at the sheet of charges near his hand, and moved it a quarter-inch. He cleared his throat softly, then moved the paper back a quarter-inch. "This court sentences you to one hundred lashes, to be executed immediately on each convict transport. Battalion and ships' crews to witness punishment."

Quint sagged a little, swallowed, then stood ramrod straight again. "Aye, Sir," he said, his voice surprisingly strong, "Thank you, Sir. God Save the King."

Flogging was such an integral part of Navy discipline that its instruments had been refined with considerable skill and subtlety. When a flogger laid onto a man's back, he was expected to take a full stroke, to make the lash whir, and make it land on the victim with a solid smack. The most common lash was a simple hemp rope, two or three inches in diameter, unraveled for about two feet. Unraveled, a common rope revealed three major strands of three strands each, the nine tails. Depending on how badly the victim was intended to suffer, such instruments were modified to inflict varying degrees of pain and damage. After the simple, unraveled rope, the next escalation of punishment was to tie knots on the ends of the tails. Knots lent density and bruised the flesh more deeply. Next was to tie metal objects to the tails. Little rings were common, which added weight and cut the flesh with every stroke. Most punishing were little balls of lead. These pulverized the flesh and chipped or broke bones.

With even the simplest lash, tails of rough fiber repeatedly striking the same place ultimately broke the skin. Leaking body fluids from the broken skin would moisten the rope, lending still more weight, accelerating further laceration. Fortunately for Quint, he was flogged with an official Navy lash, kept in a red baize bag and held for better grip by a short, red-handled stick called a "cat." Liquored up with a generous pint of grog, Quint was taken first aboard the *Lady Penrhyn*, undressed to his pantaloons, and spread-eagled against an upraised hatch grate to be seized hand and foot. As the charges and punishment were read, the ship's Marines uncovered their heads as a demonstration of respect for the King's authority.

Quint winked and grinned at an acquaintance. But when Sergeant Campion, drawing the short straw, reared back to lay down the first awful blow, Quint cringed, sucked in his breath, and pinched his eyes shut. A band of fire burned across his back and seared the soft skin under his right arm. He clamped his teeth against the next blow, then remembered Sergeant Motherwell's advice. He tried to relax, but he felt his body tighten against his will when he heard the whirring hiss and he jerked with surprise at the red band of pain. After a half-dozen blows he tried to go out of himself, to see someone else stretched across the grate, and he could do that a little, but tears came to his eyes and he wondered what he'd done to deserve such pain. When fifteen blows had been administered on the *Lady Penrhyn*, Quint was untied. He stood upright, weaved for a moment, then winked again at the acquaintance. He went down the ladder into the jollyboat two steps at a time for the short ride to the *Alexander*.

After his punishment on the *Alexander* he used his hands to help climb the ship's ladder to the deck of the *Friendship*. After the 45th blow he stumbled getting back into the jollyboat, and took longer to pull himself onto the *Scarborough*. After the *Scarborough*, he had to be supported to the side and fell the last few steps for the trip to the *Prince of Wales*. After the *Prince of Wales* he was dazed, his back a mass of welts and bruises, his broken skin oozing blood and fluid.

"One more and you're done, son," Motherwell encouraged. You're making me proud, Quint, you're making us all proud."

One eye swollen shut from an errant blow, Quint looked at him as if not seeing and nodded dumbly.

Motherwell helped him up to the deck of the *Charlotte*. A sinking sun colored the sky pink, casting a soft, roseate glow on masts and bare spars and on the tense faces of Quint's company drawn up in formation to witness the ceremony. When he saw his mates, Quint straightened, shrugged off hands, and weaved his way to his punishment. For the sixth time he laid his body across the wood and was tied. Sergeant Campion whispered in his ear as he turned Quint's head away, "You're almost done, son. Show us how a Marine can take it."

Then he stepped back and sent the 76th blow onto Quint's back.

Spying from beneath the neck of Lily the cow, Ned Thornapple watched in horror as the rope landed with a splat and sent a fine spray of blood and fluid into the light of the setting sun. Ned cringed from the next blow as if it might land on his own body. He began to whimper, but could not tear his eyes from the horrible look on the face of Sergeant Campion, a man who gave him piggyback rides. The sergeant's bare right forearm bulged with muscle and sinew, and the big artery in his neck throbbed as he smote Quint's back again and again. Ned saw blood dapples on the sergeant's shirt. Below, convicts heard Campion's grunt and the smack of the lash and the high-pitched voice of Lieutenant Creswell on the count. Her face buried in Mary B's shoulder, Seedy sobbed and jerked with each blow. "Ninety-six," Lieutenant Creswell counted, echoing each blow with a tap of his ash against a boot. "Ninety-seven. Ninety-eight." Quint's head lolled. Seedy. "Ninety-nine." Sergeant Campion stepped back, panting, and looked at the lash in his hand, sodden ends dripping darkly onto the deck. His shoulders sagged. He looked as if he would let the weapon fall from his hand. What difference did one more stroke make? he thought. What good does the last blow do? One less would be merciful, one more meaningless.

"Ninety-nine, Sergeant," Lieutenant Creswell said, rising slightly on his toes. Sergeant Campion stood in the gathering gloom looking down at the bloody instrument, at the small pools of Quint's fluids now joining at his feet. "Ninety-nine, Sergeant," Creswell said again, his voice tight.

Sergeant Campion suddenly raised his eyes to the smooth, sleek face of Lieutenant Creswell and with an expression of fury and a snarl of rage reared back his entire body and loosed one final, terrible blow, then flung the lash at Lieutenant Creswell's feet. "One hundred," Creswell said flatly, "Punishment completed, Captain. Pick up the lash, Sergeant."

"Very good, Lieutenant," Tench said, "Sergeant Motherwell, take your man below so the surgeon can tend him. Dismiss the company, Lieutenant. Double grog for Sergeant Campion."

~~~~~~~~~~

Having lost nearly fifty pounds, weak, feverish, dispirited, Malvina Thornapple lay abed and groaned. After a week at anchor, her debilities persisted, and she felt she might die. She could barely muster strength to berate Dobber. "Wasting away!" she wept, "I'm wasting away, Dobber! Look!" She showed him a surplus of cloth at her waist, "And for what? Heaven have mercy!" she moaned, and turned away.

Thus, when the Marine wives on the *Charlotte* received Arthur Phillip's permission to go ashore for an afternoon of sightseeing and shopping, Malvina Thornapple did not join them. To their delight, however, Captain Tench volunteered to be their escort and interpreter, at least so far as his Spanish could carry him in this Portuguese city. Ned begged to go along. The Captain ruffled the boy's hair. "You won't get lost now?"

Demurely shawled as Tench had advised, the half-dozen women who stepped ashore were spellbound by dozens of slaves queued at the fountain to draw water. Others loitered before hoisting the heavy jars to their heads for the journey back to their households. The wives could scarcely tear their eyes from the near-naked black bodies, the splayed feet, the women's short wooly hair, the scarified faces and arms. They clung to one another and yelped at the sight of a furtive man with his black hat pulled low hurrying past, a dark cloak drawn over his lower face, his naked sword jutting beneath.

"Just the custom, ladies," Tench reassured. But they tripped on each other's heels keeping close and couldn't help glancing nervously at the growing retinue of black faces that followed them, for the English women were also novelties. No other white women were to be seen. From Tench the wives learned that Portuguese women stayed indoors except to attend church (which they might do two or three times a day) or public festivals. Even then they hid their faces with shawls or towering mantillas. Thus the barefaced Englishwomen attracted a crowd as they made their way to the steps of the Palace, where they waited on pins and needles for Tench to register their visit.

He emerged accompanied by Captain Desvia, a Portuguese officer he'd met when Arthur Phillip presented his officers to the Vice King. Chatting in a mixture of Spanish, English, French, and Tench's scant Portuguese, the officers strolled ahead, hands behind their backs. The women trailed behind, trailed in turn by a ragged appendage of curious blacks.

Once away from the waterfront plaza, the tourists entered a labyrinth of narrow streets shadowed by long, jutting cornices and crowded with noisy traffic of vendors and slaves. In

England the women went to the market, but in Rio the market came to the women. From the heads or backs of slaves calling out their wares, the women of each house opened a specially shuttered window to buy fruit, vegetables, fish, meat, kitchen articles, flowers, wine, jewelry, clothing, china, medicines, books, religious items, shoes, lamp oil, coal, perfume, linens - whatever a household might need, including furniture. Some merchants led processions of slaves, each bearing a box or crate or chest of the merchant's varied stock. More often slaves sold directly, sometimes from a cask or box chained to their iron collars lest they sell their wares and flee with the money.

Each twisting street seemed a home of some craft or specialty - tinsmiths, cobblers, joiners, jewelers, tailors - and at every street corner busy pedestrians, slave and white alike, took a moment to bow or kneel and make the Sign of the Cross before a holy shrine or votive. The women passed hundreds of beggars, black and Indian, too old, crippled, weak, or sick to be of further use. The darkened, narrow streets, teeming with exotic, sometimes frightening sights, the cries of the vendors, the rumble of carts drawn by muscular, sweating slaves, voices haggling in gibberish, and interminable clanging church bells created a disorienting tumult.

After an hour in this overwhelming old part of town, the women were weary. But Captain Desvia led them over a steep hill toward an unfamiliar part of the bay to show them the *Passeio Publico*, a pet project of the Vice King, the still-raw beginning of an extensive formal garden that led to a broad esplanade at water's edge. As slaves winched blocks of granite into place on the project, the visitors followed their Portuguese guide along the elevated walkway. They murmured polite approval at the new paving, the large bust of Phoebus, god of the sun, and of Mercury, god of commerce. They admired the bronze railing that people leaned on to look out over the bay. Ned raced ahead to climb the railing. Fetching him with authority, Hester Thistlethwaite scolded, "Haven't you seen enough water these past months?"

Taking his charges into octagonal kiosks at both ends of the esplanade, Captain Desvia explained paintings that depicted Brazil's important industries and agricultural activities - fishing, timbering, gold and diamond mining, rope-making, the growing of cotton, sugar, indigo, and cattle. After several attempts he gave up trying to explain how a certain cactus was grown because a certain insect fed on it that in turn was squashed to make a red dye. Locals looked at the English tourists looking at the pictures, and turned to view the pictures with renewed interest.

The women were tired and quiet and ready to go back to the ship, but when they passed through the garden again Hester Thistlethwaite spied a statue Captain Desvia had neglected to describe. "Ah, she is Ceres," he said, "A beautiful woman, *nao?*"

"But who is she?" Hester asked.

Tench responded that Ceres was the goddess of agriculture, and he pointed out the sheaves and the cornucopia at her feet.

Hmph! Hester grumbled to herself. Ceres feeds the world, and all Captain Desvia finds fit to mention is her looks! Just like a man!

As they strolled back through the narrow, erratic streets of the old city, they heard an approaching hubbub of rhythmic clapping, clacking sticks, and singsong chanting. They heard people shouting and frightening hollow booms. The confusion grew as a procession drew near. The women shrank against a house, transfixed by the sight of a powerful, nearly naked slave pulling a crude, four-wheeled wooden cart on which sat a large, upright barrel. Inside the barrel, clutching its brim, a haggard old woman wailed and shook her head, crying "*Nao! nao!*" A dozen musicians in colorful neckcloths surrounded the cart, clacking rhythm sticks, stamping their feet, and honking crude horns. The man who led the taunting chant carried a heavy felling saw that he beat against the barrel, frightening the old woman with the violence of his blows.

"What was that?" the women asked after the procession passed. Tench conferred with Captain Desvia. His mouth twitched with a little smile when he turned to explain, "They're going to saw the old woman in the barrel."

"What?" the women gasped, "Why that's terrible! That's cruel! What's she done?" Tench conferred again. There seemed to be a point he couldn't grasp, for Captain Desvia uttered many a "*Nao, nao,*" and grew animated in his explanations.

Meanwhile, as if drawn by a magnet, Hester Thistlethwaite edged down the street after the procession. She held Ned's wrist in a steely grip as she muttered, "Just wait, just wait, they can't saw her in half, they can't!" She followed the disappearing procession a few more steps, then looked back to see her party still clustered near Captain Desvia. Hester decided she could turn the corner and follow a little farther.

She'd never seen a hanging, but had heard that crowds followed the tumbrel to the gallows, jeering and hooting the condemned, a practice she believed was cruel. "If I'm going to be hanged," she once said, "I should at least be hanged with dignity." But to saw an old woman in half! Why, that was terrible! How could the poor thing deserve such punishment?

A slave woman came alongside, her face scarified with a hundred tiny ridges. She also towed a boy and smiled at Hester. She pointed to Ned and to her own little boy as an indication they were much alike. Hester returned a nervous smile, and the two women followed the procession, more or less side by side, on a route that seemed to head downhill. Hester decided she wouldn't worry about re-joining Captain Tench - she could meet them all back at the jetty.

The noisy procession halted in front of a large house with a fretted balcony from which several gentlemen looked down. The man leading the chant read from a sheet of paper, directing his words to the men above. After each sentence everyone laughed, and someone boomed the barrel with the big saw, causing the old woman to shriek and cry "*Nao, nao!*" and the crowd to laugh even harder.

"What can they be saying!" Hester cried aloud, half-addressing Ned. It seemed a strange courtroom. Now the crowd shouted responses. The man would read and the crowd would repeat a response. To Hester they sounded like Catholics at their prayers. Obsessed with learning what the old woman had done wrong and what was going on, Hester stashed Ned in a doorway and squeezed through until she came to a gentlemen who stood uncloaked and bare-headed, his costume indicating a fellow foreigner. "Excuse me, but what is going on here?"

Still wearing a smile, the man turned to her with surprise. He regarded the white mobcap peeking from under her pale blue shawl, her dark blue dress and white apron. Had he seen Hester Thistlethwaite when she first crept aboard the *Charlotte* at the Mother Banks he wouldn't have recognized her as the same woman. In three months at sea she had blossomed. "Ah," he said appreciatively, "*Uma Inglesa?*"

Divining his meaning, Hester said, "Well, yes, I am English. I'm a visitor here, you see, and I'd like to know what's happening. Why are they going to kill that old woman?"

The man threw back his head and laughed. "Ah, *non, non, madame*," he said with a heavy French accent. He took her arm with one hand and waved towards the cart with the other. "This is joke, you see, a *festivalle*. They not kill old woman. *Non, non*. She not old woman, she man. She wear dress like woman. She bad woman!"

"But what has she done?"

"Ah, English lady, is joke, *non*? That *senhor*, that man at house, he give thanks, child not sick now, *comprenez-vous*? Child good now. So he make *festivalle*, *non*? He thank God."

"But what has she done? Why is she in the barrel?"

The man looked at Hester Thistlethwaite with mild confusion, then laughed again. "Ah, they say she do bad thing. They say she bad little girl. She not obey. They say she bad mother. They say she not good cook. She not let young girl, how you say, court young man. And she, how you say, talk too much - talk, talk, talk all time to *èpoux*, to husband."

"You mean she's a nag? A scold? A shrew? A fishwife?"

"Fishwife?"

"Never mind. You means she's a henpeck?"

"*Mais je ne*...ahhh, ha, ha! *Oui, oui*, she like chicken, *non*? Pick, pick, pick" and he winked and plucked her arm with his words, laughing. He turned his attention back to the scene, listened, and laughed again. "Now they say she not keep clean house. And she bad, how you say, *grandmère*?"

"Grandmother?"

"*Oui, oui*, grandmother. And she too, too, how you say, *irritaval*?" Hester looked blank. The man tried again. "She, ah, ah, bad temper, *non*? Bad temper all day. *Oui*, and she not good to little ones."

"But what are these people shouting, why are they laughing at the poor woman?" The man understood her gestures.

"Ah, they say, 'Saw old woman in barrel, Saw old woman in barrel!' Ha, ha! But is joke, *non*?" and he laughed again as he squeezed her arm.

Hester pulled away, vaguely angry. "Thank you," she said, "Thank you very much. You've been very kind." As she turned to fetch Ned, the man bobbed a bow and winked, then shifted his attention back to the scene and began to laugh again.

But Ned wasn't where Hester left him. "Oh, My Lord!" she thought. In a panic she scanned the street, the crowd, the doorways. "What have I done?" She ran back toward the street where she'd strayed from the others. "Ned! Ned! Ned Thornapple! Where are you?"

~~~~~~~~~~

Ned had gone off with Mako, who was also almost ten and also resented being towed along like a little boy. Somehow they communicated their resentment to each other, and after Hester parked Ned in a doorway and Mako's mother began a self-absorbed shuffle to the rhythms of the chant about the old woman in the barrel, the boys sidled closer and closer, making funny faces. When close enough, they clasped hands and turned to race up the street with giggles of delight. Mako led. They turned a corner, then another, giggling all the while, then walked and ran for what seemed a long while, sometimes chasing or playing brief hide and seek, until they came to a crude bridge over a ravine. Mako led the way into the ravine, yelling, "Yi, yi, yi, yi, yi!" Ned followed, "Yi, yi, yi, yi, yi!" They threw rocks into the little stream, launched debris as boats, caught frogs and grasshoppers, had a long sword fight with sticks (during which each died a number of times with great drama), and waded in search of treasure (Ned having first to take off his shoes, which Mako put on to strut about with a proud chest). Tiring, they then played more quietly, sometimes separately and sometimes together, making their own private play sounds which the other might begin to imitate. They exchanged a few words. Ned taught Mako "shoe," and they exchanged words for "mouth" and "nose." They both thought "pee" was very funny.

As the light turned dusky, Mako suddenly pointed to his mouth, made chewing motions, and scrambled up the side of the ravine. He paused at the top to yell, "Yi, yi, yi, yi, yi!" Having to put on his shoes, Ned sat down and yelled back, "Yi, yi, yi, yi, yi! Wait for me!" but when he emerged from the ravine Mako was gone.

Ned had no idea where he stood. Thankfully, after looking around he spied the gleam of the setting sun on a fortress atop a steep hill that he'd seen from the *Charlotte*. He knew the fortress was near the waterfront, so he pointed himself in that direction and hurried along the narrow, darkening streets.

He couldn't travel in a straight line. Streets turned this way and that, ending at a church or a little square or a dead end. Shutters banged as people closed up their houses for the night. The church bells fell silent. A muffled figure hurried past. Ned heard the slapping feet of a tardy slave. He caught one last gleam of the hilltop fortress, then completely lost sight of his destination. He began to whimper, and when he saw four watchmen with long lances marching up the street, he hid under a bush until they passed. Weary, hungry, not knowing what to do, he saw frightening visions of goblins and ghosts and bugbears and crippled beggars. He cried quietly into the crook of his arm. This was unfair. He felt angry at Mako, at Mrs. Thistlethwaite, at his mother.

He must have fallen asleep, for the next thing he knew he was raising his head in the dark to listen to a far-away lowing. Lily the cow! He scrambled from beneath the bush to turn his ear this way and that. Aye! there it was again! Hurrah!

He set out on light, hopeful feet along the dark streets, encouraged by corners lit by weak, flickering pools of votive light. He hurried along, trotting as quietly as he could, pausing to listen for Lily's periodic moo. As the mooing grew louder his spirits rose. Now he heard other cows on other ships and knew he would soon see the water. One last corner and he found the bay stretched before him. But it was empty of ships.

He stood dumbfounded. Had his parents sailed off without him? Was he left here alone? Across the water he saw the vague outlines of familiar steep peaks. And there, looming like a mountain, was the hill with the fortress! But where was he? He heard the lowing again, coming from nearer the water. Leaving the last of the buildings behind, he trotted toward the bay and in a few minutes came to a long, low, half-walled shed nearly at water's edge. The mooing came from the shed.

Its stone walls rose a foot or so higher than his eyes, and Ned scrambled up so he could see over the top, and tumbled back with a cry of fright when his nose touched the broad nose of a horned cow. Collecting himself, he climbed the wall again and spoke to the cow as Lieutenant Creswell had taught him, "Sooo, cow, sooo." He heard other cattle getting up, blowing, shitting, pissing. He heard the click of their hooves as they ambled over, curious about the visitor. Ned scratched between the eyes of one, then another, and another. After he'd scratched several, the animals became insistent, and thrust their heads forward to be scratched still more. He scratched until his fingers cramped, then found a stick and scratched with that, all the while talking quietly, giving names to the barely discernible faces, telling them his troubles, enjoying their warm, vegetable breaths, their heat, their smells. After a time he grew sleepy and made a nest in a nearby pile of hay. In a few moments he roused himself to call out "Good night, cows," then curled up again and fell asleep.

Shouts and bellowing cattle awakened him. Disoriented for a few moments, he lay still, then scrambled up in the gray light of the predawn. In the shed nearly naked black men yelled and thrust at the cattle with lances. He saw two men with huge muscles swinging broad-bladed axes. He ran to a nearby fence and stood on a rail better to see. Axes rose and fell. Blood spurted into the air. Bawling with fright, panicked cattle jumped half over the wall to escape. The cattle crowded into corners, trying to hide behind each other, but the lances jabbed and beat them apart. The animals slipped in gore, stumbled over each other, fell heavily. They shook the earth. A cow he remembered thrust its head over the wall, its eyes rolling with terror, then an axe bit, nearly severing her head. Riveted by the carnage, trembling with fear, horrified, Ned watched the axes rise and fall and rise and fall as blood gushed and cattle bellowed for what seemed an endless time. But at last the killing ended. The shed quieted except for the occasional, futile rousing of an expiring animal and the subdued, foreign voices of the butchers. Rio had its meat for another day.

Ned climbed down and walked to the water's edge, numbed by his experience. Morning sunlight sparkled on wavelets. Hundreds of gleaming gulls wheeled and screamed in anticipation of their morning offal. He stood looking over the clean blue water, bright with morning light, then unconsciously turned to the left and followed a rutted track along the shore of Boqueirao Beach toward the fortress on the hill. He walked flat-footed, eyes to the ground, his memory spewing bloody images of slaughter, of Quint's cruel punishment, of shrouded bodies slid into the sea. In a surprisingly short time he rounded the fortress hill and saw the familiar ships of the English fleet. On the jetty near the pyramidal fountain Captain Tench waited with a swollen-eyed Hester Thistlethwaite and Dobber, Ned's henpecked father.

"Yi, yi, yi, yi, yi!"

Stork turned to see Ned hanging over the afterdeck rail trying to catch John Coffin's attention. Since his disappearance and rescue a couple weeks earlier, the boy had sought out Coffin, the only black convict aboard the *Charlotte*. "Yi, yi, yi, yi, yi!" Ned called again. Coffin reluctantly tore his rheumy eyes from the English slaver making its way to anchor and sent a brief, pre-occupied smile and wave to the boy, who waved back with a wide grin.

"A slaver! A slaver!" rippled through the *Charlotte* when someone identified the pennant. People crowded to watch the drab little ship's progress. To most on the *Charlotte* the world of slavery was as remote and exotic as the world of royalty, for slaves no longer lived on England's soil. In truth, most on the *Charlotte* had never thought about slavery one way or another, just as they'd never considered the world beyond their own villages or neighborhoods. What did England's endless foreign dithering and wars mean to them? Its discoveries and quarrels, its commerce and disasters and scandals and endless parliamentary debates? All were the dim business of the ruling class. Aye, till that dim business was made real by a new tax or impressment or another reality. Like the slave culture of Rio.

"You never saw anything like it," the wives exclaimed when the to-do over Ned's disappearance settled down, "Why, there must be ten slaves for every white, even children. They do everything! Fetch, cook, carry, clean, dig! They bear the country on their heads and backs! Why, we saw poor old blackamoors so crippled by bearing they couldn't stand, their legs twisted this way and that. Lord! And the way they just turn them out when they're no good anymore! Beggars on every corner, in every doorway, a more pitiful sight you never saw! It's a wonder they aren't sent to slaughter when they're too broken down! What kind of a Christian country is this?"

"Indeed, no different than your own," John Coffin might tell them. "It was you English who bought my mother on the Slave Coast and hauled her to Jamaica." Aye, his mother and innumerable others who cleared the Empire's land in America and the West Indies, who drained its swamps, grown its tobacco and sugar, its cotton and indigo and rice, who gave suck to its children, put rum in its glasses and marmalade in its pots. Multiply by almost 200 this little slaver coming into Rio, and that was how many English ships still hauled slaves to the colonial outposts of Portugal, Spain, Germany, France, and the Netherlands. Aye, John Coffin might say, "Jolly old England, your Christian home and mine. No slaves on England's righteous soil, but the world's biggest trafficker in slaves!"

The shackled convicts followed the progress of the slaver to its anchorage near the Isla de Cobras. "Well, thank God we're not on that ship!" James Martin said. "I hear they crowd three and four hundred on a ship not any bigger than this old tub. Aye, and starve them too, so they're too weak to make trouble. And if they lose one out of four they still turn a profit."

Will Bryant spit a gob. "They got chains, we got chains."

James looked at Will's impassive profile and shook his head.

They watched the disgorgement from the English brigantine. Even from a distance they could see the emaciation of the naked cargo. They stumbled often, enfeebled by starvation and crowded confinement. Wearing thick, iron collars, the slaves were looped together on deck in strings of ten or twelve by chains run through the collars. Prodded into jollyboats, they were rowed close to shore, then made to jump into the surf and wade, some so weak waves repeatedly knocked them down. To the accompaniment of curses and laughter, sailors jerked them upright again by their hair or iron collars. After the slaves staggered ashore and onto the quay, they were plopped on the ground under the eye of an overseer. The patch of black bodies spread like a spill as more and more emerged from the maw of the slave ship. The women on the *Charlotte* gasped when they saw the females, some carrying children, many naked except for their iron collars. Although unchained, the women too struggled through the surf to shore. One carried a small child on one arm and tried to keep another's wailing head above the waves. She could scarcely drag him along, and the little slave choked, cried, coughed, and fought until a sailor plucked the half-drowned child from the water by an arm. The jollyboats shuttled to and fro.

John Coffin knew what would happen next. When the ship emptied, the slaves would be herded to shelters at the edge of town, where an abundance of fruits and vegetables waited. At first from hunger, later by threat of force, they'd gorge themselves on an endless supply of bananas and oranges and plantains and watermelons and cassava until they regained strength and muscle lost during months of scant food. Then came the auction block. Some would resist by trying to escape.

Some would commit suicide. After being sold, some would toil in gold and silver mines, others hack out plantations in the rainforest, still others pull carts, push sugar cane presses, pick coffee, wash clothes, cook meals, sew, and vend goods in the streets of Rio. These last were the lucky. In the mines and rainforests life expectancy for a slave was five or six years. Although this aspect of its economy was not depicted in the kiosks of the *Passeio Publico*, Brazil required a constant re-supply of slave labor brought by European ships running the slave ship shuttle.

The slaves were formed into a line to begin a doleful procession toward the feeding sheds. A whistle blew on the *Charlotte*, sending John Coffin and the convicts below. Reluctantly he tore his eyes from the black bodies disappearing behind a warehouse. Perhaps a relative moved among them, a nephew or niece.

"I'm saying good-bye to my family," his mother said when he'd heard her feverish lips forming syllables strange to his ear. Tired and worn and old, soon to die, her fingers plucked the worn coverlet in her little room off Master Hoare's kitchen. Years earlier, when John Coffin was perhaps four or five, Master Hoare told her it was time for her little boy to learn to serve. That's when she'd first intimated to John why his life would be different from then on.

As he grew older, John pieced the story of his heritage together - how men with guns had surrounded the village of his father's family, killing or capturing almost everyone in the compounds, using the butts of their muskets to cave in the heads of the old and the sick. He learned something of his mother's confinement aboard a slaver, how those who died were simply tossed overboard like refuse. He learned something of her stoop labor in the cane fields of the West Indies and of how, when her belly grew big with him, her precious fruit of Africa, Master Hoare brought her into the house to cook and clean.

After Master Hoare made plans to take her to his estate in Exeter, she'd begged him to take her new baby too, and he acceded, for he was not ungenerous. "But I'll have no nigger talk in the house. He'll speak English." Perhaps that's why his mother never told John Coffin how he came by his name. She'd named him Koa Fena, after his father, but having trouble with the foreign sounds, Master Hoare announced that "Coffin" was close enough and that "John" was a good Christian name. That was one evening after he'd taken his pleasure with her.

At age five John Coffin began a life of liveried servitude, complete with lace ruffles and wig, serving but not seen, and assumed ignorant. When his coffee-colored sister reached age five she began servitude by helping his mother in the kitchen. Some time later Master Hoare informed his mother the law was changed and she and her children were no longer slaves. They were now servants. Her pay would be a shilling a month, and he'd allow John and his sister to work for their keep.

John Coffin proved to be a good servant. After a half dozen years of anticipating the needs, desires, and whims of Master Hoare, he moved with assurance through the polished brass and oiled walnut of the aging man's elegant house. But when John's mother died and his sister took over the kitchen, Master Hoare sought to use the daughter as he had the mother, and John Coffin rebelled. The plan he devised for a new life was neither good nor practical, for he thought that by filching a few pieces of Master Hoare's silver he could secure passage to Boston for both him and his sister. But betrayed by the fence, he was instead sent to Botany Bay. Master Hoare promised the court that he'd look to the welfare of John Coffin's sister.

"See any brothers out there?" Will Bryant asked with a snort. John Coffin stopped, murder in his eyes. James slid between them. "Easy, John," he said, putting a hand on Coffin's shoulder. Coffin shrugged off the hand, his eyes unwavering. After a seemingly endless, silent, staring contest, Bryant said, "Ah, never mind," and slid into his compartment.

~~~~~~~~~~

Aboard the *Sirius*, Arthur Phillip penned the final words of a letter to Lord Thomas Sydney, Home Secretary. "*I hope to sail tomorrow, as I only wait for the accounts to be settled with the contractor....*" He sat back and massaged his neck. This letter should make them sit up and take notice. For all the sniping in England, he was proving his critics wrong. So far only sixteen dead. He realized now he should have put in at Port Praya for water and fresh food, but with the winds at the harbor entrance such cat's paws he feared ships might get entangled in the little bay. With better luck he could have made up for his error and taken on water at that damned dot of an island called Portuguese Trinidad in the middle of the south Atlantic. But they were unable to find it, and when they reached Rio they discovered their reckoning was off by 25 miles.

He sipped sherry and looked over his letter. "*I have the pleasure of saying that every assistance we have wanted in this port has been most readily granted.*" He still didn't trust the Vice King, however - a sleek, portly man, immensely rich and immensely greedy. As Don Luis de Varconcellos had extended a pudgy, bejeweled hand for Arthur Phillip's obeisance he'd announced, "*Ah, Vossa Excelencia, o meu pais fica enriquecido por causa de sua presenca,*" which Arthur Phillip interpreted to mean "You're going to pay through the nose for what you need."

But referencing when Portugal paid him handsomely for his services in Rio, Arthur Phillip returned: "*Ah, e eu, Vossa Excelencia, teno side enriquecido au estar aqui.*"

The Vice-King glanced at his secretary, and after a moment's consideration, rose from his chair laughing, and advanced to shake Arthur Phillip's hand. Thereafter, credentialing letters or not, Arthur Phillip was treated with the honors due a colonial governor, and found himself saluted with dipping flags and trailed by an honor guard whenever he set foot ashore.

~~~~~~~~~

As Marines lounged on deck enjoying their last rose-gold sunset over the hills of Rio, Porter asked Quint about his book.

"Blimey, " Quint blurted, "I never thought it'd be so hard!" And as if he'd been waiting for just such an opportunity he eased his still-injured body to his knees, then his feet, and began to lecture. "Blimey, Porter, did you ever think about where a book begins? It's not easy. We started in Plymouth, right? But we only went to Portsmouth! So is Plymouth the beginning, or Portsmouth? Portsmouth is when we really started for Botany Bay."

Several Marines abandoned their gossip to cock an ear towards Quint's earnest declamation. "So suppose I was to begin my book with Portsmouth. But suppose I never get to Botany Bay - you know, disease or an accident or something."

Porter grinned and said, "No more coney for old Quint!" A burst of laughter.

Quint grinned. "Right! But listen, that's what's hard about writing a book, trying to figure all that stuff out! I never thought just thinking was such hard work, but sometimes my head just pounds!"

Porter favored the group with a foolish grin. "Me, too! Sometimes my head pounds trying to think a hard-on to pound on." The Marines laughed roundly.

Quint's smiled was perfunctory. "Gentlemen, I think my book should be about what we see when we get to Botany Bay - the savages and animals and birds and the things that grow there. It should be about what we do after we get there, about how we make a new land, like they did in America."

Someone said, "That's good, Quint."

"But there's something I haven't figured out, mates, and that's who my book is about."

"Quint!" Porter cried, "You said me and Stork would be in the book, that we'd all be in it!"

"Well, yes, but now I'm thinking, who's going to make this new land? You, Porter, or you, Stork? Any of you? No! As soon as my time's up I'm heading home. And you are too, right?" Heads nodded.

"Well, then," Quint asked, indicating the below-decks, "Is it them down there? Those thieves? Those felons? Is that who's going to make this new land?" There was thoughtful silence. "I don't know," Quint said, "How can my book be about them? It just doesn't seem right." Forgetting his injured back, he started to shrug, then winced with pain before he shook a finger at his audience. "But I know this, mates - maybe I don't know who my book is about, but I'm not going to start till we get to Botany Bay! That's the real beginning of my story. Botany Bay."

Cape Town
1787

*I*n the early evening of September 8, four days out from Rio, Mary B squatted over a blanket, red-faced and sweating, her lips bloody, pushing at the child within and fighting to free herself of its symbiotic mass. Kneading Mary B's lard-greased abdomen, Liz Bason encouraged and instructed. "You're doing fine, Mary B, real good! Big breaths now, big breaths. That's right! Tell me when you feel it coming. Do you feel it? Yes? Push! Push! That's right, push! You're doing good, Mary B, he's coming, you're working good, Mary B!"

Liz inserted her fingers. "I can feel his head! That's good! He's coming the right way! Big breaths now, big breaths! Is he coming again? Yes? Then push! Push!" From her compartment above, unconscious of gnawing her lips until they bled, half-sick with fear, Katie was fascinated by the scene below. She couldn't help thinking of a snake swallowing a frog.

"He's crowning!" Liz announced, "He's crowning! Give him another good push! That's right! that's right! Good, good, good! There now, love, he's crowned. You can rest a minute. The worst is over." Liz rested her hand on the protruding crown to prevent an explosive birth. A dozen women crowded the lamplight to murmur comment and approval. Breathing in frantic, raw gasps, Mary B grunted at the red pain, more searing than any she'd ever imagined. She felt Seedy press a wet cloth to her brow and nodded appreciation.

"Give her some more caudle, Seedy. Drink, Mary B, you need to keep your strength up." Mary B sucked at the cup of spiced wine. She'd begun sipping the potion early that morning when her waters broke and Liz Bason sent word to John White. Weeks earlier the surgeon had knelt to place a hand on her swollen womb, pressed one ear to her abdomen to listen, and asked how she felt. Not knowing how she should feel, Mary B said she felt as good as could be expected. John White looked up askance. "It'll probably be a normal delivery," he pronounced. "I think you'll do fine by the women, but let me know if you need me." Mary B nodded, not sure what to say. She feared John White and his cold, metal instruments. She found his breezy self-assurance irritating. What did he know about carrying a baby? About being a prisoner of this swollen thing within, of this thumping life that broke her sleep and sent bile to her mouth and made her leak? How could he tell her baby seemed normal? What if it was misshapen or had no fingers? Or no nose! One woman whispered of seeing a newborn with a hole for a face, and that night Mary B dreamed that rats gnawed her belly and ate the face of her baby. She waked in terror and buried her face in Katie's bosom.

Another night she was sure the baby was dead because she couldn't recall feeling movement for several hours. Trembling, she roused Liz Bason. "He's just asleep, love," Liz reassured. "Even babies inside have to sleep. Just wait, he'll stir soon enough." She stroked Mary B's forehead. "Come in with me and Little Jacob for the night." Later when Mary B felt the baby's familiar bump, her heart soared. She could have kissed Liz in gratitude. How many times Liz had quieted Mary B's fears!

Weeks earlier Mary B waked Liz, convinced she was going into labor. "No, love, it's not your time yet. It's what they call a false labor. It's your body getting ready." Liz giggled. "I thought the same with my first. I guess we all do. How could we know?"

As the day of Mary B's deliverance crept closer, the prison buzzed with the women's stories:

"You remember that Irish giant that came to Bristol town? O'Brien? Aye, more than seven foot tall he was, and with a prick to match! Well, I heard a preacher say they took him to the convent. Aye, to breed the nuns! Aye, because the Pope said! He wants Catholics to be giants and take over the world! Aye, he does! But the nuns all died. Aye, from his big prick. Aye, and then they had a secret burial in the convent. Aye, at midnight. But one nun had a big cunt and she didn't die. Aye, and she got so big they had to haul her in a hand barrow. Aye, and when she came to term they cut her open. Well, the baby weighed two stone, don't you know! Aye, can you imagine, two stone! Aye, and now they keep him in the convent! Only two years old and he weighs five stone! Aye, the preacher said!"

"Well I saw a birthing once where the poor woman was in labor for two days and worn down to a nothing, mind you. Wanted to die! Well, the surgeon had to come from the next town, and when he rushed in, Heavens! What do you think? He'd forgot his knives! Well, the baby was stuck, he said, so big he'd never come out, so what do you think? Why, he took a poker from the fireplace and poked in and broke the head! Aye, like a pumpkin, mind you! And all this time the poor woman was thrashing around so it took six of us to hold her down. I had her left arm myself, mind you. My, like a bucking horse she was! Well, he had to do it, see? The head was so big. He had to break it. We all heard...crack! We did! I swear I felt that 'crack' in these very hands! Lord, the sweat just poured off the poor man's face, and we all heard it...crack! Well, there was no way that thing could've come from any normal woman! Aye, and then he went in with the fireplace tongs and started pulling pieces out, little red pieces, like chunks of pumpkin rind. Aye, he did! And then he just broke off more! Well, he did! I saw a piece of arm myself, and then a leg from the knee down, and maybe a piece of foot. Lordy! It was awful! And the poor thing suffered two days more before she died. Bless her! From the bleeding, mind you! A blessing, we all said, truly. Truly. A suffering saint, she was. A martyr, mind you. But at least the vicar came to pray before she died. And then, wouldn't you know? her husband married again that same summer. Aye, took a mere girl, mind you!"

"Well, when we was in Bristol jail didn't that woman from Cornwall throw a black monster? A head like a calf! And a tail! But with arms and legs, you know. We saw it, didn't we, and it cried like a bleating calf. 'Beeaaahh, beeaaahh!' And then they took it away and burned it. We could smell the stink from the hair and all. It wasn't human hair, you know. And when that Cornwall woman smelled her little monster burning she went stark mad, didn't she? Bashed her head against the stone till she killed herself. Aye, bashed the back of her head till it was mush. And her blood and brains started oozing out. And then the turnkey's dog licked it up, didn't he? Aye, licked up her brains. My, that was the smartest dog!"

Mary B tried to avoid these horror-story sessions, and would move away to think of pleasant things - her memories of a sunny sheep meadow or of childhood games with her sister Ann and Little Bill. But Katie couldn't stay away. Perched nearby, her pouty mouth tight with fear, her green eyes inwardly picturing the bloody scenes, she unconsciously squeezed her swollen belly, her narrow hips.

Other discussions were more clinical. Should Mary B take to her bed like a gentlewoman, or go about as always? And how should she deliver? Abed on her back, or squat, or sit like the

French did, between two chairs? Someone told of a midwife who made her women stand and hold an overhead rope.

As Mary B hoped, Liz Bason took charge. Liz had borne two earlier babies besides Little Jacob, and knew enough to quiet Mary B's fears of labors that failed to deliver, of a surgeon's cold knives and cruel clamps, of bloody deaths, evil spells, and monsters. While not a midwife, Liz had attended many birthings as a gossip, a trusted friend of the delivering woman. Gossips not only assisted, but bore important witness, for sometimes when a woman was distraught by the burdens of too many pregnancies and too many children, fearful of another mouth to feed or of superstitious signs, she might snuff out a life just delivered. But attending gossips could protect an innocent mother from accusations of infanticide. More important, with one of ten women perishing in the birthing process, a woman wanted the support of close friends during this frightening, dangerous experience. "They say men are bound by battle," Liz Bason said, "but none are bound like women in a birthing room."

Birth was seen as a natural event, and interventions were uncommon. A midwife who knew the stages of pregnancy and the process of birthing served primarily as a coach. She instructed, encouraged, and comforted as the laboring woman's body carried out its excruciating but natural task. Amidst fear, agony, and blood, the midwife had to know when to send for help, usually for a surgeon with his instruments, but sometimes for a preacher with his prayers.

Despite their sibilant stories rooted in ignorance and superstition, the women in Mary B's birthing room were now a better group. Arthur Phillip had transferred the Bristol clique of troublemakers to the *Friendship*, and brought over a half-dozen of that ship's best behaved to the *Charlotte*. He believed in separating good and bad eggs. For one young officer on the *Friendship*, stories of whoring, drunkenness, cursing, and brawling had become his titillating stock-in-trade. "By God, I was glad to see that Dudgeon bitch flogged," Lieutenant Clark grinned at a Saturday night mess in Rio. "She asked for it, and we gave it to her too, two dozen on her bare back, and the bitch screaming like you never heard, awful, filthy curses. God! The women on that ship, the way they carry on! London stewmeat! Turn your back and they rut like stoats." His eyes glittered. "Aye, but she got what she deserved, she did, that whore!"

"All right now, Mary B, push again! This little man wants to be born. It's almost over. You're doing good. Again now, push! That's right! His shoulders are coming. Good, good! I've got his arms! Good, Mary B, good! Push again! We're almost done! There!"

And suddenly the baby was inside her no more. Mary B vaguely felt its absence as she collapsed against Seedy, her body still contracting. She went with its paroxysms. As from a distance she heard, "It's a girl!" and then a keening little cry, and she felt Seedy's fingers squeeze congratulations and then the welcome, cool cloth again.

"A beautiful little girl, Mary B. You'll have her soon." Liz snipped the umbilical cord with her sewing scissors. Relieved the terrible pain was over, a part of Mary B didn't care. She became aware of Seedy pulling at her wrists. "You can let go now, Mary B, it's over." And gradually she did let go, releasing the stanchions she clamped with fingers frozen into claws, the red mass of pain that had throbbed larger and larger, and a thousand anxieties and fears. Her head lolled. Seedy put the caudle to her lips, "You did good, Mary B, real good!" Others echoed Seedy's praise.

And there was comment. "Aye, it's hard, they say, the first." "I remember my ma. She said a man would never know such pain." "My ma, too!" "But we do it, don't we? We always do it." "I hope I do as good as you, Mary B! You did good!" There was another chorus of approving agreement, and Mary B, slowly coming back, nodded and smiled a little. Katie listened, silent.

In a few minutes Liz Bason had wiped and swaddled the infant. She offered her to Katie, but Katie shook her head and withdrew. She hadn't spoken for some time. After another woman took the bundle Liz bent to massage Mary B's abdomen. Crooning and murmuring words of comfort and encouragement, she worked until Mary B felt another powerful contraction, went with it again, and sensed a soft, mushy mass leave her body. Liz picked up the bloody placenta and held it to the lamplight. "Good," she said, and deposited the tissue in a slop bucket. She rinsed her hands and bent once again to swab Mary B's loins and genitals with John White's all-purpose elixir, oil of tar. "Raise up a little, love," she said, and bent to place a cloth beneath Mary B. "There, that's it for now. Just lie back and rest, and give me the little girl." Liz took the swaddle and peered into the squalling red face, checking it once more.

"Here she is," she said, presenting the tiny bundle of child. "A beautiful baby girl, the first born on the *Charlotte*, and after all these many months! Good for you, Mary B!" A patter of applause congratulated Mary B as she took the bundle and regarded the clamant little face. "You'd best give her suck. She'll settle down then, and you'll heal up faster." As Mary B kissed her baby she smelled the infant's sweet, slightly acid odor and saw the tiny mouth make sucking motions. She poked in her purplish nipple and felt little tugs, surprisingly strong. "Just relax, Mary B, and let your milk come." Liz winked at Seedy, "And I'll try a wee bit of caudle now."

Mary found herself smiling at the little face, and when she saw a dribble of translucent milk leak from the minikin lips she smiled up at Liz to a chorus of soft ohhh's and another spatter of applause. The terrible, searing pain was already bowing out of her consciousness, retreating to some deep recess of memory. Led by Liz, the women finished off the caudle with toasts to Mary B and her new baby.

Next morning, John White came down to inspect the newborn and pronounced her as healthy as he'd ever seen. As word spread that a fine baby girl was born the night before on the *Charlotte*, a wordless satisfaction pervaded the prison transport. James accepted Will Bryant's congratulations with an involuntary grin. "At least I won't have another namesake to deal with," James winked. But the pride in his voice betrayed his comment. Nonetheless, deep down he felt like a baited worm tugged this way and that by hungry fry. He imagined what it would be like to have his wife's little James to play with, to tickle and make laugh. But he'd acknowledged his obligation to Mary's baby too. And what if he never saw his wife again? What if he never saw England again?

During Mary B's convalescence, Seedy took over as their group's mess matron and passed on congratulations from James. Days slipped by. Several times James thought of asking to see Mary B and the baby, but he never did. Three weeks later, taking his turn at the pumps, he saw her on deck in her familiar gray cloak. She carried a little bundle. James waved to her, his tanned face wearing a huge smile. Mary B smiled widely in return and raised the baby so he could see. James thought Mary B had never looked so lovely.

~~~~~~~~~~

Tench stripped a film of sweat from his brow with a finger and regarded Will Bryant's letter with mild dismay. He'd stuffed the letter between the pages of Boswell's *Tour*, then forgot to send it on a homebound English whaler just putting into Rio. Now the letter would have to wait until the fleet reached Cape Town. But Tench knew it wasn't unusual for a letter to take three months or a year to reach its destination from a distant port. Most writers sent two or three versions of their correspondence by various ships, never sure which, if any, might finally arrive, or when. Tench wondered if he'd ever censured a letter from the convict they lost overboard that morning. He couldn't remember. He wasn't even sure he remembered the convict's face - a man named Brown.

Will Bryant's folded sheet was addressed in a surprisingly practiced hand: "George Bryant, Trewardreath, Cornwall." As he began to read, Tench was again reminded of letters he'd censored in the American War - the language so simple, so direct, as if there was no time for flourishes and formalities. Were they premonitory?

Bryant wrote that he'd turned matters over many times since leaving Plymouth and now knew he might never return. "We have been almost four months at sea and are not yet half way to Botany Bay. To say a place is the end of the world doesn't speak to the matter. Unless you try to go there, you don't know what you say." Tench raised an eyebrow and read the passage again. A felicitous insight. Bryant went on to say they were treated fairly, but found the idleness hard. Then, "George, dear brother, I want you to try to get my boat back. If you can get my boat, Ann will never be on the parish, for many would fish who only lack a boat, and she would do well by them, and also the children. Do what you must. You have my blessing. Perhaps she should marry again. At least then she will have a life." A few remembrances and blessings and a curiously embellished signature, "Wm Bryant."

Tench reread the last few lines. Was Bryant married? Was he saying his wife should take up with another? But maybe, despite the children, they weren't really married. Or perhaps she was a relative. Shaking his head, Tench folded, initialed, and sealed the letter. For some inchoate reason, Will Bryant's letter saddened him.

He wiped another finger of sweat and sprawled in his chair, unconsciously balancing against the ceaseless lifting and plunging in house-high swells. Outside, a cold, westerly wind shrieked. He tapped Bryant's letter on the little table while his eyes roamed the familiar knots and shadows of the whitewashed bulkhead. The candle twitched at each tap.

Why Bryant? he wondered. Almost every fisherman along England's south coast smuggled. Fishermen smuggled, merchants smuggled, officials, gentry - anyone with half an opportunity. Smuggling was the rule. So why exile a small-time smuggler when the First Fleet sailed past a hundred others in the Channel? What sense was there in sending one smuggler to Botany Bay? What sense was there in leaving a wife and children without a provider?

But perhaps there wasn't sense so much as machinery, a huge, clanking machine of law and punishment. Something happens - an untoward event, a moment of weakness, of passion - and suddenly a man was caught by the scruff and carried aloft by the machinery of law, helpless as Don Quixote. It was a matter of being found out. Some got caught and punished, while a hundred other events and moments of weakness and passion went unpunished. Of course, those who pulled levers and engaged gears and hanged and imprisoned and flogged and exiled thought they meted out justice. But, Tench had come to realize, they only stirred the stew.

Cause and effect! Do bad and you get punished; then you will no longer do bad. Very rational! After Quint's flogging the Marines certainly attended to business with alacrity - for a couple days! Then it was back to the usual. And what did Quint learn, that a man was built for fucking and not for flogging?

Tench stared at the whitewashed deckbeam, its every crack, crease, and shred so familiar he could draw them from memory. Of ten thousand acorns dropped from a tree, one had taken root and survived. For a century, perhaps two, the little oak grew, changed with the seasons, fed squirrels, deer, birds. Then someone chose this particular tree to cut down, square up, and ship to a warehouse to lie for years until chosen for this ship to become this beam, this grainy, whitewashed metaphor. What a marvelous mystery, how things came together! If he could know why this particular beam had come to be here, he might know why Brown was lost overboard that day.

Tench wiped his brow again and regarded his wet finger. Shit! He sighed. Was the voyage getting to him? Here he was, almost thirty years old, and still losing himself in metaphysical

111

swamps! Did he expect some monkish vision would suddenly gleam from his candle, some astonishing insight?

Tench moved Bryant's letter close to the flame. What if he'd failed to discover the letter in Boswell? Was misplacing it part of some larger plan? some predestination to delay its delivery? Was finding it tonight predestination? And what if he burned it? Would that be the same as not finding it? Or was finding and then burning it part of a larger design? Or what if he started to burn it, then changed his mind? Or what if he started to burn it, then changed his mind but was too late and the letter burned anyway? Was all predestined? Was he predestined? Was there such a thing as predestination?

Tench snorted and tossed the letter aside. Bryant! How absurd that they didn't have the convicts' records! They'd get to Botany Bay and know nothing about him or any of the others. Was this oversight also part of a larger design?

Tench poured himself a glass of canary and drank it off thirstily, shuddering. He lit the gimbaled lamp by his bed, blew out the candle, undressed to his shirt, and stretched out. He poured another glass and drank it off at one draught. He felt lonely and vulnerable. "Goddamn ocean," he said aloud. He picked up Boswell's *Tour*, hoping the book would carry him away.

When he riffled the pages a dried flower fell out. He recognized the token, and sniffed it. Hardly any scent remained. Like his dreamy infatuation with Esperanza, the scent was but a faint memory. Riffling the pages again, he noticed Novembers, Octobers, Septembers, Augusts. He looked to see if Boswell had an entry for this day, September 19. He felt a little spurt of excitement when he found one. The subject was wives. More predestination? He read, *At breakfast Dr Johnson said, 'Some cunning men choose fools for their wives, thinking to manage them, but they always fail. There is a spaniel fool and a mule fool. The spaniel fool may be made to do by beating. The mule fool will neither do by words or blows; and the spaniel fool often turns mule at last: and suppose a fool be made do pretty well, you must have the continual trouble of making her do. Depend on it, no woman is the worse for sense and knowledge.'*

Tench re-read the passage, then laid the open book face down on his chest, and closed his eyes. "Pompous ass!" he said aloud.

112

He listened to the shrieking wind and felt the *Charlotte* shudder as it plunged into the dark sea and struggled free. How long since he'd been abed with a woman and felt safe and warm and satisfied? He thought of the placid widow in Dock and how the familiar pre-dawn light would slowly give shape to her bedroom, the canopied bed, the bureau against the far wall with the cracked mirror and dried flowers. He remembered the feel of her warm flank beneath the coverlet, her sleeping profile with her lace nightcap askew, her soft snore. Pleasant. Some eight years older, widow of the late Captain George Bridewell, at first she'd hinted that she and Tench might be more public. Instead of always entertaining him in her rooms, she thought they might go for walks and see plays and attend concerts. But Tench wasn't inclined to public demonstrations of attachment; even his visits to the widow were almost clandestine. She must have wondered what he sought to conceal, or avoid, and perhaps didn't want to admit it was deeper involvement or more commitment. In any event, she'd come to accept less, realizing it was more than she might otherwise have.

Tench suddenly waked shivering. His feet felt like ice. How long had he slept? He blew out his lamp and scrambled under the blankets, shivering violently, then sat up to snatch his greatcoat from a peg and spread it over himself. He curled in a little ball, his icy hands between his warm thighs. Gradually his shivering slowed and the faces of men and women began floating across his view. They floated and eddied in a disembodied procession of strangers. Where did they come from? The street? Casually examined portraits? Plates from unremembered books? Were they created in some wellspring of dream world imagery?

Tench settled more deeply into his nest. His mother's face floated near, forever young in a portrait done before she died giving birth to him. In three-quarters pose, she sat in a white dress trimmed with pale blue. In her lap half-hidden hands held a small bouquet of forget-me-nots. Wide browed, she looked out with level gray eyes. Her rich, reddish-brown hair was unpowdered and loosely pinned beneath a small, blue-ribboned cap. Her small mouth wore a faint smile. So young! One day Tench had been struck by the realization that his mother and his laundress bore a resemblance in their brows and eyes.

Tench was unable to get warm. He felt a foot. Still ice. Was he getting sick? He thrust out a hand to pull in the bottle of canary, uncorked it with his teeth, and drank straight from the bottle, one, two, three large mouthfuls. He shuddered, recorked the bottle, and kept it under the blankets. Its cool, smooth shape felt good on his groin. "Maybe I'll have to nurse myself," he thought with a smile. "Here, Wat, have a tug on this teat!" An image came to him from a few days earlier. He'd happened upon Mary B nursing her baby, her features softened in the pale prison light. At the sight he drew in his breath. Her eyes lowered in serene gaze, her face a milky glow, she seemed so incongruously at peace in the odorous dungeon he'd stared for several moments, wondrous. When she looked up and noticed him, she smiled and wordlessly pulled the baby from her breast, its lips leaving the nipple with a faint popping sound. Smiling himself, Tench held out a finger that the baby grasped with a surprisingly strong grip and tried to suck. Tench asked if the little girl had a name. "No," Mary B smiled, "I'm still thinking," and put the baby to her breast again.

Something about the convict woman drew his interest, something beyond a resemblance to his mother, but he was unsure what. "It's not her learning," he said aloud under the blankets. "Why, no, sir, 'tis not her learning," he answered himself in what he imagined to be Dr. Johnson's stentorian pomposity, "But, sir, learning is not all, as beauty is not all. But together a little learning and a little beauty are much, sir. Depend on it!"

Tench chuckled under the covers. "You got that, Boswell?" He squeaked out the cork again and drank, spilling some of the wine. He sucked the spill from his shirt, then had an idea. He giggled as he tore a strip from the bottom of his shirt and stuffed it into the mouth of the bottle. He lay back and put the rag end to his mouth. A trickle of sweet wine came to his tongue, a grown-up's sucking bottle. He positioned the bottle handy to his mouth and put his icy hands between his thighs again. "Fool," he said aloud, "I hope you don't die tonight." A few minutes later he spoke again under his covers, "Maybe it's predestined."

When Lieutenant Creswell peeked in Tench's cabin next morning he found his friend mumbling deliriously and entangled in a jumble of clammy, wine-soaked bedclothes. He fetched John White, who was a little shaky himself. After a brief examination, White pronounced that Tench too had come down with the ague, prevalent on the *Charlotte* since Rio. He said he'd have his servant Broughton look in on Tench from time to time. Creswell nodded and stooped with some effort in the cramped quarters to retrieve Tench's sucking bottle from the floor. He looked at the sticky rag-end with mild curiosity before he placed the bottle on the table and gently replaced the flower in Boswell's *Tour*.

For the next three days, sometimes sweating, sometimes shivering, Tench lived in a dream-like world in which imagination and reality were one. In his delirium he lectured, laughed, wept, pleaded, and protested so convincingly that at times young Broughton wasn't sure whether it was the real Tench or his helpless charge. Then the fever broke, and Tench's mind began to clear. As if he'd only napped, he immediately looked for the rag-stuffed bottle, and felt immense relief to see it sitting innocently on the table. Later, when John White looked in, Tench thought he saw a veiled, knowing look cross between Broughton and the surgeon, and Tench wondered how he could determine what was dream and what was real, for he held a vivid memory that John White had come into his cell and found him sucking at the rag-end of his bottle and John White had taken off Tench's shirt and then undressed himself and crawled into bed with him and that young Broughton had brought a scented candle and produced a bucket of warm, soapy water with which he bathed the two naked men and sang softly of two orphan brothers who lived in the roots of a old oak tree. In the dream, young Broughton breathed an anise breath and was very gentle when he washed their private parts and raised their legs and washed their bottoms with languorous pleasure and rubbed them all over with warm oil that smelled faintly of pine and wrapped them in one warm blanket so their foreheads and noses touched and they could taste each other's breath and tongues and could feel each other's heartbeat and feel the hot pulse of each other's penis while young Broughton stroked their heads and crooned a song of an orphan boy who lived in a cave by the sea and one night saw the seals come onto the shore in the moonlight and take off their skins and become beautiful young women who danced on the shore and when they discovered the boy watching they rushed to put on their skins and he begged them to take him too, but they swam out to sea and he tried to follow and swam so far he could no longer see land and was never seen again. And young Broughton wiped salty tears from the cheeks of the two men, but now it was

Mary B who lay with Tench, and then young Broughton lay down too and put his arms around them both so their bodies melted into each other with a sweet flowing and Broughton put his lips to their lips and fed them warm, sweet wine from his soft lips that sent delicious warmth spreading though their conjoined body like ripples in a pool and made their body melt away and melt away until there was nothing left but eyes, which grew heavier and heavier and finally closed in sleep.

Tench studied John White as the surgeon smiled an odd, crooked smile. White put his hand on Tench's forehead, then laid his ear on Tench's chest, and listened to his heart while looking vaguely into the captain's eyes. Tench thought, he's acting as if nothing happened!

White sat up and held Tench's hand between his own. The surgeon's forehead was beaded with sweat and a little shudder shook his body. "Wat, you should feel much better in a few days, good as a newborn." Tench's eyes flicked to the wine bottle. Had they both sucked from the bottle? Had they really lain together in bed, naked, melted into each other? Had White brought Mary B to his bed, too?

Round-eyed, Tench stared at White, who still smiled his odd, little, crooked smile. "Now it's up to you, Wat, the worst is over." The surgeon wiped his brow with the back of his thumb and looked at the sweat. "Shit," he said, and pushed himself up. He wavered. "Broughton? Broughton, I...." He stretched a hand toward his servant, then his eyes rolled up in his head, and he collapsed like a heap of empty clothes, striking his forehead on the corner of Tench's table.

Upset, the wine bottle rolled to the edge of the table. Broughton made a grab for it. Too late. The bottle fell, and dealt the surgeon's forehead a second blow, a bony clunk.

~~~~~~~~~~

Will Bryant waked in pre-dawn darkness and listened to the snores and sighs and deep breathing of men asleep. The air was foul. Many suffered from persistent, feverish diarrhea that sent them to the slop buckets in a constant, groaning procession. He heard a ruffle of grunting and wondered who was getting buggered. The boy? Or Tartop? James stirred next to him. Will regarded the dim outline of the sleeping man's face. Suddenly Will realized why he'd awakened, for the idea was before him startling clarity.

"Jimbo," Bryant said aloud, "I'm going to have your Mary B. Yes, I am, Jimbo. I'm going to have her."

Later that day Tench nodded acknowledgement of Will's request to speak with Mary B about marriage, then turned away in thought. A half-mile distant the *Friendship* plowed through heavy seas. For three weeks the fleet had climbed the southern latitudes from Rio and now sailed almost due east for Cape Town. After days of favorable winds and running seas, expectations ran high that water rations cut soon after Rio would be restored. But while the progress of the fleet was good, conditions were not. Heavy seas pitched the ships mercilessly, sending cascades of seawater over coamings and down hatches to soak bedding and clothes. Bone-chilling rain and cold kept people in the below-decks for days on end, huddling for body heat and savoring warm food. But today the sky had cleared, the temperature risen, and the decks sprouted convicts and Marines blinking in bright sunlight. They dried clothes, aired bedding, and moved in the luxury of fresh air and open vistas.

Wearing his thin, blue kersey convict jacket, already threadbare at the elbows, Will shivered in the brisk wind and shifted from foot to foot while awaiting Tench's decision. He knew Tench would want something in return. At last Tench faced him. "Aren't you already married? Don't you already have a wife?"

Bryant gave a little start of surprise, then laughed, "Oh, you mean my letter! Is that what you mean?" Tench said nothing.

Will shrugged, "That's my sister. Her husband drowned sometime back." Tench nodded, waiting for more.

Will studied the rigging for a long moment. "I never much liked him. I thought he wasn't good enough for her."

Tench nodded. "And what about the counterfeit coins?"

"The what?"

"You know what I'm talking about, the counterfeit coins in Rio."

"Well, I didn't have anything to do with that!" Bryant's voice was suddenly whiny.

"You know about it."

Bryant studied the rigging again before answering, "Well, I suppose it doesn't matter now." He was unable to suppress a little smile. "It was the pumpkins."

"The what?"

"The pumpkins. Remember the pumpkins we picked up in Santa Cruz?" At Santa Cruz convicts had lugged aboard hundreds of little pumpkins, one of the few vegetables available to Arthur Phillip. "Well, some of the boys figured they make a perfect casting form because if you cut them in half you can line the two halves up just right." Bryant demonstrated by cupping his hands together, one atop the other. "Then they took sand and ashes to make a molding sand and melted old buckles and pewter. They cast the coins in the pumpkins. Of course they had real coins to make the impressions."

Tench shook his head in mild wonderment. What ingenuity these convicts brought to their criminal behavior! "And when you all came tramping down to look for the makings, why, there they were, scattered here and there, big as life, those little, dried pumpkin shells. And you never figured it out!" Bryant laughed so hard he bent at the waist and held his stomach.

Tench couldn't help smiling at Bryant's amusement as he remembered his fruitless forays into the men's prison. No sooner had the *Charlotte* dropped anchor in Rio than fruit vendors were alongside with their wares. Several convicts sent up money to buy the welcome fruit, but one seller, biting a coin, hurled it back with an angry cry and a stream of abusive Portuguese. Chagrined, Tench tried to mollify him by explaining that the culprit was a convict and that such behavior was not typically English. To placate the aggrieved vendor Tench purchased three large bags of oranges and gave them to the women. Six pence well spent.

Still weak and out-of-sorts from his illness, Tench huddled in his shawl and gazed past Will at the convicts milling in the pen as he thought: all the work these convicts went through to cast half-crowns, risking hard punishment if caught, then having their phony coins found out anyway so they ended up with nothing for their labors.... What kind of mind works to such ends? After a long silence, Tench said with a small smile, "Well, score one for you."

Bryant grinned back but said, "Not me, Captain, not me."

They stood face to face, perhaps three feet apart, measuring each other. Finally, rising on his toes and sinking back again, Tench asked, "Tell me, Bryant, why are you here?"

Bryant's eyes shot to the side. He barked a brittle, humorless laugh. "Well, Captain, sir, I just tried to protect my people." When Tench said nothing, Bryant added, "I'm a fisherman, and I just tried to protect my people."

"Are you a smuggler? Is that why you're here, because you're a smuggler? That's what I heard."

"God, no! I'm no smuggler! I told you, I'm a fisherman! In fact, I'm a fourth cousin to Lord Marlborough's secretary, and I was just trying to protect my people!"

"Well, how? Get on with it, man!"

"I crossed an excise man!" Tench nodded, but looked dubious. "But it's not simple, Captain. There was this old lady's house, see, a safe house with a big cellar where they'd hide goods just off the boats. Well, the excise man shows up and starts nosing around. So I take him in my boat up the coast on a wild goose chase, and while we're gone they get the goods out of the old lady's cellar, see." Bryant paused, thoughtful. "But somebody snitched, some spy. The excise man heard about the goods getting moved while I had him in my boat, so he had me arrested."

Tench looked puzzled. "But why? What evidence did he have?"

Bryant laughed his brittle laugh again. "Evidence? Evidence? I'll tell you what he had, Captain. He had this!" Bryant reached a hand near Tench's nose and pretended to tweak it.

"What?"

"I swear to God, Captain, I don't know why, but when he said he was going to arrest me for taking him on a wild goose chase I just reached up and gave his big, greasy nose a good twist!" Then Bryant told how he'd gone to trial and been sentenced to seven years and how the excise man had seized his boat and was using it to chase smugglers.

The town was good about seeing to his welfare while he was confined in the Launceston jail and later on the hulk in Dock, bringing food and money and looking after his affairs. But appeals failed, and his famous cousin, Lord Marlborough's secretary, declined to get involved. "But to the town, to my people, I'm not a criminal, Captain. I'm a hero, and they'll never quit telling how I tweaked that bastard's nose! No, only the law sees me a criminal. To my people I'm a hero."

119

Arms akimbo, skillfully braced against the motion of the *Charlotte*, Bryant's mouth twitched with a faintly arrogant smile. From a sudden squawking, Tench knew Ned Thornapple was bedeviling his chickens again. That boy! The captain studied Bryant a few moments, then turned away.

He found Bryant unlikable, and was put off by his boasting and self-aggrandizement and sneakiness. He wondered what part Bryant had really played a few days earlier, when Tench flogged a hulking lag who'd begged off his turn at the pumps because of illness, the first convict punishment administered on the *Charlotte*. Still weak from his illness, Tench went below to investigate a clamorous prison and found the malingerer engaged in vehement argument with Bryant. Something snapped. Suddenly Tench was fed up with the lazy, conniving, lying cargo he safeguarded, with their wheedling, petty intrigues and cheating and foul-mouthed ignorance. He ordered the malcontent to receive a dozen lashes in summary punishment. Within ten minutes of hauling him out the Marines sent back down a chastened, bruised convict. And for the rest of the day the convicts were eerily quiet.

Tench faced Bryant again. The convict remained with his arms crossed, but the arrogance in his eyes was clouded by doubt, and he shivered in the nippy wind. Tench's voice was so soft Bryant had to strain to hear. "All right, Bryant, I have no objection at this point if you talk with her, but I want to talk with her first."

"Oh, thank you, Captain, thank you," Bryant said through a spasm of shivering. "I was just thinking of how the boys all say they're glad you're on this ship, Captain. Yes, sir, that's what they say, God's truth." Tench made a face and walked away to look out over the taffrail. Bryant raised his voice, "And does the Captain think he'll talk with her soon?" Tench paused, the wind flapping his shawl at Bryant. He didn't bother to address Bryant when he said softly, "I'll see."

Bryant's words tumbled through his shivering, "What, what did the Captain say? What?"

Tench ignored him and stared at the wake of the pitching ship. Bryant remained rooted, shivering, his resentment growing. Finally turning to go, he sneaked a half jerk of his fist in an obscene gesture.

"Aye, that's the Table Rock, and to the left, that's Devil's Peak." The bewhiskered sailor pointed with his pipe. "And over there's Lion's Head. And that smaller hump this way? They call that Lion's Rump!" To Mary B, gawking with a knot of women, the flat-topped mountain crowding Cape Town's harbor looked foreboding. She held up her baby to see the famous Table Rock, looming more than 3,500 feet. "See there, little baby? Your first mountain, and I saw my first but a bit ago!"

For almost three centuries Table Rock had served as a landmark for Europeans plying the tea, silk, and spice trades. Situated midway between Europe and the spice ports of the Far East, Table Bay was a haven from the Cape of Storms, a message center where seafarers left letters under specially marked rocks to be picked up by passing ships and carried on. Otherwise, this barren, wind-swept bay, devoid of permanent settlement, held little attraction. Nomadic Hottentots sometimes bartered their cattle and fat-tailed sheep with ships that put in, but fresh meat did nothing to relieve the agonies of scurvy, which might fell half a crew by the time a ship reached Table Bay from Europe or the Far East. Indeed, this haven was a lonely resting place for the dead long before it became a permanent settlement for the living.

Settlement was part accident. In 1648 a homebound vessel of the Dutch East India Company was blown ashore and grounded. During the months they were stranded, the Dutch harvested impressive crops of vegetables from seeds salvaged from their wrecked ship. After rescue, their glowing reports prompted the Dutch East India Company to establish a port of call to tend sick seamen and re-supply ships with fruits and vegetables. Strictly business, the settlement they christened Cape Town restored sea-ravaged men and ships, at a profit.

As the First Fleet skirted Robben Island to enter the harbor, Arthur Phillip spied more than a dozen ships at anchor - most of them flying the flag of the Dutch East India Company, but he also saw the flags of England, France, Sweden, and Denmark. Once again he felt relieved. Not only had he arrived without major mishap, but Cape Town appeared at peace, never a certainty here. England and France both coveted this strategic little Dutch city, for whoever controlled the Cape controlled the sea lanes to the Far East, and European powers jockeyed constantly for advantage in that lucrative trade. Small wonder that Cape Town's governor was a little nervous when

121

he watched the cloud of English sails heading into his harbor, but he relaxed when they were identified as the convict fleet. He knew they were coming. In fact, he'd received secret instructions not long before: don't make things too easy for the English. After all, whatever success the English had at Botany Bay would only diminish Dutch influence in the Pacific.

The *Sirius* was no sooner anchored than Arthur Phillip dispatched an emissary to inform the authorities of his mission, then hastened ashore. Time was growing short. To arrive too late to get a crop in and harvested before Botany Bay's upside-down winter season would be a serious blow to his plans. No harvest would mean an extra year of living on handouts that had to be hauled halfway around the world.

He found Cape Town's provisioners eager to meet his immediate needs. A small supply of fresh meat, vegetables, and bread was delivered later that day to the convict transports. But for his other requirements, Arthur Phillip had to contend with the Council of Policy, a committee of Dutch East India Company officials that controlled virtually every aspect of Cape Town's economy. The quantities of flour, livestock, and seed he sought required the Council's careful consideration, he learned. "I'll see what I can do," Mynheer De Witt told him, a major provisioner in Cape Town. "But there was drought, you see...." He let his sentence hang. Arthur Phillip shrugged. He gave the provisioner a list of his needs and assembled his principal officers for an official call on the Governor.

Lieutenant-Colonel Cornelis Jacob van der Graaff would be the last of Cape Town's governors to live in such splendor. Along with a princely salary, he had use of three stately residences, fancy carriages, a stable of fine horses, dozens of servants and slaves, first rights on produce from the Company farms and gardens, and a commission on almost every tax levied (of which there were many). He could afford to be generous, and he received his English visitors in a sumptuous palace situated on the grounds of the extensive Company gardens. Of course he'd do what he could to speed the English on their way, he assured Arthur Phillip. But the Council of Policy was very busy and a long drought had seriously cut the colony's harvests. Meanwhile, he hoped the English would take time to enjoy the sights of the countryside and the comforts of the town before resuming their difficult voyage.

He introduced Colonel Gordon, Scottish-born and twenty years at the Cape, a botanist without peer in his knowledge of the area and its produce. Colonel Gordon would advise them on seeds and cuttings, beginning with a tour of the Company's famous gardens.

Governor van der Graaff smiled. "Our gardens are no longer just fruits and vegetables, you know." Was Arthur Phillip aware that Cape Town's magnificent oaks had sprung from England's acorns?

~~~~~~~~~~

"See here," Malvina Thornapple snapped, "I'm a freeborn Englishwoman, and I will go ashore!" Her feet planted firmly before the ship's ladder, her hair disarranged, her bonnet slightly askew, dressed in voluminous badly wrinkled black (her Sunday best), Malvina punctuated her speech with a furled parasol. "I'm not a Marine, I'm not crew, and I'm certainly not a convict! I have every right to go ashore, and that's where I'm going! Now stand back, Dobber!"

Their confrontation had attracted a circle of the curious. In a tumbling exodus from the ships, every officer had hurried ashore to take lodgings in town, and Malvina was proving more than a match for whatever authority remained aboard the *Charlotte*. Porter caught Quint's eye and twirled a forefinger near his temple. "I see that, you silly monkey!" Malvina cried. "I'm not mad! You're all mad! Mad to stay on this death ship! You're no better off than those awful criminals! Yes, Dobber, you! Look at you! What are you, a sheep? And look at me!"

Malvina clutched a handful of excess dress and shook it at Dobber, a gesture that was now a motif of her lament. Dobber would not meet her eyes, but stood just beyond jabbing range of her furled weapon, his look fixed on the tips of her shoes. "I've lost six stone. I can't eat, I can't sleep, I'm dizzy all the time, and I can't walk more than two steps without knocking into somebody, and this...this ship stinks like a sewer! And I have no place to myself! None! It's killing me, I tell you, and I'm getting off before I'm dead! Now stand back and let...me...be...free!" This last was accompanied by emphatic pokes of her parasol. She teetered as she stooped to pick up her valise and backed to the ladder in a sword-fighter's retreat.

"But, Malvina," Hester said calmly, "You can't leave your children. They need you." With that, half-hidden behind Hester's skirt, little Nell popped a thumb from her mouth, tilted back her head, and let go a fearsome wail. Hester sheltered little Nell with one arm as she extended the other,

"Let's talk, Malvina. I know it's hard, Lord knows! But let's brew up. We'll sit and have a good chat, just you and me."

Malvina was beyond persuasion. Her eyes bright with a feverish vision, she would not look at Hester as she fumbled for the rope of the ladder. Below lay the safety of a waiting scoot that would take her ashore, a water taxi. "No," she said, "I'm getting off this boat, and that's that!" She took another step back. "You there!" she commanded the boatman below, "I'm coming down! Take this!" and she tossed down her valise. Dobber took a tentative step. "No, Dobber!" she cried, whirling to jab him in the stomach, "You must give me this! If you try to stop me I'll never speak to you again! Never! Now leave me alone. I'm going ashore! I've got to save myself. I'm dying!" She backed down the ladder a step and paused. "Nell, sweetie, I'll send for you as soon as I get settled. Mother won't leave her little Nell. Come here, darling, and give us a kiss!" But little Nell would not. She ducked her face behind Hester's skirts. At this something seemed to go out of Malvina. She sagged.

But she straightened again after a long moment and backed down another step. "Well then, Dobber, if that's the way you're going be, so be it! The children will stay with you for now. Come to your senses, you poor man! Don't you understand? And where's that boy? You don't even know, do you? He could be drowned!" She backed down another step and appealed to her audience, "I'm not mad! How can I be mad? Look what's happening! We're sick, we're wobbly, we quarrel all the time! No, I'd sooner live as an alehouse drudge than live one minute more on this awful boat!" And with that she made her uncertain way to the waiting craft. Those above peered over the bulwark as she gathered her dress, settled firmly on a thwart, and set her chin for the Cape Town pier. When the scoot had pulled several boat lengths away Nell's sudden howl caused Malvina to turn back to the row of faces still staring down from the *Charlotte*. A muted whimper rose in her breast.

"Sweetheart," Dobber Thornapple implored, his nasal term of endearment the first word ventured since his wife's breakout began earlier that morning, "Sweetheart, eh, will you be back for evening mess?"

As a chuckle ran through the little throng, Malvina rolled her eyes and searched for Dobber among the faces. "Dobber, you poor fool! You'll have bind and gag me before I set foot on that boat again!" She was sorry to see that her words only

124

amused the crowd. Porter mimed a trussed-up, suck-cheeked prisoner hopping about.

"Oh!" Malvina huffed, squaring her shoulders once more towards the beckoning shore. A few minutes later she turned to look again at the shrinking shape of the *Charlotte*. "Well, that's that then," she said as she turned her back on the ship. And a few moments later repeated softly, "Aye, that's that."

~~~~~~~~~~

After having lived ashore for almost a week, Tench was compelled to return to the *Charlotte* with the expedition's chaplain, the square-jawed Reverend Richard Johnson. A wave of weary disgust engulfed the captain as he mounted the ladder with reluctant feet. He found the smells, the smothering confines, and the human cargo of the little vessel disgusting. He had to force himself to respond pleasantly to the greetings of his men.

The big story, of course, was Dobber Thornapple's wife. But after talking the matter over with the private, Tench and the Reverend agreed that a few days ashore might be all she needed. "Give her a little room, Thornapple. Let her keep her feet on solid ground for a few days. She'll come around. We all know it's been hard for her."

Never one with much to say, Dobber nodded, then saluted at his implied dismissal. But he suddenly turned back to seize the captain's hand and bent to kiss it, then caught himself and instead shook it gravely, once, never raising his eyes. Tench and the Reverend exchanged amused looks.

"Now, then, Reverend," Tench said when Dobber was gone, "You want to visit the sick and attend to the baptism of our new baby girl, eh? And perhaps raise everyone's spirits?"

But oddly enough, while Tench was still weak and dispirited from his illness, those on the *Charlotte* were regaining their spirits on their own. Despite having been confined aboard for five months, they now enjoyed fresh meat, vegetables, and soft bread to eat, and moderate rations of wine to fortify their blood. And while many were still half-sick from the fever and dysentery that had followed them from Rio, none were deemed ill enough to send ashore for hospital care.

Those confined to the ships were entertained by a succession of vendors who pulled from ship to ship selling varied wares. While the season's early produce was meager - onions, peapods, chives - it was cheap enough, and for those with more to spend an abundance of goods was held up for their consideration - tea, sugar, spices, cloth from the orient, tobacco, spirits, souvenirs, old newspapers, used clothing, sewing necessities, pets, caged fowls, trussed and tethered animals. Some chose to commission sailors. On daily trips to fill the waterbutts or bring back the ships' provender, an enterprising sailor might secure almost anything Cape Town had to offer. But for most of the penniless convicts, the sights of the port were their only treat.

After spending just two hours on the *Charlotte*, Tench happily returned to shore.

As he made his way across the rough parade ground near the fortress that guarded the anchorage, slaves scythed and raked the over-grown field in preparation for the militia's annual exercises. Others set up the standards for the militia companies. While the mandatory day of drill was supposed to be an annual affair, this would be the first such assembly since a mobilization several years earlier, when farmers, craftsmen, and petty officials were mustered to oppose a rumored English invasion. But many lacked heart for their militia obligations. Those living on the fringes of the settlement were more concerned with Hottentot raiders than English invaders. No, to abandon their homes and kin and cattle to defend Company property in Cape Town was not a priority.

A handful of swag-bellied, blue-uniformed officers overseeing the preparations saluted Tench, "*Hallo, Engelsman!*" They held a bottle aloft in invitation. Tench smiled as he returned their salute but pointed towards the town. The paunchy officers laughed and turned back to themselves.

Unlike so many English towns whose streets followed old cow-paths and water-runs, Cape Town was laid out with the regularity of a chessboard. The streets and whitewashed stone houses and shops and warehouses possessed a dreary sameness. Ornamentation was almost unknown. With the exception of the church steeple, buildings were roofed with heavy tiles to withstand the Cape's violent winds. None rose higher than two stories. But the interiors of many burgher houses were a different story. Thanks to the influence of an earlier time, when luxury-loving French *sabreurs* had lived among them in a brief military alliance, residents of the Cape now indulged their long-deprived appetites freely. Imported French furniture, plate, fashions, music, and a gay social abandon prevailed in many households.

Tench's destination was the house of Mynheer Johann van Jaarsveld, where he and John White shared an upstairs room. Crossing the fountain square, one of Cape Town's few cobbled thoroughfares, he glanced up at the remnants of three black-skinned corpses hanging in chains from a high gibbet. Raucous birds wheeled and croaked as they visited the carrion. Seemingly oblivious to the gruesome scene above, lounging burghers and visitors in town for the militia exercises traded gossip around the fountains or gawked in shop windows.

Tench himself had paused to ponder an array of sugarcoated pastries when he heard his name called. He knew it was a brother officer, Captain Meredith, who was probably drunk again. Pretending not to hear, Tench ducked into the baker's shop.

~~~~~~~~~~

A grim-faced Arthur Phillip pretended to look out the window of the provisioner's warehouse as he spoke sotto voce to John Shortland, his commissary agent. "Something's going on, John. I don't know what, but we're being swingled. What time is it?" Ripe with experience that served Arthur Phillip well, Shortland nodded as he fished out his watch. Apart from the fresh supplies provided for the ship-bound, Arthur Phillip had received no satisfaction from the Council of Policy. Days of waiting had stretched into a week. A couple days earlier he'd sent a second appeal to the Governor. Now he mused aloud, "Is it something we don't know about, something back home?"

Shortland cleared his throat. "I don't think so, Sir. I should think Colonel Gordon would tip us off if there were any real danger. After all, 200 good English Marines are not to be taken lightly, even with the militia in town. And I can't imagine the Dutch wanting to deal with 500 convicts as well."

A faint, responsive smile indicated agreement as Arthur Phillip chuckled, "We could threaten to turn the women loose too, I suppose. A plague to go with their drought."

The two were laughing at Arthur Phillip's joke when Mynheer De Witt puffed in from the street, wiping his round, red face with a large handkerchief. "Ah, I'm glad my English guests find something amusing," he said in wheezy, heavily accented English. Large-bellied, wearing a tense smile, he perspired heavily from his hurried return from the offices of the Council of Policy. "I offer humble apologies for my delay," he said with a very short, awkward bow. "Is it something you could share?"

Plumed hat under his arm, Arthur Phillip smiled and inclined his body in graceful return. "Of course. We were discussing our cargo and wondered if perhaps the Council of Policy would consider a trade."

Mynheer De Witt thought seriously for a moment, then saw the joke and laughed loudly, his belly and chins jiggling. "Ha, ha, ha! you English! So always making jokes! A trade, ha! ha! I must tell my friends of your trade, ha! ha!" Arthur Phillip and Shortland joined politely in his laughter. Then the Dutchman stopped laughing, his belly stopped jiggling, and he did his best to pull a long face, difficult given his round countenance. "Ah, my friends! I feel so bad I bring you no good news." He made a show of looking warily around, then lowered his voice. "I cannot learn why the Council takes so long. There is talk of shortage, you see, but here it is always too little or too much. If 'too little' will turn a guilder, it is 'too little,' If 'too much' will turn a guilder, it is 'too much!' Heh!" He wiped his face again as Arthur Phillip and Shortland joined eyes. "Heh!" De Witt said again as he studiously folded his handkerchief in the long silence.

Arthur Phillip straightened, again inclined his body in a courtly bow, and thanked the contractor. "I'm sure we'll work something out, Mynheer," he said, extending his hand. De Witt quickly wiped his own hand and shook Arthur Phillip's vigorously. "Of course," he said, pumping away, relief in his voice, "We will work things out with our English friends. I will work very hard to work things out. *Ja, ja*, of course!"

When he opened the door to see them out, he looked carefully up and down the street. Then he confided, "Perhaps you will try our little sugar cakes tonight, heh? They are very good, and our sugar cakes are not always to be had." The Englishmen looked blank.

After they were on the street, Mynheer De Witt called again, "*Ja*, and not always to be had."

Some distance away Shortland wondered aloud, "What was that about?"

"I don't know," Arthur Phillip said, "But I think you should find out."

~~~~~~~~~~

Tench paused when he heard Broughton's shriek and White's giggle. "Hold still now, damn it!" White's voice had a good-natured gruffness. Broughton shrieked again. "Damn it, Broughton, hold still so I can get it in!"

"But I can't help it! I'm afraid you're going to hurt me!"

"I'm not going to hurt you. Take another drink." Broughton giggled again.

Tench wondered, what are they doing in there? Should I knock? Should I leave?

White's voice: "All right. Now hold still so I can get it in. Hold still. Don't flinch. That's right. That's right." Broughton giggled nervously. "All right, I think I've got it. Just ease.... Goddamn it!"

The sound of a slap, Broughton's whimper, then his voice, whiny, "Well, I can't help it, I'm afraid you're going to hurt me."

"Shit, it can't be that hard. Put your coat back on." Tench held his breath. He heard the clink of metal and the sound of White's boots crossing the room. His imagination kaleidoscoped wild fantasies. Impulsively, Tench knocked.

Flush-faced, White's slender servant opened the door. "Ah, Wat," White said somewhat distractedly, "You're back, good! How're things on ship?" Coatless, he was examining a large, metal, pincer-like instrument.

"Fine, fine," Tench said, his eyes darting about the bare room - two narrow beds, one rumpled. White's chest. A plain bureau. An oil lamp. A table. Broughton stood nervously by the door, his hand to his cheek.

"All right, Broughton, you can go now," White said. "I'm sorry about.... Here!" He tossed Broughton a small silver coin. Broughton snatched clumsily but only succeeded in batting the coin to the floor, where the three men watched its fateful wheeled progress to the dark space beneath White's rumpled bed. The servant rolled his eyes and rather mincingly got down on hands and knees to retrieve the coin. As he was maneuvering himself to reach under the bed, a ripping sound betrayed splitting pants. Broughton's hand flew from under the bed to his behind. Tench and White exchanged amused looks.

At last, his coin safely in hand, his face crimson, Broughton bowed out with averted eyes.

When they were alone White said, "I can't seem to get the feel of this damned thing." He tossed his instrument onto the bed and sat down to pull off his boots.

"What is it?"

White picked up the instrument, plumped his pillow, took his ease, and studied the simple mechanism. "Forceps," he said, working the jaws experimentally. "I had him make a fist in his sleeve to see if I could get a hold and pull it out, but he kept jerking away. I don't have a feel for these things."

Tench put his parcel on the bureau, stepped over, and held out a hand. "Are you going to need them?"

White passed the forceps. Hands behind his head, he said, "Probably. You know that little woman that's due, the one with the bad foot...Katie Prior? I think her baby's a big one."

Tench returned the instrument, doffed his coat, stretched out on his own bed, and put his hands behind his head. "And you've never used those before?"

"Christ, Wat, I'm a Navy surgeon! I cut things off and sew things up! Oh, I watched once or twice." He paused. "Well, there can't be that much to it! Just reach in and clamp on and pull. What can be hard about that?"

Tench looked over and smiled. The ugly bruise over White's eye was faded to a greenish-yellow patch. "Not much," he finally answered. They were silent for a long while, when Tench asked about White's ascent of Table Rock. That the surgeon had recovered so quickly from his illness and was able to climb Table Rock perplexed Tench.

"Ah, Wat! Too bad you weren't along! It was hard but fun! But we never thought to bring water! Christ, we were so thirsty when we got to the top we drank from puddles! Can't you just see the chief medical officer drinking from a dirty puddle! What an example!" They laughed, and then White told of the magnificent view from the mountaintop. Tench in turn told of Malvina Thornapple's flight from the *Charlotte* and shipboard gossip of a boozy feud going on between Captain Meredith and a ship's surgeon named Arndell. "What a couple of boozers!" White commented, dismissive.

"Ah, well," Tench said, feeling somewhat defensive, "You know Meredith's been melancholic since he was passed over for promotion. Without a promotion he feels himself a failure."

"Poof!" White said, "A man may drink because he feels himself a failure, then fail all the more because he drinks."

Tench opened his mouth, "Or, well...." then remained silent, suddenly feeling the talk was trivial.

The room stayed wordless as the soft gloom of late afternoon deepened. His eyes distant, White began clamping a knee with the forceps, pulling it towards him, releasing the forceps, letting his stockinged foot slide back on the bed.

Vaguely ill at ease and out of sorts, Tench sensed something building. He wanted to talk about his dreams - those confusing memories from his febrile delirium aboard the *Charlotte*, memories so real Tench still wondered at times whether they were dream or reality! For weeks he'd been unable to get the details out of his mind as he wondered, what part of me dreams such things? such unlikely events, such unspeakable acts? Now he even dreamed about dreaming those dreams. But whom could he talk with? Not White. White was part of the dreamscape that might be rooted in reality. But there was more to it than that.

White had begun to embarrass Tench, and Tench was starting to feel uncomfortable about their closeness. About their too-closeness. Perhaps what Tench sensed was only increasing familiarity and trust, a lowering of barriers, but at Rio there'd been an incident in a hospital that made Tench squirm. While touring the facility White insisted on demonstrating an amputation technique now in vogue in England. No matter that he'd sail away in a few days and leave his hosts to deal with the consequences of his interference. The impropriety of his act seemed never to have occurred to the English surgeon.

And women! True, Rio had a certain reputation for sexuality, and men in Rio practiced that surreptitious public foreplay so common to Latin culture. But more than once White shocked Tench by winking up at women who looked down from their balconies, or by brushing against women in church, or by bandying with tittering, virginal novices through convent gates. And Tench would never forget the embarrassing evening when White fingered the unplaited hair of Captain Desvia's wife.

After they'd drunk much wine and joked and laughed with great familiarity over coffee and port there came a moment when White expressed skepticism about Desvia's claim that Rio women grew their hair to marvelous lengths. The Captain called his wife into the room to prove his point. Her face averted, her eyes lowered, Desvia's wife unpinned her hair and shook loose an ankle-length cascade of shiny, black, silken splendor. With an exclamation of surprised delight, White sprang to his feet and ran his fingers through it. Even gave it a tug to test its authenticity. Even bent to sniff its perfume. And murmur words that neither Tench nor their Portuguese host could hear before he turned to them with a too-bold grin. "Ah, Captain Desvia, you have me," he said.

But it wasn't just women. White now exhibited a familiarity with Tench himself that Tench found at once flattering and disquieting - a hand on the shoulder, a squeeze of the arm, an open, disarming smile directly into Tench's eyes. It was as if the farther England receded, the closer White came.

Suddenly, surprising himself, Tench blurted, "John, do you ever think of marriage?"

White paused in his forceps play and thought. "No," he finally said to his raised knee, "No, I don't." A long pause. "Do you?"

Tench roused himself and swung his feet to the floor, his back to White. He sat hunched on his bed, hands braced on either side. He answered quietly, "No, not much."

"Well, that surprises me, Wat. Someone like you. Don't you want children? An heir?"

"Ahhh, I don't know. A child. An heir. I suppose I don't have any great investment in perpetuating myself...that way." This last he muttered to himself.

"What?"

"Nothing."

"I thought you said something."

"I, oh, shit, it doesn't matter!"

"Is something on your mind, Wat? Are you upset? Are you upset with me?"

Tench half turned to glare over his shoulder at White's puzzled face with such intensity his friend showed mild alarm.

"Jesus! Wat, what is it! You look frightening!"

"Frightening? Frightening? I'm frightening? Christ, I'm the one who's frightened! I'm so frightened I have no idea what I'm going to do in the next ten seconds!" He suddenly swung his legs over the bed, sprang to his feet, and stood over White, his fists clenching and unclenching. "What do you want from me, John? What do you expect from me?" He shook a fist in White's startled face, "Do you want me to be your Broughton? Is that what you want? Well, I'm not going to! I'm not going to, hear?"

White scrambled to his feet on the opposite side of his bed, his expression confused, his hands half-raised in defense. "Jesus, Wat, what's going on? Are you drunk? What're you talking about?"

"You know damned well what I'm talking about!"

"I don't, damn you! I don't know what you're talking about! All I know is you come in calm as can be and the next minute you're yelling at me! Well, I'm getting out till you calm down. I don't know whether you're drunk or deranged, but I'm not sticking around to figure it out! You can go hang!" He snatched up his coat and boots and headed for the door. Tench stared at White's empty bed.

"Wait," Tench called, as White reached the door, a plaintive appeal, "Please wait. Don't go, John. I'm sorry." He turned, his arms limp at his sides, "Please don't go. I'm not drunk. I'm just, ah, not feeling well. And I...."

White waited, but Tench said nothing more. Finally, his voice solicitous, White asked, "What is it, Wat? What's bothering you?"

His eyes on the floor, Tench gave a dismissive shrug, "Oh, you know...when I was sick." White remained silent. Finally raising his eyes, Tench said, "I dreamed, John, really mixed-up dreams, and you were in them." When White opened the door Tench was afraid he was going to leave, but the surgeon called for Broughton to bring up a light. In a minute Broughton appeared with a taper. White crossed to the bureau, lit the lamp, and adjusted the wick. A warm, yellow glow leapt like a cat to the planked floor.

"Sit down, Wat, and tell me."

So Tench, after twisting uncomfortably as if he were considering escape from between the beds, finally sat, elbows on his knees, his head in his hands. He began shaking his head slowly in negation, as if denying a voice within. White turned down the lamp, padded over, and sat opposite, but Tench never raised his eyes above White's stockinged toes. Tench sighed loudly a number of times as he continued to shake his head slowly. Finally, tentatively, groping for words, he began.

At first he was vague, skipping from one matter to another, and didn't tell all the things in his dreams, all the things he'd felt. But as his confession spilled forth, more and more detail filled the space between the men. Tench went back again and again to certain parts of his dreams and told them over, adding still more details, sometimes not sure whether he was telling his actual dreams or his dreams about his dreams, or drawing on some other corner of his thoughts that only seemed like dreams. He even told of Mary B. He spoke without interruption, pausing only to correct himself with a better word or phrase until finally, exhausted, he felt empty of words. Silence. Then his shoulders rose high as he drew in a tremendous breath, and slumped as he expelled it in a long, long sigh. He sat depleted.

They were quiet for some moments. Then John White leaned over and softly kissed the top of Tench's head. Tench seized the surgeon's hands and held them to his lips, his downcast eyes tear-filled. Shoulders shaking, Tench wept, and White kissed the top of his head again.

"Ahhh," Tench said at last as he twisted away. He wiped his eyes with a sleeve, a sheepish half-smile on his averted face. He couldn't bring himself to look at White. "So now you know."

"Aye, now I know."

"Well?"

"Well, what?"

"Well, what do you think?"

White didn't answer, but after a moment stood and moved to the bureau. He turned up the lamp again. "Well," he said at last, "I think you had holy dreams."

Tench stared. "What?"

"Holy dreams. I think you had holy dreams."

"What do you mean?"

"I don't know. I just think there's something special about some of our dreams, something spiritual. You know, holy."

"You mean they weren't about, you know, us?"

White laughed. "No, I don't think they were about, as you so charmingly put it, 'us.' Not that way. Besides, what about your laundress, Mary B? Were they about her? Do you have feelings for her?"

Tench started to nod, then caught himself. He'd kept silent about his odd sense of connection to Mary B, about his feeling that something bonded him to the convict woman. Not just their shared experiences aboard the *Charlotte*, but something else, perhaps an accident of history or ancestral Cornish blood. Whatever it was, whether out of jealousy or protectiveness (he was unsure which), Tench had not yet told her that Will Bryant wanted to discuss marriage.

Tench's attraction to Mary B wasn't physical, an oddity that made his feelings even more difficult to understand. He could understand a physical attraction for the convict woman. White could too. But this other attraction? Across class and culture, across a vast gulf of learning and experience? It made no sense. Yet there it was, a feeling - no, a reality - as real as his dreams.

Tench shook his head decisively. "No, I don't have feelings for her. But then, what do you think my dreams mean? If they're holy?"

White waved off the question with a laugh, "How should I know. They're your dreams. I'm no Moses. As a matter of fact, I'm hungry. What's in here?" He pointed to Tench's parcel, the grease stains on the paper indicating something to eat lay within. "It sure smells good."

"Sugar cakes."

But Tench wasn't ready to let the subject go. "Listen, John," he said, "I've got to know - no, I don't have to know, but I want to know - do you, do you have, ah...?"

"Feelings for you? Is that what this is all leading to, what you want to know?" Tench started to shake his head, then nodded. White let the silence grow as he fingered a corner of the parcel reflectively. Tench waited with bated breath. White bent to look for the knot. "Sure I do," he said.

Tench gulped and stood with an open-mouthed stare. Seemingly oblivious to the effect of his words, White began untying the string as he asked, "So what are you going to do about it? We aren't schoolboys, you know." He spread the paper and bent to inhale the sweet aroma with eyes closed, "Ahhh!" He turned to Tench, "May I?"

Tench nodded dumbly. White raised a sugar cake in mock salute and took a large bite. He closed his eyes and chewed with obvious pleasure. Tench finally found his voice.

"I...I...well, I guess I'm going to, to, ah, do nothing."

White opened his eyes. They twinkled. "Well, then," he said, admiring the latticed texture of the sugar cake in the lamplight, "That's that then, isn't it? There's really nothing more to say, is there? Here, have a bite," he said, proffering his sugar cake. "They're quite delicious. No? Mmmm, I wonder if it's tea-time?"

~~~~~~~~~~

Anna Labonne was the last of the original Huguenots, a frail relic from an era already forgotten in the Cape colony's busy history. Her parents were one of several families to receive Dutch sanctuary from Catholic persecution in France, with the tacit understanding that their sanctuary was actually 6,000 miles away on the desolate, wind-swept Cape. Born on the voyage to her new home, Anna Labonne was only two months old when her parents came ashore. Although educated aristocrats, they arrived penniless, and it was only through the generosity and friendship of Cape burghers that Monsieur Labonne rebuilt their shattered lives. Now, more than ninety years later, sole survivor of that Huguenot family, nearly blind, Anna Labonne lived in a small house near the waterfront with the help of a one-time slave, now her companion, friend, and eyes - an old, black woman some fifteen years younger.

So close to death, Anna Labonne had little interest in the world's dramas. News of the famous, of war and rumors of war, of discoveries, fires, floods, earthquakes, storms, of maritime disasters and famines - all such news was old news to the old woman, just the millwheel of fortune grinding out its endless trickle of human vicissitude. While Cape Town burghers hurried to the waterfront to learn the latest from ships putting in, Anna Labonne lived oblivious to the affairs of the world on the dwindling remains of her family's fortune, trying to return the good works her own parents had received so long ago. What better way to spend her last days?

That she chose to bestow her good works exclusively on women stemmed from a gradually awakened awareness during her long, solitary life, a realization that a woman's lot was basically unfair. When younger, Anna Labonne upset many a man with her stubborn assertion of this new insight, but she hardly cared; it was all their doing anyway! Thus, long before she grew so old that she became simply an old, old woman, Anna Labonne had become something of a Cape Town pariah. But while some of the older inhabitants remembered her as an unsettling female provocateur, the rest of the town (if they knew of her at all) saw her as just a gentle, harmless, and mostly forgotten *ancienne*.

Anna Labonne knew she could do little to amend a woman's lot in this world but try to deal with its sad consequences, like a patient mother who comforts a runty child tormented by playmates. Although so frail she sometimes spent much of the day in bed, she kept an open door for the occasional female supplicant who might arrive battered and bruised and frightened or simply too exhausted to go on. In her little stone home with scarcely a trace of whitewash remaining she provided a corner to sleep in, a warm kitchen, and a parlor that offered more silence than conversation, for Anna Labonne was also quite deaf. Conversation required considerable effort. And it was to Anna Labonne's little house that Malvina Thornapple was directed by a woman in a dramshop.

Despite her intention to "get settled" in Cape Town, Malvina actually had no chance of putting down roots. For a man to get settled was one thing. A man could be pressed into service as a soldier or sailor or contracted out as labor. But an obviously distraught, slack-skinned, lower class, unskilled, unschooled, married woman jumping ship was something else. Too bad she was married, or she might quickly find a husband. Too bad she was unschooled, or she might teach. Too bad she was white, or she might sew clothes, bake bread, milk the cow, comfort the sick, feed the fires, fetch the water, swill the pigs, churn the butter, attend to birthings, sweep the floors, air the bedding, go to market, cook for the master, launder the linens, weed the garden, polish the silver, feed the chickens, light the candles, or do any other women's work that domestic servants performed in England. But in Cape Town these were the tasks of the dark-skinned. Really, what could she do? Cape Town had no manufactories, no wigmakers, lacemakers, dyers or weavers. It had no inns or theaters. It even lacked a convent. But it was business-like. Even the orphanage was an item on the Company's balance sheet, supporting itself by selling off the sea chests of sailors who died in the Company hospital.

In short, nothing was landed or shipped out without Company approval, including Malvina, for Cape Town was not a community, but a large business, and except for the likes of Madame Labonne, run for profit. The fate of Malvina was fixed before her determined foot even stepped ashore.

She'd been sleeping in an alcove off Anna Labonne's kitchen for almost two weeks when Tench paid his first visit. Anna Labonne's dim, threadbare parlor had not been graced by such a dashing figure since the days of the French *sabreurs.* Admitted by Malvina, his eyes sun-dazzled, Tench stood awkwardly just inside the door until he made out a tiny figure seated across the room. He stepped over to Anna Labonne, bowed, and lifted a translucent hand to his lips. He spoke so she could hear. "*Pardonnez la liberté que je prends, Madame Labonne, mais je m'appelle Capitaine Tench, et je suis parti d'Angleterre pour faire votre connaisance.*"

"*Ah, mon Capitaine,*" Anna Labonne laughed lightly, leaning forward, her milky eyes taking on a glow, "*Bonjour et bienvenue, monsieur l'Anglais.*" Her voice was so faint he could scarcely hear. She patted the cushion beside her, "*Asseiez vous, restez ici.*" Tench sat.

As they talked, Anna Labonne looked straight ahead with her useless eyes as she concentrated on Tench's words. Malvina remained by the door, intensely curious but unable to understand much of what passed between Tench and her venerable benefactor. "*Je pense que vous parlez le français mieux que moi, non?*" The two seated figures laughed softly.

Then Anna Labonne asked in English, "Are you married, *mon Capitaine*? Is *Madame* your wife, *votre épouse?*"

Tench chuckled and patted Madame Labonne's hand. "*Mais non, Mademoiselle*, I could not marry until I came to Cape Town to see *Mademoiselle Labonne.*"

"*Oh, Capitaine, vous êtes très charmant, très charitable, vraiment un galant.*" They laughed again and her laugh turned into a little cough and Anna Labonne dabbed her mouth with a lace handkerchief. She made an indistinct sound in her throat, almost a moan, and her milky eyes lost their attention. Except for that occasional throaty moan she remained silent for long moments. Tench listened to the loud tick-tock of the mantle clock, then looked questioningly at Malvina. She only shrugged. Finally he leaned close, "*Madame?*" Madame Labonne jerked slightly and touched the corner of her mouth again with her handkerchief and smiled.

"*Ah, monsieur, je m'engage dans une rêverie, pardonnez-moi.*" She sat straighter. "Your wife, she is Charlotte, no? and you have many children. *C'est bon, les enfants. Les enfants des enfants sont la couronne des viellards, oui?"*

Tench looked at Malvina. "*Oui, Madame,*" he said to the ancienne, "My wife is the *Charlotte,* and I have many children."

"*C'est bon.* Be good to them, *parce que vous savez, les pères sont la gloire de leurs enfants.*"

"*Oui, Madame.*" Tench leaned closer. "*Et leurs mères?*"

"*Leurs mères!*" Almost creaking, Madame Labonne turned to face Tench more directly. She might have been studying him. He wondered if he appeared as more than a dim silhouette through her cataracts. When she spoke her voice was stronger. "*Les hommes insensés méprisent leurs mères, et les femmes!*" She sat back, sighed, and after a moment rapped her cane on the stone floor. Her slight, wispy-haired companion entered to shuffle slowly across the room. Tench stood. Without a word the two women clasped bony hands and, leaning back, the black woman slowly pulled Anna Labonne to her feet.

"*Eh, bien,*" Anna Labonne groaned when she stood, "*Encore je me lève, Dieu merci,* always up, always down." One hand on the black woman's arm, the other on her slim cane, she inclined her head to Tench, "*Excusez moi, mon Capitaine, mais je suis trés fatiguée. Adieu, que Dieu vous bénisse ainsi que votre épouse.*"

Tench bowed, "*Adieu, Madame Labonne, et merci beaucoup.*"

As the two ancient women made their slow, painful way from the room, Tench rocked up and back on his toes with his hands clasped behind his back as he regarded Malvina. When they were alone, he said, "Well, then, you're looking well, Malvina. You're getting back the healthy bloom of a true born Englishwoman." Although she'd scarcely set foot outside the door, Malvina raised a hand to her cheek. She suddenly felt hot, and wondered what Tench and Anna Labonne had said about her. Hesitating, she replied that she thought she felt a little better, but could rarely eat more than soft bread.

The two stood in awkward silence, then both began to speak at once, stopped, then stepped on the other's words again. They laughed politely at their clumsiness as Tench gestured for Malvina to speak first. "I was going to offer tea," she said, touching the back of her hand to her forehead. Heavens! she was hot!

Tench said he'd love to take tea with her, and surprised Malvina by following her to the kitchen and seating himself at the scarred table, chatting all the while about trivial events on the *Charlotte*. As Malvina moved between cupboard, fireplace, and table, Tench told how Lily the cow had given birth to a fine heifer calf, how the Reverend Johnson was making regular trips to the *Charlotte* to hold services, how Corporal Baker's foot now seemed completely mended, how Hester Thistlethwaite's husband was bedridden with fever, how little Ned had caught a strange, reddish fish.

At mention of her son Malvina halted as if yanked by a rope. She turned to the captain, her face haunted. "She's gaunt," Tench thought. Embarrassed by his slip, Tench twisted in his chair and hurried on to tell of the livestock and supplies being loaded. Whatever matters had caused such delay by the Council of Policy, the issues now seemed resolved, and preparations for departure were nearly complete. They'd sail in a few days. He stopped talking when he heard the screech of the kettle swung into the fireplace and saw Malvina's shoulders begin to shake.

He crossed to the fireplace and turned her gently around. She kept her hands over her face as she sobbed silently. "Shhh, shhh, shhh," Tench said as he enfolded her. She stiffened, then let go a little, still shuddering with silent sobs. Even through his uniform she felt very warm, and Tench felt moisture through her dress as he patted and rubbed her back and made his shushing sound. He felt her slowly relax as his head slowly bent until his cheek touched the side of her head. Ever so slightly their bodies began to sway side to side in warm communion. His eyes closed, Tench became aware of a faint tightening in his groin. Lord! he thought, giving himself some distance as he continued to caress Malvina's back, the woman is wasting away to nothing!

"I'll die if I go back," she whispered. Her shoulders began to shake again with quiet sobs, "I might as well drown myself as go back."

"Shhh, shhh, shhh," was all Tench could find to say.

She pushed back and looked up at him with teary intensity. When she spoke she almost hissed, "If you make me go back I'll kill myself!" She turned to the fireplace and poked at the fire, then said in a matter-of-fact voice, "I'm not a good mother anyway. No one will miss me. Dobber will just find some other woman. Some other silly woman!" Still standing, Tench heard her words but was thinking of the awkwardness of the situation and of the need to get her back on the *Charlotte*. Time was growing short.

"It's all right," he said. "No one's going to make you go back." He turned to pick up his hat from the table, "I'll see what I can do about getting your children ashore. I'll see if Dob... Given the situation I think your husband can resign. I'll help find you all a place to stay till you can catch a ship back home. But there's not much time. Not more than two days, I'd say. You'd best have your things packed. "

"Oh, Captain! Do you mean it? Do you mean we can all go home?" Malvina rushed over to throw her arms around his neck. "Oh, thank you, thank you, thank you!" She kissed his cheek with a smack. Suddenly embarrassed, she seized his hand and covered it with salty kisses. She looked up, smiling brightly through tear-sparkled eyes. "Oh, thank you, Captain, you're a saint on earth!"

In the early dusk two days later, Captain Tench again appeared at the door of Anna Labonne's small house. Again he charmed the antique Huguenot, who excused herself to go to bed. Again he accompanied Malvina to the kitchen. Again he sat at the knife-scarred table while Malvina bustled about preparing tea as she blathered about the day's sunshine, the sweet air, the early flowers, barely able to control a swarm of questions about her children, about Dobber, about getting the rest of her things off the *Charlotte*, about how soon their family might return to England. After she poured tea and they were settled in their chairs, Tench reached into his side pocket to produce a small parcel of sugar cakes. He opened the wrapper and held them for her to smell. "Mmmm," Malvina said, her eyes closed as she breathed in the rich aroma, "All of a sudden I'm starving. Here," she said, jumping up, "I'll fetch a proper plate." It was then that Tench shook a shower of tinctured opium into her tea.

As they drank tea and nibbled sugar cakes, Tench asked Malvina about her early life, about her family, about her town. Enjoying the tea and pastry, basking in the captain's attention, ebullient over the good turn in her fortunes, Malvina was talkative, expressive, and increasingly animated. Tench grew worried. After spiking three cups of tea and seeing no change in her lively chatter, he thought, My God, it isn't working! At the next opportunity he dumped the rest of the vial into her cup. Only after she drained that cup and one more did Malvina's speech become lethargic. Her eyes grew glassy, her shoulders bowed, and with a yawn and a deep sigh she pushed cup and saucer away with careless gesture, lay her head on her arms on the table, and slept.

Tench studied her for a few moments, reached over to raise a limp finger, saw it fall with a faint plop on the table, then rose and lifted the bar on the rear door. He clucked his tongue into the evening. Almost instantly Sergeant Motherwell hurried through the door, out of character in a dark frock coat, followed quickly by Stork, Quint, and a stolid Marine named Bill Mitchell, all dressed as sailors. Stork and Quint each carried a stout pole, and Mitchell a length of sailcloth and a blanket. The four newcomers blinked in the kitchen-light before becoming aware of Malvina sprawled across the table. "Blimey," Quint breathed, hearing her snore, "Gone as a goose."

Finger to his lips, Tench beckoned Mitchell. He pulled aside the curtain of the little alcove containing Malvina's cot and pointed to her valise and parasol, ready to go. Within moments they'd fashioned the poles and sailcloth into a litter on which they gently deposited the unconscious Malvina. Then, Quint in front, Stork behind, they hoisted their sleeping burden, sharing a quick smirk of relief for her melted figure. Tench bent close to listen to her breathing and arranged her hands over her bosom with some care. He draped Sergeant Motherwell's blanket over her body and loosely secured her chest, middle, and legs with loops of rope. At Tench's nod the procession quietly bore Dobber's unconscious wife into the pale moonlight. Tench slipped a note from his pocket to prop against the teapot with a guinea coin (for which he would claim reimbursement). The note informed Madame Labonne that Malvina had decided to continue on to Botany Bay with her husband and children.

On the street the covert group was joined by a Dutch watch-officer. "Without a hitch," Tench smiled. Chuckling quietly, the men eased down the dark street to the pier and deposited Malvina in the *Charlotte*'s jolly boat. Tench shook hands with each of his men, then bid them goodnight to head for a wineshop with the Dutchman, satisfied with the night's work and already mulling in his mind how to make an entertaining story of their adventure. White would find Malvina's resistance to the opium interesting.

After the officers disappeared into the shadows Quint spoke urgently as he motioned back up the street. "Sergeant, I think we dropped something back there, something that fell off. I better go get it," and in a blink he'd vanished into the darkness.

"Quint, Goddamn you!" Motherwell stage-whispered, "Get back here!" But to no avail. The sergeant scrubbed his face and jaw with a meaty hand, then rubbed the back of his thick neck. To no one in particular he said, "Jesus, someday that kid is going to get in big trouble!"

In a few minutes the anxious Marines saw Quint emerge from the night carrying some hapless burgher's small pot of blooming geraniums, snatched from a doorside. "Quint!" Motherwell groaned, that one word expressing such perplexity, irritation, wonder, and relief that the men laughed softly. In the darkness Quint's grin gleamed around the black gap of his missing tooth. He whispered, "Well, when a man has to leave his lady he should bring her flowers when he gets back."

Later, when he presented Seedy with the potted flowers, remembering the awful punishment of his flogging and foregoing a hurried opportunity to make love to her, he whispered, "My heart says 'Aye,' love, but my back says 'Nay.'" Seedy placed a hand on his proud chest, looked into his eyes, and nodded her understanding. She too had fallen in love.

~~~~~~~~~~

"*Sirius, at the Cape of Good Hope, Nov. 10, 1787*" Feeling miffed, Arthur Phillip was making his report brief. Two days earlier a packet of the British East India Company had put into Cape Town with letters from England, but none for him. None! Damned bureaucrats! Damned sniping fools! But he knew what would happen weeks hence when his report from Cape Town reached London. Some Home Office functionary would hold his letter by a corner as if he smelled the convict stench and say "At last!" and twitter that Arthur Phillip must have dawdled at the Cape colony. In reality, after the Council of Policy finally granted his requests Arthur Phillip had pushed day and night.

He turned in his chair to look out the galley window at the massive structure of the fortress. Had he really guessed right? That it was that slippery Spaniard in Santa Cruz who'd made the mischief? What other reason could explain the Council of Policy's odd behavior? He dipped his quill, bent to his paper and continued, "...*I was informed that the crops of corn having failed the year before last the inhabitants had been reduced to the greatest distress, and that I could not be permitted to purchase any flour or bread. I, however, obtained an order for three days' bread for all the ships...."* He paused. How much should he tell of events that finally led him to call the Governor's bluff? He decided he couldn't write about the sugar cakes - the Home Office would think him batty. He continued, *"...and as I found on enquiry that the last year's crops had been very good, I requested by letter to the Governor and Council permission to purchase what provisions were wanted for the Sirius and Supply, as likewise corn for seed, and what was necessary for the live stock intended to be embarked at this place."*

Arthur Phillip laid aside his quill, sprawled in his chair, and rubbed his eyes, which complained of their soreness. Then he sat with his face in his hands and listened to the distant squeal of winches and faint calls of men yelling instructions and the bellow of cattle being hoisted into ships. How odd events followed at times. Sugar cakes! When Shortland asked Tench if he knew of any special meaning sugar cakes might have, Tench had fixed the Naval Agent with such a stare that Shortland thought his question had given deep offence. Shortland hastened to explain the Dutchman's enigmatic suggestion. And Tench, rolling his eyes back in his head to think in that odd, irritating way of his, had actually cried, "Eureka!"

That was when Shortland learned that Tench had brought sugar cakes to his dwelling, which John White praised over tea to their landlady, who informed her English boarders how lucky they were to find such delicacies in the shops, because during times when grain was in short supply it was against the law to bake such fine-flour pastries. Thus, "Eureka!" Tench had hit on the answer to Mynheer De Witt's hint about a connection between sugar cakes in the shops and the posturing of the Council of Policy.

Even after Arthur Phillip sent a second letter and made a personal entreaty to the Governor, the Dutch authorities remained obdurate, insisting that drought had depleted their grain stocks and that they couldn't possibly spare so much flour for the new English colony. Mulling over the impasse, Arthur Phillip finally dispatched Major Ross to the Governor with a hamper. He instructed Ross to present himself in full dress uniform and to say the hamper was a gift from His Majesty's government. The hamper held three bottles of Canary sack and a half-dozen sugar cakes. The very next day the Governor dispatched a letter granting all their requests, along with an invitation to dinner for Arthur Phillip and his principal officers. At table everyone had gotten on very well, and spoke not of a word of sugar cakes or shortages.

Arthur Phillip snuffled behind his hands. *Omnia tempus revelat.* Sometimes these silly intrigues were amusing, but not now, not with so much at stake. Rousing himself and returning to his task, Arthur Phillip finished penning his report, skimming over the events of the past two months in a single paragraph. He scraped back his chair and moved to the window. A lighter approached with three brood mares. Pregnant mares, pregnant cows, pregnant sheep, pregnant pigs - for the long, final haul to Botany Bay his fleet was filled with animal life in vitro. Two for the price of one. Vessels within vessels. So far there'd been only one human birth, a baby girl born on the *Charlotte* and named for the ship. Perhaps one day that baby girl would bear her own children in the new colony. There lay the real hope, Arthur Phillip believed - to get the land broken and things born. As he gazed out the galley window, his vision of neat rows of cottages and gardens came to him again, of green pastures and harvest-gold fields of grain, of contented sheep and cattle.

He enjoyed his reverie for some moments before he shrugged himself back to the present and returned to his table. Why no word of his wife? The silence puzzled him. Was she already dead? He decided not to write her, but bent to make a note about convict marriages. How was he going to handle marriages among the convicts?

Botany Bay
1787

"*M*other of God! I'm dying!" Katie whispered. She'd gone into labor as the *Charlotte* weighed anchor in Cape Town a day earlier. Now, drugged with caudle, frightened, and writhing in pain from insistent, fruitless contractions, she felt herself slipping away. "Hannah, Hannah, will you pray for me? Oh, Jesus! It hurts!"

Hannah knelt close and took Katie's hand to press on the soft, worn cover of her Bible. She began whispering sibilant prayers, rocking to and fro. Her knees numb from the long vigil, Mary B patted Katie's other hand and exchanged teary looks with Seedy and Liz Bason, all of them haggard with fatigue from vain efforts to assist Katie's labor. They'd encouraged, massaged, squeezed, pushed, walked, and propped Katie in a half-dozen positions, but still the child would not descend.

"Sweet Jesus!" Katie gasped when still another powerful contraction took hold. As the paroxysm grew, her back arched, her chin reached upward; the cords of her neck were sharp, white ridges. She abandoned the Bible to grip Mary B's hand so hard Mary B thought her fingers would break. "Ah, no! No! No! No!" Katie rasped.

When the contraction at last subsided, Liz Bason exchanged another look with Mary B and pushed to her feet. She swayed, hobbled to the hatchway, and rapped with a dipper on the ladder. "Quint!"

Startled, Quint half-raised his musket, then knelt to the grate. "Blimey! It's awful, those screams. How is she, any better?"

Liz shook her head. "Best get the surgeon."

John White was playing chess with Tench when Broughton tapped at his door. Seeing Sergeant Motherwell behind his slender servant, White grimaced, "Does duty call?"

Motherwell nodded. "It's that Prior woman."

White looked at the board as he stood. "I'll have you this time."

Tench snorted good-naturedly and watched White roll up his sleeves. "Can I help?"

"Get my apron," White told Broughton. To Tench he said, "Thanks. I'll need a table. Same as for the wounded." The instructions surprised Tench. Was a surgeon's workspace the same for battle and birth?

"Roust a couple of convicts," Tench told Motherwell. "No, get James Martin and somebody else dependable. Set up a place in the women's quarters."

White took a bottle from his bag, poured some of its contents into a vial, and handed the potion to Broughton. "Give this to that Bason woman and tell her to put some in the caudle." He busied himself with his bag of instruments.

The area at the bottom of the ladder where Mary B had given birth was crammed with Cape Town freight - barrels, crates, bags of grain, bundles of hay, Lily the cow's new heifer calf, and two pregnant sheep. Within minutes, pausing in their work at the horror of Katie's screams, James Martin and Will Bryant cleared a space and wrestled a couple barrels into position. They spanned the barrels with planks while Motherwell dispatched Quint to fetch sailcloth from the piles aft.

"We need more light," White announced as soon as he arrived. "Jesus, why do these things always happen at night?"

"Get more lanterns," Tench ordered. White doused his hands with oil of tar.

When Porter appeared with lanterns, Motherwell positioned Porter and Quint on either side of the makeshift table. Quint aimed the bulls-eye of his lantern at Seedy. "The light of my life," Quint announced, his grin tense.

"Shut up, Quint," Motherwell growled. "You two get lanterns, too," he told Will and James. Katie screamed again, a piercing cry that set the calf mawing and the sheep bleating and Charlotte and Little Jacob squalling.

"Jesus!" White said, "Let's get going before the whole ship starts bawling. Wat, stay, will you? I might want you to wipe my brow." He grinned and winked.

As soon as Katie saw the surgeon and the makeshift table she began to pant and try to break free, but Liz propelled her forward, "We've got to do it, Katie, we've got to do it."

"Get her up here, damn it!" White snapped, his brow already beaded.

Perhaps it was the monstrous pain of another contraction that suddenly ended Katie's struggles, or perhaps the opium-laced caudle, but her body stiffened until she stood rigid as a post, then she suddenly wilted, and the women eased her onto the table. Liz stepped in front of White to arrange Katie's dress, now drenched with sweat, amniotic fluid, and blood. She raised Katie's knees and slid her hand beneath her dress and crooned as she rubbed Katie's swollen womb. Katie lay still.

White watched and listened, then bent to Liz Bason's ear. "Step aside. I'll make this as painless as I can."

He put his hands beneath Katie's dress and groped, one hand on her belly while he insinuated the other into her vagina, slowly easing into the birth canal. He looked into her half-closed, stuporous eyes. Porter shifted a little to see better under her dress.

From his studies White knew he should be able to feel the baby's head with either his right hand inside or his left hand outside. If it was his right hand, now several inches into the birth canal, the news was good. If not....

Katie went into another contraction. Again she stiffened, then began writhing and panting, her power surprising the surgeon. "Jesus!" he gritted, "Hold her down!" Liz Bason and Hannah each had a leg, Mary B and Seedy her arms. Katie bucked and twisted, moaning "Ah, no! No! No! No!"

When the contraction subsided, White continued his analysis. He was uncertain, but he thought the head was presented partway through the pelvic rim. Perhaps an arm was in the way or the upper rim of the pelvis was too small. Or the baby too large. He looked at Liz Bason, a question in his eye. Almost imperceptibly she nodded.

Shit! White thought, his shoulders sagging. He withdrew his hands and wiped them on his apron, studying Katie's face. In his mind's eye he saw the words in one of his reference books: "Newborn heads are soft and quite flexible." Should he give her another dose of opium? He decided no, and reached in his bag to withdraw a flask of brandy. He squeaked out the cork with his teeth, smelling on his upraised right hand the famous flowers from the garden of female delights. He drank a healthy draught and offered the flask to Tench, who shook his head.

White put the flask back in his bag and took out the forceps. He raised an eyebrow to Tench before turning back to Katie. "Now, listen," he told the women, "The baby's stuck. I'm going

148

to go up and get it. If you hold her still it'll take only a couple minutes. Are you ready?"

The women nodded frightened faces. "All right, then," White said, and experimentally worked the jaws of the forceps. He put his hands under Katie's dress again but suddenly withdrew, fished oil of tar from his bag, and wiped the forceps with the antiseptic. He turned back to Katie, but withdrew his hands once again, this time to uncork another big swig of brandy. "All right, then," he said half to himself, and squared his shoulders.

He was ready to insert the forceps when he paused, wondering blankly what would happen if she began another contraction. He couldn't remember any mention of that eventuality. He regarded her face. She lay passive, breathing shallowly. He decided to go ahead.

He was surprised at how easily the forceps slipped up the birth canal. He felt his instrument bump against pelvic bone. With his left hand on Katie's bulbous belly just above the pelvis, he closed his eyes and tried to imagine the curved ends of the forceps, the position of the head, and was pleased that he could actually picture the movement of the baby as he worked his left hand to maneuver the head into position. He felt confident enough to smile slightly as he turned the forceps to grasp what he knew were the sides of the skull. He had a solid hold and was reaching with his left hand to begin his pull when Katie began another contraction. She writhed under the restraining hands. "Hold her, hold her!" White cried, an edge of panic in his voice, sweat suddenly pouring from his brow. Katie twisted with unnatural power as White struggled to keep his forceps engaged with a steady pulling pressure. "Hold her, Goddamn it, hold her!" The women bore down desperately, laying their weight on Katie's limbs.

White wasn't sure what was happening. Was the head through the pelvic rim? Was it still locked in his forceps? Katie gave another violent twist, and White felt something slipping. He re-clamped the forceps and pulled forcefully. Katie screamed one long ear-piercing screech after another, flooding the air with her pain, setting off a crescendo of cacophonous wailing, bleating, and screaming of animals and humans.

"Shut up!" White cried, "Shut up!" He felt a surge of adrenaline and pulled at the resistant mass, twisting the forceps one way and another as if pulling a tooth. Katie continued screaming, even after Mary B put her head next to

Katie's, covered her mouth with her fingers and begged her to be still, knowing that Katie couldn't hear and would not heed, but desperately wanting an end to the awful screaming.

"Shut up!" White cried again, the screams and barnyard uproar unnerving him. Suddenly amidst the tumult he heard a new scream as the creature at the end of his forceps began emerging, and he glanced up to see Mary B holding up her hand with a dumbfounded expression, blood pouring from the end of her little finger. She began to shake it as if to shake off her wound. Katie had bitten Mary B's finger nearly through at the first joint, then fainted.

"Jesus Christ Almighty!" White said, flinging back Katie's dress. He stared at the bloody, vaguely humanoid mass clamped in his forceps. Blood seeped steadily around a gory, bluish umbilical cord snaking back into Katie.

Tench heard Porter retch and saw the light from his lantern fall away. He thought the moment seemed frozen in time, when every color, every sound, every smell, and every detail would remain etched forever in his memory. He saw Liz Bason reach down, rip off the bottom of her dress, and stuff it between Katie's legs. She took Hannah's hand and pressed it hard with her own against the wound of Katie's sex. He saw Will Bryant thrust his lantern at Seedy and take Mary B's bloody finger and clamp it tightly in the tail of his shirt, put his other hand on the nape of her neck, and urge her, "Hold on! Hold on!" He saw the surround of shiny, white, panting faces stare in shocked silence at the gruesome scene on the table, and other mute, wide-eyed faces crowd the edge of light. He saw the wildly rolling eyes of Lily's calf and heard its front hooves scrabble in panic at the side of its makeshift pen. He saw the sag-shouldered figure of John White look blankly at the inert creature he'd just pulled from Katie's body, the thing itself wondrously small to have caused such suffering. He became aware of his own hands extended towards White as if in supplication, spattered with a woman's blood.

Katie's moan seemed to thaw the tableau. With mechanical deliberation White released his forceps. He gagged and turned away when he saw shreds of what were probably Katie's interior membranes clinging to his instrument, knowing suddenly that in his panic he'd seized her flesh along with the baby's head and ripped it off. For a baby born dead, or killed in the borning.

"We've got to stop the bleeding!" Liz Bason cried, "She's bleeding too much!"

Her call to action stirred White, and he turned back to his patient with a glare in his eye. "Bandages! From my bag! Wat, in my cabin, a box marked 'Women'...have Broughton get it! Hurry! Sergeant, get some more laudanum into her. Not too much! No, wait, Quint, you do that! Sergeant, see what you can do about that finger. Take it off!" He fished scissors from his bag and tossed them to Motherwell, "Dunk the stump in oil of tar!"

So White reassumed control, and the scene quieted as he snipped off the umbilicus and shifted the bloody infant to a barrelhead, then extracted what he hoped were the remains of the placenta. He packed Katie's wounded interior, and inserted and secured a pessary from the 'Women' box. During this entire time Katie's gory nether parts remained in full view, but in the horror of the event she'd become disembodied. Even Porter ceased to stare at Katie's naked flesh and looked anxiously to her blood-drained face for signs of recovery.

Finally White stepped back and let go a huge sigh. He was wiping his hands on his bloodied apron when Katie began gagging. Raising her heaving body upright, he swung his arm and delivered a heavy blow, then another, and a third, before a large chunk of half-digested salt meat popped onto her blood-soaked lap. With a roll of his eyes to no one in particular, White picked up the meat between thumb and forefinger and held it towards Tench. Tench knew the surgeon was about to make a joke when the scene was again frozen by a faint but distinct, eerily feline little mew. Puzzled faces looked at each other, then almost as one their eyes swung to the tiny, blood-streaked body on the barrelhead, fluid leaking from its umbilicus. No one moved. No one breathed. Every heart stopped. Then they heard it again, a fragile, tentative voice announcing to their lantern-lit world that a new life had begun.

~~~~~~~~~~

In the early morning of January 21, 1788, Arthur Phillip scrambled atop a sandstone outcrop with scant dignity to stare in disbelief at the *Sirius* gliding through the headlands of Botany Bay. How was this possible? At the news of an approaching sail his first thought was that Dutch ships had come to challenge the new English colony. But it was the *Sirius* all right. Arthur Phillip shook his head in dismay. Two months earlier, fed up with the poky, erratic sailing of the convict transports and cargo ships, he'd moved to the nimble *Supply* and loaded his four fastest ships with Marines and convict artisans for the final 6,000 mile dash to Botany Bay. Over such a distance he hoped to gain two or three weeks before the laggards arrived, time enough to explore and evaluate the land, to choose a settlement site, to build temporary storehouses and workshops, and get the gardens started. But now, less than forty-eight hours after his "flying squadron" dropped anchor, here were the slowpokes!

"There's the *Prince of Wales*," Shortland said excitedly. He studied another ship rounding Point Solander. "And there's the *Charlotte* right behind! And the *Lady Penrhyn*! All the convict transports are here, thank God!"

"Very good, Lieutenant. Signal the *Supply* to send a pilot to the *Sirius* and get her anchored. Do you see the cargo ships?" With nearly all the supplies for the new colony loaded on just three ships, Arthur Phillip worried about what he'd do if one or two or even all three never arrived. He'd be little better off than a shipwreck!

Shortland knew the supply ships were a sore point with Arthur Phillip, but the Navy Office insisted that Shortland not put any of the colony's food supplies aboard the transports lest convicts seize a ship and sail away with provisions for years. He remained silent as he studied the gap between the headlands, his arms trembling from the telescope's weight. Shortland endured several more minutes before he announced with a quiet sigh of relief, "I think I see the *Fishburn*."

But even with the safe arrival of the rest of the fleet, all was not well. Arthur Phillip had already discovered that Botany Bay offered poor opportunities for settlement. Edged with saltwater marshes, it revealed an abundance of sandy, unproductive soil and a paucity of fresh water. Its anchorages were difficult and exposed to heavy southeast seas. He'd already decided to explore alternatives, but with the early arrival of the transports he'd have to hurry, for despite his need to get the convicts off the crowded, unhealthy ships he couldn't just land them anywhere. While not overtly hostile, the natives were clearly not friendly. Curious, yes, and shy, but ready to defend what they considered theirs. And this land was theirs.

"I think that's the *Golden Grove*," Shortland said.

Of course Arthur Phillip didn't understand the gray-haired native who'd greeted their ships with a brandished spear and a vigorous speech shouted across the water, the bone through his nose lending a fierce aspect of defiance: "Beware! This is our place! We have lived here forever. Our spirit ancestors us gave this place. Go away! This place is filled with sacred things. I saw you before. You took our things! Now you have come again. Beware!"

Turning to his companions, the old man warned, "I saw these creatures long ago! We should kill them! We should make them go away! Who will help me kill them? Who?" Awed by the incomprehensible human figures on the moving islands, his young companions did not respond, and after a few more threatening gestures toward the invaders the old man stalked away.

Approaching the cluster of natives in his cutter, Arthur Phillip had made signs he wanted to drink. After conferring among themselves, the natives pointed with their spears farther down the shore and from a safe distance followed the progress of Arthur Phillip's boat. Since that first encounter, however, no one could induce the natives to come any closer.

"And there's the *Borrowdale*! Good show!" Shortland lowered his telescope and snapped it shut. "That's it then, Sir. Congratulations!" Shortland's smile was broad as he extended his hand, "Thank God, sir, you did it!"

Arthur Phillip's smile was perfunctory as he accepted Shortland's handshake. That all eleven ships safely made the 14,000-mile journey was remarkable, but....

"Thank you, Mr. Shortland. Take my cutter to the *Sirius* as soon as she's anchored. Extend my compliments to Captain Hunter and bring him to the *Supply* at his earliest convenience. We've got to revise our plans."

~~~~~~~~~~

"Thank God," Mary B breathed when she heard the anchor pay out. Overhead she heard feet scurry and excited voices exclaim at the sights. They called across the water to the other ships, "Yes, we're well, we're well, we're safe come." Blue sky and puffy sunlit clouds smiled through the open hatchway. The air smelled of land and growing things and faint woodsmoke.

Aye, safe come, Mary B thought, a part of her angry despite her relief at their arrival...safe come only after the worst two months of her life. Two months that had seen monstrous seas and violent winds, a sailor swept away in spumy swells higher than the yards, spars snapped, sails shredded, rigging carried away, cloud-wrapped lightning and incessant thunder. Two months that had seen fog and icebergs and teeth-chattering cold. Two months that had seen the women huddled in their dank prison, frightened beyond words, fever and sickness rampant. Two months that had seen Willard Thistlethwaite and a convict named Thompson die of galloping dysentery. At times Mary B thought they were all being carried along some horrible, storm-tossed, endless seaway to Hell.

And yet through it all her baby had thrived, oblivious to everything but Mary B's warmth and milky breasts, her murmurs and kisses and sometime desperate hugs. Even Katie's frail John Matthew clung to life, his misshapen skull bearing crease marks from the forceps, while Katie herself, attended to with unusual assiduity by John White, made a slow but steady recovery.

Safe come, the women wore irrepressible smiles of relief. They congratulated each other and chattered excitedly as they prepared for release from their shipboard dungeon. In Cape Town still more women had been transferred to the *Charlotte*, and for the past two months they'd all lived as cramped for room as any in the fleet. Those like Mary B who'd come aboard in Plymouth had lived in this dark, crowded, stoop-low prison for 319 days. Oh, how she longed to get away from so many people, to stand erect and feel the earth beneath her feet, to walk more than a few steps before having to turn around, to drink water until her stomach might burst. How she longed to feel the warmth of a fire, to watch the progress of the sun and moon across the entire span of the sky, to mingle in the flux and flow of life on land. She wanted to chew a blade of grass, to show her baby living things - leaves, flowers, ants, bees - wanted to watch her learn to walk, to teach her to talk.

And more. Somehow, come at last to Botany Bay, this land of improbable story, rumor, and endless speculation, this wilderness where she'd live out the rest of her sentence and for all she knew the rest of her life, Mary B hoped she might finally understand why, in that silly, unthinking act by the Citadel, the truth of which she tried again and again to reconstruct, its details obscured by drunkenness and shock, the telling and retelling, by her dreams and the passage of time - that she might finally understand why all this was happening to her. Perhaps here she would at last come to understand her unfolding fate.

She'd been imprisoned for two and a half years. She'd lost her family, her maidenhead, most of her hearing in one ear, the tip of her little finger, and control of her life. She'd thirsted, fed on wormy food, suffered extremes of heat and cold, known terror and utter despair. But she'd survived. And borne a baby girl and made new friends. And now was ready to pick up the pieces of her life and go on. And deal with James.

She had to deal with James.

But Mary B was not alone in having man problems. Those women whose bellies swelled with new life needed to settle matters with the sailors and Marines who'd gained their favors during the long months at sea. Would they marry? Perhaps. Rumors were that Marines and sailors who wished to settle in the new colony would get land, as would convicts who completed their sentences. With land to farm, a man and woman could build a life in Botany Bay far more promising than awaited most of them back in England. In truth, Arthur Phillip was not alone in his daydreams of cozy thatched cottages and fields of ripening grain.

Of course the thought of never seeing one's family again was heart-rending. And in truth most of the women would actually lead lives of poverty, drudgery, childbearing, sickness, and mistreatment. Some would give up and kill themselves, some go mad, some wallow in drunkenness and befuddled debauchery, and some would suffer sickness and early death. But as the anchors splashed down in Botany Bay on that bright, sunny day the women's relief was palpable, their prospects dawning, their thoughts hopeful, their spirits high.

Mary B's face brightened in the festive atmosphere. She swaddled Charlotte in a clean length of blanket and cooed, "Aye, we've come to Botany Bay, haven't we? Aye, you too, you've come to Botany Bay too, haven't you? Aye, you have, you've come to Botany Bay too, haven't you!" Charlotte rewarded her with a happy, three-cornered smile. Mary B picked up her baby and in the low confines of the prison swung her in a stooped little dance, making up words to an old familiar tune:

Oh, we've come to Botany Bay-ay,
We've come to Botany Bay-ay,
We've come to Botany Bay-ay
And now we go ashore!

"Ouch!" Mary B cried when she straightened too high and struck her head. But she laughed good-naturedly, and began an improvised chorus

And now we go ashore!
And now we go ashore!
Oh, we've come to Botany Ba-a-ay...and now we go ashore!

Other women joined her song around the sunny hatchway, laughing through round after round of the simple little song. Even Katie smiled through her pain and gingerly swung John Matthew in time.

Mary B's prospects were brighter than for most, for Tench had informed her a couple weeks out of Cape Town that Will Bryant wanted to talk with her about marriage. James was no longer on the scene. He'd been transferred from the *Charlotte* to Arthur Phillip's "flying squadron."

Tench was officious when he informed Mary B. "It's against regulations for the two of you to converse, but he keeps pestering. I guess in fairness to the child...." He paused. "But I feel I must tell you...." He paused again and looked away for a moment. "Will Bryant may already have a wife."

Mary B felt something go out of her. Will Bryant too? Was this all just a game for men? "I left a letter from him at Cape Town, but whether it was to his wife I don't know. He said it was to his sister...are you all right? Are you ill?"

Tench studied Mary B's suddenly pale face, her confusion and hurt plain to see. *Why does she enter my dreams?* he wondered, and again felt the draw of an inexplicable connection.

Her mind a muddle, Mary B looked down at the sleeping baby without seeing her. What did it mean if a man had a wife? Was she a real wife? Were they married in a church? Or was she like Liz Bason, a common law wife who bore children and kept house? Liz was already planning to marry a man named Hatherley, the *Charlotte*'s carpenter.

"Well, why not?" Liz said, "That bastard Bill left me rotting in jail, didn't he? We won't see Bill's face in Botany Bay, will we? And I've got Jacob here, haven't I, and that's all I've got after seven years with bastard Bill! Well, fair's fair and I've got to take care of myself!" This time, however, Liz Bason would make sure her marriage was a real marriage, with a preacher and proper papers.

Mary B shot a glance at Tench. He stood with hands clasped behind his back, rocking up and back on his toes, his gaze fixed on some distant place. She wondered, why does Will Bryant want to marry me? What can he know about me except what James might tell him? She shuddered and her stomach tightened at the thought that James might have blathered about her wanting to be married. Damned James! What did she owe him? He'd nearly jumped out of his skin at the thought of marrying the mother of his child, this sweet little girl! Oh, how he'd whined!

Oh, James, she thought, you didn't even try to see your own baby!

Aye, what did she owe James?

The answer wasn't simple, for Mary B carried fond memories of James. In the Exeter jail and on the hulk he'd treated her with kindness and made her laugh and feel special. She'd come to love him for his generosity and companionship. Although she couldn't remember the incident, he swore she'd been the first of them to smile in the tiny courtyard of the Exeter High Jail. Soon after he'd presented her with a limp daisy, bowing extravagantly and telling her it was all that remained of a huge bouquet he'd picked but was commanded to give up by a duchess in a carriage. After that he brought her tidbits acquired through from his contract labors in the sawpits. He hated his sawpit work. But when he dragged his shackles through the iron-bound inner door, stumble-weary, sawdust in his hair and coating his face and ragged coat, a glimpse of Mary B at the pump or near the copper would brighten his countenance and straighten his back, and he'd flash a smile that made her heart dance.

During her first hard weeks in Exeter he made jokes of matters that seemed overwhelming. "Why, don't you worry, Mary B, Sunny Jim will make it right." And he had, many times, whether they nuzzled beneath a stairwell or shared a bit of meat or an onion or a rare swallow of throat-searing gin. He shared his pittance earnings with Mary B because he liked her and wanted to make her jail life easier. He liked her firm grip and frank, friendly eyes, her wide smile and straightforward manner. In return she was feminine, pliant, and companionable.

Perhaps to preserve this growing mutual trust, there came a day when he made a confession to Mary B. He told her he had a wife but that after his arrest her family had shut off all communication with him. He'd written two letters, both returned unopened. "Maybe that's the way she wants it too," James said, scuffing the dirt, "Over and done."

After his confession Mary B and James became even better friends, and their physical contact awakened feelings in Mary B long dormant...thrills down her back and tingling in her breasts and groin. Perhaps inevitably there came a Sunday afternoon when something new and urgent stirred in them both. Their laughing died. They sensed mutuality and nurtured the stirrings. Their kisses beneath the staircase grew more passionate, oblivious, and James fumbled with clothing and explored with fingers until their bodies sank to the floor and Mary B felt a sharp pain, an insistent intrusiveness and fluid warmth. She heard his groan of ecstasy, "Ah, Mary B!"

The pain gradually slid away. She let her body seek its natural rhythms as he covered her neck with kisses and caressed her face. Wonderful. Eyes closed, she felt proud of herself and complete in a new way. James grew more insistent, more demanding, and he gripped her hard, and Mary B felt herself losing her rhythms and being used. Suddenly James was plunging, his breath rasping, an alien presence that suddenly grew rigid as a statue, then sank to Mary B's breast with a groan. James lay panting in exhaustion. Mary B held him, wondering what happened.

Over time James learned to be a better, more patient lover, and after their transfer to the hulk in the Hamoaze he purchased time for them in the cubbyhole. Their "little house" they called the closet-size shack rented out by Blackpool's turnkeys. There, for a brief time almost every week, they enjoyed an interlude of make-believe domestic normalcy, and privacy. They talked of acquaintances and news from town and sometimes shared a sweet from the women's quarters or a treat from town. And they made love.

So that was what Mary B owed James Martin - a good friendship and a prison love that helped to keep her going. Did it matter that he was married? She couldn't really say. If his wife wanted nothing more to do with him, was he really married? Why wouldn't it be all right then for him to love Mary B, and for her to love him back? And what was marriage anyway but a lot of promises? Oh, it was easy to promise, to say "I'll do this and I'll do that," but those were words. It was easy to say, "I'll do what's right," but another thing to know what's right. And what might be right for one might not be right for another.

No, all Mary B really knew was that news of his newborn son and the long voyage to Botany Bay had interrupted their affair and perhaps broken it. Until they could talk things out, she wouldn't really know whether there was anything to salvage. When she lived in a real "little house" on Botany Bay, she realized, it might not be with James.

"Captain Tench, Sir?" Again the soft 'zzz' that Tench found so pleasurable. Mary B's eyes were downcast. "Will Bryant...," she took a huge breath and expelled it, "It's only right that I talk with him, for the baby's sake, if you'll permit it, Sir."

Tench nodded. His feelings were mixed. He wanted to lift her chin and explore her eyes for her real feelings, but the hood of her cloak hid her downturned face. Why do I feel jealous? he wondered, what can I be jealous of?' "Very well," he said, "The Governor will be wanting marriages. You might keep that in mind."

Mary B only nodded, unable to meet his eyes.

~~~~~~~~~~

"Keep hold my hand and don't be afraid," Tench said. Little Ned tightened his grip on his little wood sword and squeezed Tench's fingers. Pre-occupied with the boat, the natives had not yet noticed Tench's party. The black, muscular men were the most exotic humans Tench had ever seen - bushy, matted beards, wild hair. Behind stood two or three sag-breasted women, one burdened with an infant. One of the men held the hand of a small boy. "Steady," Tench cautioned his party of Marines. He felt Ned's fingers squeeze more tightly and heard him begin to pant. His own breathing quickened as Ned sidled more behind him. "No trouble now, Sergeant, we want no trouble. Ground your arms." He heard Motherwell's quiet command and the thump of muskets butting the ground. "Blimey!" Quint muttered, "Straight out of Hell!"

Discovering the Marines, the band of natives began a surprised, excited jabber as the women dashed into the underbrush. It took the natives a long minute to re-group and begin a cautious approach. They were a fearsome sight with their scarred faces and cicatrix scars running the lengths of their arms. Two were bedaubed with whitish smears in vague patterns and one had painted his face completely white. Completely naked, their penises seemed of a size with their broad noses; black knobs jutting through tangles of pubic hair. Fierce-faced and grizzle-bearded, the foremost wore a large, white, curved bone poked sideways through his nose, its ends protruding upward like tusks. He carried a shield of bark and several spears with jagged tips. Thrust through a thong at his waist was a heavy club. The others also wore bones through the septums of their noses and bore shields and spears, or brute-heavy clubs.

When they approached within a dozen paces the grizzled elder rapped on the face of his shield with his spears and shook his weapons at the Marines, crying in a rough voice something that sounded like "Warra, warra." The others took up his cry as they beat weapons on their shields and made feinting motions towards the strangers.

With Ned in tow, Tench took two steps forward, the pale, open palm of his right hand raised towards the natives, evocative of some idyllic depiction of Bradford or Penn greeting the natives of the New World. "Good morning," he said through their threatening babble, "My name is Captain Watkin Tench of His Majesty's Royal Marines. I come in peace."

The natives fell silent. The grizzle-beard elder slowly raised his spears so they pointed to the sky and the two parties faced each other in the quiet calm of the early morning. How do I explain? Tench wondered.

Sent to look for fresh water, his party was one of several exploring Botany Bay's shoreline and marshy reaches. Indeed, the transports had scarcely let fall their anchors when Arthur Phillip ordered the wheels of progress to begin turning. Parties were deployed to cut grass for the livestock, scout for water, seine for fish, dig sawpits for lumbering, cut firewood, and explore the upper reaches of the bay. The pace was energetic, heady, and exuberant. They'd arrived! This was their new land!

Now the unofficial godfather to the boy at his side, as Tench had nosed along in the *Charlotte*'s longboat, the swells gentle, the water lapping at shoreline rocks, he'd become aware of the peaceful quiet. Like a blossom unfolding in sunlight, his awareness enlarged as he took in the variegated greens of the forest and the undulating, silver-green marsh grass and rushes. He saw patches of pale, green grass scattered among the rocks and the darker green of towering, sentinel trees and the nearly black green of distant forested hills. Here and there a splash of yellow-white cliff peeked through. A sunbeam slid through an assembly of trees, creating a ghostly slanting column of spirit-mist from the rising vapors of the night's rain. An occasional puff of air brought odors of rich vegetation to his nose. Tench inhaled deeply, intoxicated by the heady aroma of green-life and fishy smells. His Marines responded too, and if they began their excursion nervous about the threat of encountering the seemingly invisible natives, their apprehensions soon faded. They sat silent as the boat slid forward with the slow, clock-clock rhythm of the oars, sometimes turning to look at a sun-splashed outcrop of rock, a fire-scarred tree, or follow the flight of startled waterfowl flickering light and shadow against a pristine sky. Sometimes one would nudge a neighbor to share a special sight. So still, so peaceful and secluded.

Strongly moved, Tench thought, How beautiful! An entire land to build on, an entire land to explore! He felt privileged to be one of the first! At heart a man of books and newspapers and conversation, always ready to speculate about the latest discovery and the latest disaster, the latest breathy turn of events in Moscow or Madrid, he now sensed a profound significance in the moment, a moment with meaning far beyond the daily blather of table-talk. This was a rare, deeply personal moment. He wanted to savor it. Leaving the rest of Motherwell's squad to follow in the boat, he decided to walk hand-in-hand with Ned on an inviting stretch of black-sand beach. A few Marines trailed behind. And now the spell was broken by their meeting with the natives.

His right hand still raised, Tench said, "We mean no harm." He reached into a side-pocket and extracted a handful of beads. Ned in tow, he took three slow steps towards the natives, proffering the beads. "Aren't these pretty? See how pretty they are? Don't you think they'll look good on you? Here, you can have them, take them."

A rough word from Grizzlebeard stopped his progress. The reek of rancid fish and woodsmoke emanating from the natives was over-powering. The elder pointed with his spears at Tench's feet, indicating he should place the beads on the ground. "All right," Tench said, bending to deposit his offering, "If that's what you want. But I assure you we mean no harm. We're just looking for water." He cupped his free hand and made drinking motions. "Water," he said, "Water, good water."

"Warra! warra!" Grizzlebeard challenged, again assuming a threatening pose.

"No, no, no, no," Tench answered, raising his palm, "Waa-ter. Waa-ter. See? See?" He took off his hat, made a scooping motion, and pretended to drink from his hat.

When Tench bared his head, the natives exclaimed in surprise. They jabbered excitedly. Tench realized the idea of headgear was apparently unknown to them.

Grizzlebeard once again raised his spears for silence, but several moments passed before his band obeyed. Planting his spears in the sand, Grizzlebeard made motions to his head, indicating he wanted Tench to put on his hat.

"Oh, yes, of course," Tench said. He put on his hat, took it off, put it on, turned to the side, struck a pose. "See? See? I take my hat off to you." He turned and doffed his hat in a courtly bow. Marines sniggered. At their sniggering, some of the natives sniggered too. Then one stepped forward and mimicked Tench's bow to perfection, saying with remarkable precision, "See? See?" At this a couple natives laughed and began bowing to each other, causing the Marines to laugh at the natives making fun of Tench. Tench himself smiled and nodded, then doffed his hat again and again to Grizzlebeard, who joined the laughter, the gap of his missing foretooth black amidst strong white teeth, his laughter releasing the rest of his band to mill around, bowing and cavorting and crying "See? See?" The boy slipped loose and danced up to Ned, where he jerked bows like a pecking chicken, laughing, "See? See?" He carried a sharpened stick, his spear.

As their antics grew more and more exaggerated and silly, Tench and his Marines withdrew into a nervous reserve and watched with fixed grins. Grizzlebeard's command brought the natives to a sudden halt.

Ned's heart stopped when he saw the ugly elder point his spears at him and say something to Tench.

"What?" Tench asked, "Do you mean this boy? This boy?" He pulled Ned forward and placed him squarely in front.

Grizzlebeard eyed Ned while conferring under his breath with the nearest men. The boy's shoulders trembled beneath Tench's hands. "Easy, Ned," Tench murmured, "They mean no harm." Finally the elder called forward the boy in his own band. Positioning the child before him, he began a speech in which he gestured repeatedly towards the child's genitals, squared his own shoulders, threw out his chest, and stamped his feet authoritatively. Then he made intercourse motions with his fingers and gestured to Ned. "See?" he said.

Non-plused, Tench kneaded Ned's shoulders. At last with a tone of discovery, he said, "Ahhh!" Then, "Ned, I think they want to see what you are." He unbuttoned the two buttons of Ned's wooly shirt. "See?" he said, exposing a patch of the boy's thin, white chest, "See, Ned's a boy, a boy!" The sight of Ned's milk-white skin raised more shouts of astonishment, and Grizzlebeard nodded and smiled at little Ned. Again Grizzlebeard made a speech, sometimes pointing at the dark child before him, sometimes at Ned. He gestured that Ned and the boy should join hands. With this, the other natives began to shout encouragement and joke among themselves. One of them pantomimed rear intercourse with another. Tench felt a shock of recognition. In a voice of disbelief Tench said, "Ned, he thinks you're a girl."

"Argh!" Ned cried, twisting free, "A girl!" He faced Grizzlebeard, "I'm no damn girl!" He fumbled with his breeches. "See? See?" he said, his breeches around his knees. He raised his shirt and thrust his hips at Grizzlebeard, his little white penis an accusing finger, "I'm no girl! I'm a boy! a boy! I'm a boy!"

The natives stood in shocked silence, then erupted in laughter. Again they began to mimic, this time Ned's hip-thrust, growing more and more animated until they were weak-kneed from mirth, a few belly-laughing so hard they sank to their knees and held their sides in howling, helpless laughter. No one laughed harder than Grizzlebeard, who stamped his feet, pounded his shield with enthusiasm, threw his head back to the sky, and yowled.

At first the Marines laughed with them. But as the natives' behavior grew more and more extreme, the Marines again grew uneasy, and when Grizzlebeard observed the white men no longer participating, he too quieted, and his people grew silent behind him. He again regarded the strangers and turned to confer for a long minute. He faced Tench and pointed to Tench's groin with his spear. "See?" he said.

Tench sputtered, "Why! Why! Good Heavens, man! Absolutely not! I don't care who you are, I'm not going to put on a show!"

Reading Tench's demeanor, Grizzlebeard grew serious, drew himself up, and shook his spears. He began a long speech in which he gestured to the ground, to the water, to the woods, to the distant hills, to his followers. As he spoke, Tench nodded agreement, filling the pauses of the naked man with, "I know, I know, absolutely! Yes, yes, you're right. You're absolutely right!" When Grizzlebeard ceased he stood quietly for a moment, then again pointed at Tench's groin. Tench motioned the nearby longboat closer and turned to his men, indecision on his face. He was about to order them to present their arms when Quint stepped forward.

"I'll show the buggers, Captain. If they think we're a bunch of women, I'll show them a spear!" With Tench's relieved nod, Quint handed his musket off and stepped up to Grizzlebeard. He towered over the native by half a foot. Wearing a close-lipped smile, Quint slowly undid the string of his breeches, slowly unbuttoned his cod-flap, slowly pulled out his shirtwaist. The natives stood riveted, expressing quiet wonder and puzzlement at the mechanics of Quint's complicated coverings.

"Hum!" they exclaimed when Quint revealed his sex. They crowded close, peering at his genitals, and began to explore gently with fingers and noses the textures and smells of his red coat and crossed belts, his breeches and leggings. They fingered his buttons and the leather of his cartridge pouch and his bayonet scabbard and the laces of his boots. "Hum!" they repeated as they discovered each new aspect of his attire. When one put forth a hand to touch his penis, Quint jerked back. "No, no, no!" he said with a broad grin, "I'm saving that!" and began to button himself up.

When the natives saw Quint's grin, they shouted in astonishment and jabbered excitedly. Grizzlebeard took Quint's hand and led him over to Captain Tench. Pointing first at his own mouth, then Quint's. He drew himself up and thrust out his chest. He spread his bearded lips in a grin that revealed his own missing foretooth. When he motioned to his followers, three stalwarts spread their lips also, each revealing a missing foretooth. "Well, I'll be damned!" Motherwell said.

Tench murmured, "They must take a tooth for manhood."

Grizzlebeard began to knead Quint's brawny arm as he lectured Tench. He placed a spear and throwing stick in Quint's hand and indicated that Quint should throw it. "Me?" Quint said, hefting the long, reedy spear, uncertain what the throwing stick was for, "Me, throw this?" Grizzlebeard nodded encouragement as he addressed first Quint, then Tench, urging Quint. But when Quint and Tench stood uncertain, the elder called a stalwart near and pointed to a whitened driftwood stump half-buried in beach sand perhaps sixty yards distant. The natives excitedly parted to either side of the flight path. Quint's competition, broad shouldered and with cicatrix scars on his arms and a bone piece in his hair, looked into Quint's eyes and smiled as he fitted the butt of a spear on his throwing stick. For a moment he balanced his weapon between thumb and forefinger, then turned, eyed the stump, took two quick steps, and launched his spear in a fluid blur.

Thirty pairs of eyes watched the arching, slightly tailing flight of the spear speed towards its target and saw and heard it sink dead center in the stump with a thunk, the shaft wagging in slow eccentric circles like a friendly, expectant dog. "Blimey!" Quint breathed, his words drowned in the excited shouts of the natives and the surprised exclamations of Marines. He was about to try his own throw when Tench held his arm.

"Shoot it," Tench said quietly.

"What?" Quint asked, looking confusedly at Tench, then at his spear, thinking that somehow he was supposed to shoot the spear.

"Shoot the stump!" Tench said. "Don't try to throw that damn thing. You'll look a fool!"

Quint studied the unfamiliar objects in his hand for a moment, then wordlessly exchanged them with Motherwell for his musket. The natives rumbled, their faces puzzled.

Quint was not a good shot. He was fast, yes, and he could load, lock, and fire quick as any. But a little near-sighted, he'd gotten by in practice because most targets in his military training were only twenty or thirty yards away and most military shooting was in massed volleys.

He took a big breath. The stump had definite edges only if he squinted. He glanced at Sergeant Motherwell, who winked and nodded encouragingly. When Quint returned his attention to the target Motherwell rolled his eyes Heavenward.

Quint wondered if he should ask someone else to shoot. He'd never seriously tried to hit a target at this distance before and he was not a confidant Marine as he took another big breath, checked the priming of his weapon, and set the cock. He raised his musket and squinted along the barrel.

The natives murmured as they watched the clumsy stick with its spearpoint waver, settle, then suddenly belch fire and a cloud of black smoke and make a thunderous noise. Some of the natives dropped in fright, holding their ears and covering their faces. Some bolted for the woods. As the report of Quint's musket died and the smoke thinned, the confused, fearful eyes of the natives still remaining turned to follow the intense stares of the white men, who suddenly burst into cheers. Lucky Quint! The small explosion at the stump was Quint's three-quarter-inch ball shattering the native's spear. Quint lost his breath as Marines pounded his back.

Other than seeing that most of their stalwart's spear had disintegrated in an instant of noise and confusion, the natives had no idea what happened. A now-confident Tench patiently explained to Grizzlebeard several times how Quint's weapon sent a heavy round ball to destroy the spear.

At last Grizzlebeard began to understand. He asked by sign if the musketball could also strike him. At Tench's self-assured nods the elder became very thoughtful. He tested the tip of Quint's bayonet and fingered the hole of the muzzle, as if by touch he could divine the secret of its power. He studied the musket for some time, Tench's babble of explanation lost to his concentration. Other natives waded out to the boat to feel its strakes and gunwales, its tholes and oars, while others mingled with the Marines, again exploring their clothing, weapons, skin and hair. The Marines wrinkled their noses at the smell of the natives and admired their firm, muscular bodies. Porter feasted on the naked flesh of the women who now peeked hesitantly from the woods some distance away, their broad-nosed features not unlike his own. To one side, Ned and his native counterpart tussled and romped and laughed as they mimed the events of the morning.

Suddenly Grizzlebeard stepped back from examining Quint's musket to shout a gruff command and stalk back down the shore toward the stump. There he stood at half-face, posed with his shield and spears held aloft. One by one his group pulled themselves from the strangers and straggled back to cluster near the elder, their eyes trying to read his thoughts, to understand the reason for his sudden withdrawal and sullen demeanor. But Grizzlebeard said nothing, and after his group reassembled he set off in the direction they'd come from.

The Marines watched, baffled. As the last of the natives disappeared around the bend, the boy in tow, Ned raced out a few steps to send one last, shrill communication to his new playmate, "Yi, yi, yi, yi, yi!"

~~~~~~~~~~

Terra australis incognita, the unknown southern land. For centuries theoreticians had speculated about a landmass in the Southern Hemisphere that balanced the weight of the continents in the north. Not discovered in the South Atlantic, it was thought to lie somewhere in the Pacific, an ocean so distant from Europe that even reaching it was a hazardous enterprise, so vast that its islands and cultures were discovered and then lost, often for generations.

In the early 1600's, a Dutch trader looking for a faster route to Djakarta sailed due east from the Cape of Storms and went too far. He fetched up against a barren land not shown on charts. Later ships following his route also touched along the vast shoreline, and by mid-century several expeditions from Djakarta had sought some utility in this desolate land. But no one found anything worth having. The Dutch marked the discovery on their charts as "New Holland" and forgot about it. Thereafter, only ships poorly navigated or blown off course touched on New Holland, whose size remained unknown.

After the Dutch lost interest in New Holland, more than a hundred years passed before England dispatched Lieutenant James Cook to Tahiti, ostensibly to observe the transit of Venus across the sun in efforts to sharpen astronomical calculations. But after completing his Tahiti mission, Cook opened secret instructions that took him to New Zealand, where he proved by charting its coastlines that it was not part of *terra australis incognita*. Determined to discover and chart the eastern shores of New Holland, he sailed west, and a thousand miles later nicked the southeast edge of the continent

167

whose eastern bounds had eluded searchers for decades. He followed the coast north for more than a week before putting into what came to be called "Botany Bay."

Cook would report finding abundant fresh water, friendly natives, and soil that seemed rich enough for planting: in short, an anchorage eminently suited for habitation, a judgment shared by two important passengers - Joseph Banks, a rich, young patron of exploration and natural science, and David Solander, a Swedish botanist. On a later voyage Cook circumnavigated the south polar region, proving that no great continent projected into the South Pacific. Now the best minds of the time were satisfied that no significant landmass remained undiscovered. *Terra australis incognita*, the sixth continent on the globe, was now just *terra australis*.

Between the time of doughty James Cook and the landing of the First Fleet twenty years later, not a single European of record touched on New South Wales, which was the name bestowed by Cook on his discovery. Arthur Phillip knew no more about Botany Bay and New South Wales than what Captain Cook and his passengers had observed two decades earlier - observations blurred by time, distance, and politics. A careful observer, Joseph Banks had noted that much of New South Wales was "barren," but in the enthusiastic rush of politicos to promote Botany Bay as a penal colony, Banks swung to the position that an English outpost there could be self-supporting in two years and would open up important new trade opportunities for English manufacture.

But such addled thinking had not hindered European powers in the past. For three hundred years their explorers had sailed the globe in small, leaky ships, not knowing where they were, what they were seeing, or how to find it again. Nonetheless, they landed here and there to plunder what they could and claim the rest for distant sovereigns. When rival powers staked rival claims, even to lands largely unexplored, they sometimes drew a boundary line on a fanciful map to settle the matter, and thus determine by where the line ran whether the aborigines learned Dutch, Portuguese, Spanish, French, or English, and whether they entered the Heaven run by Catholics or the Heaven run by Protestants.

All according to form, of course. Arthur Phillip's commission from King George III established his territory as encompassing all the land and islands from the northern tip of Cape York (10° 37' S) to the southern tip of Van Dieman's Land (43° 39' S). The

168

boundary extended west to the longitude of 135° E. From his clutter of tents on Botany Bay, Arthur Phillip would rule by virtue of this commission a territory of one and a half million square miles, roughly half the continent. But George III had no idea how much land this was, and probably didn't care, for he was going quite batty at the time. In any event Arthur Phillip's domain was roughly 15 times the size of the English monarch's own adopted island nation. Whether only half the continent would be enough was not yet an issue. For the nonce, His Majesty's government would not quarrel with the Dutch Stadholder over his claim to the other half, another million and a half square miles of western wasteland, give or take a few.

But the Dutch were not England's only competitors. The Spanish maintained a firm grip on the Philippines to the north, and guarded a Pacific lifeline from Manila to Acapulco for their treasure ships. And the Portuguese still maintained enclaves in the Orient for their share of the spice trade, while the French laid claim to numerous Pacific islands and were actively exploring further opportunities. Even individual traders looked for lucre with armed expeditions sent to seize what opportunities they might. Keeping a close eye on all these actors, its bleak coastlines bounding much of the Pacific's northern rim, was Russia. But the Dutch, weakening under the constant pressure of England's empire building, presented the strongest rivalry to the English.

Thus, when dawn lookouts glimpsed two large ships approaching Botany Bay only three days after the *Charlotte*'s anchor splashed down, the English recognized an incongruity that set their hearts racing. No one had seen a strange sail for two months. The nearest trade route was thousands of miles distant. Probably no European ship had touched at Botany Bay for twenty years. Yet here on the heels of their own arrival two strange ships were joining them! Excited and apprehensive, curious, they watched the mystery ships struggle in the distance against contrary winds and strong currents, gradually losing way. By midmorning the ships had slipped out of sight behind Point Solander, unable to gain the English anchorage.

Who were these mysterious strangers? Had they come to challenge the English for possession of this savage wilderness? Was one of them perhaps the English breadfruit ship with a companion vessel? Were they the van of the Second Fleet, more convict ships?

Arthur Phillip acted with a sense of urgency. Absent from England for eight months, he realized wars might have started he knew nothing about. Lacking information about the world's affairs, deception and ambiguity were prudent weapons. He ordered the English flag raised so it was clearly visible from outside the headlands of Botany Bay and sent the *Supply* out with colors hoisted to identify the mystery ships.

Actually, Arthur Phillip had already decided to abandon Botany Bay for a vastly superior site a few miles up the coast. But whoever the intruders were, he didn't want them to know his intentions. He forbade everyone except senior staff from communicating with the strangers. He recalled all land parties and ordered all ships readied for early departure next day.

But the *Supply* returned from its reconnaissance almost as ignorant as when it went out. The mystery ships had retreated too far. They were clearly not the Second Fleet, however, and the English ship sent to Tahiti to collect breadfruit trees wasn't one of them. Indeed, they weren't English at all. Perhaps they were French or Spanish, perhaps even Portuguese.

Unknown to Arthur Phillip, Lieutenant William Bligh's soon-to-be-famous breadfruit ship, the *Bounty*, was nowhere near. Indeed, Bligh was no closer than Santa Cruz in the Canary Islands and months behind schedule. The *Bounty* wouldn't reach New South Wales until Arthur Phillip's colony was six months old. By then Bligh would be so far behind he'd not put into Botany Bay at all, but head directly for Tahiti. There the mutiny awaited that would link the fate of the *Bounty* to Mary B.

Perhaps the mystery ships pushed Arthur Phillip a bit over the edge. At dawn next day a stiff east wind sent in the huge ocean swells that made Botany Bay an undesirable port. The swells pitched the anchored ships like woodchips. Nonetheless, Arthur Phillip gave signal to hoist anchors for immediate departure. Agitated, impatient, pacing, he witnessed the *Supply*'s fruitless struggle to escape the bay and near collisions of several transports floundering in the swells and adverse winds. "God Almighty!" he yelled as he beat the compass box in frustration. Officers blanched. But his anger vented, Arthur Phillip calmed. He signaled the fleet to anchor again and retired to await ebb tide.

Word of Arthur Phillip's tantrum spread quickly to Marines and convicts. "Well, it's nice to see he's human," James Martin remarked to no one in particular. He thought a moment before he spoke again, more publicly. "I expect if we took his pants down he'd look like us." He paused again. "Well, maybe." The men roared.

Early next morning, Mary B heard groaning capstans once more rouse the *Charlotte*. The seas seemed a little calmer. But in a few minutes she heard warning shouts, a flurry of orders, then distant cries of alarm. Sailors above voiced exclamations of dismay. Coming about in the still-unfavorable wind, the *Friendship* had collided with the *Prince of Wales,* splintering its jib-boom and destroying its victim's mainsail. Continuing to come about, the *Friendship* now plowed in front of the *Charlotte*, just getting underway. Again shouts of warning, sounds of running feet, then a horrible shuddering and the awful groans of rending wood. The *Charlotte* had collided with the *Friendship*. Its bowsprit was locked in an embrace of wood, chains, and rigging with the *Friendship*'s larboard shrouds. The agony of repeated collision and ripping timbers persisted. Screams of fright and alarm filled the below-decks as the entangled ships drifted toward a rocky outcrop pounded by ocean surf. The winds that had frustrated their departure now pushed the convict ships toward disaster.

Sailors hacked frantically at the tangled rigging, the shock waves of their thudding axes palpable to the convicts below. Nervous superiors exhorted, "Cut us clear! Cut us clear!" Sensing catastrophe, Marines and convicts scrambled to gather their belongings in a pandemonium of curses, orders, cries, and entreaties. Some sat dumb, awaiting Fate.

After a lifetime of terror but only minutes of frenzied effort, the ships at last broke free and swung apart. Sailors swarmed aloft to bring them under control.

The *Friendship* recovered first and dragged itself out of harm's way. But nearly lifeless, blown broadside, the *Charlotte* continued its fateful course toward the rocks. A voice just above the women's hatchway talked to the ship, "Come on, come on, come on, you can do it, come on, come on, come on, you can do it."

Mary B found herself mouthing the encouraging litany. For anxious minutes the little ship remained unresponsive, the voice above coaxing, entreating, praying.

At last a tremor of life.

A swelling cheer rose from sailors and Marines as the *Charlotte* roused herself with a shudder, dipped her bow, and began to answer the helm. Then she was heeling sharply, coming around with authority and bearing away from the ominous black rocks with only yards to spare. Prayers of thanks followed cheers and backslapping, with the silent meeting of eyes, head-shake smiles, embraces, and tears of relief.

Mary B hugged her baby and sobbed, at the end of her endurance. When the *Charlotte* at last put into Botany Bay, the promise of relief from long confinement had raised the women's spirits to giggling, silly heights. They sang, they primped, pulled together their belongings, chattered about their favorites among the men, and what they first would do ashore.

--"I'm going to eat leaves and grass and every green thing I can get my hands on!"

--"I'm going to find the Marine with the biggest jug! 'Come here, Soldier boy,' I'm going to say, 'Annie's got a big thirst and a big itch!'"

--"I'm going to plant my feet in the ground and grow roots. I won't go to sea again till they chop me down and drag me out!"

But their euphoria was short-lived. When they learned no settlement awaited them at Botany Bay and saw the unrelieved wilderness of marshland, rocky outcrops, dark hills and primeval forests, their spirits plunged. Even worse, they heard Arthur Phillip wouldn't let the women disembark until he found a new settlement site and it was properly prepared. That might be weeks!

Now this narrow escape from possible death drained the last of Mary B's strength. It was that afternoon, while the *Charlotte* plowed towards Arthur Phillip's new destination, abandoning the poetic promise and mythic opportunities of Botany Bay, that Mary B decided she would marry Will Bryant.

"I don't have a wife," Will reassured her as Tench stood nearby, listening to the convict's breathy, impatient argument. "I don't have anyone. I'm not going back to England, even if I could. My fishing boat's gone and there's nothing back there for me anymore. Besides, I don't think any of us are going back. How can we, at the end of the world like that? We'll be stuck in Botany Bay forever, you and me, stuck there together, and we'll have to make the best of it."

Charlotte in her arms, Mary B tried to follow the rush of his words, the wind sometimes whipping them away. She found it hard to keep his eyes, so smoky and penetrating and insistent. "I thought you were special the first time I saw you," he continued. "You know that, don't you? I thought about you and watched you. I knew you were special, and I wanted to see what you were like. Sure, I knew you were Jim's, we all knew that, we all knew you were Jim's woman. But that didn't stop me. I could've told you about Jim. But I didn't. And you know why? You know why?"

Mary felt herself falling under the spell of his insistence, his eyes. She looked away, her mind awhirl.

"I'll tell you why. I wanted to be fair, see? I knew you were Jim's but I wanted to be fair, that's why. Jim's a good sort. But it's no good with him, is it? Jim's got a wife and he's a Catholic and now he's got a little boy back there. And remember that day in Portsmouth? Aye, I asked you to be my Mary B, didn't I? You remember, don't you? Sure you do. I knew then Jim wouldn't do right by you - not with that news about his boy. And I knew you were going to have his baby then too."

Mary looked up at him in amazement.

"Aye, I knew you were going to have his baby, Jim's baby, but I said, 'Jim, are you going to marry that woman?' That's what I said, aye. And you know what he said, what Jim said? I'll tell you what he said. Nothing! That's right, nothing." Will stepped back and flexed as if to loosen knots in his shoulders and neck, then bore in again, his eyes dancing over Mary B's downturned face. He put his hands on her shoulders and leaned close. "Mary B! Mary B, I've watched you from the day you came onto the hulk. I've watched you for almost two years! I know who you are! I know what I want. Mary B, I want you to be my wife! I want you to marry me! I want you to live with me so I can take care of you and your baby. As if she was my own! I will! As if she was my very own. I like young ones around me, see? And I hope we have our own! I promise and I swear to God Almighty. Look!" he commanded, stepping back to raise his right hand, "I swear to Almighty God that I want to marry you and that I will care for you forever and take care of your baby as if she were my very own! I swear to Almighty God! See? That's what I promise."

Mary B felt giddy. No one had ever spoken to her like this. His words were so compelling, so unreal. She felt buffeted, overwhelmed. When she sagged a little under the burden, Will grasped her again. He spoke more slowly, gently. "I remember the night of Katie's birthing. I remember like it was this morning. I watched you then too. I never took my eyes off you. And when Katie chewed off your finger..." he raised her hand to his lips and kissed the rosy, tingly end of her foreshortened little finger, "I never felt so close to anyone, not anyone." He released her hand and stepped back.

"I know you don't know me, Mary B, not very well. But I'm a good man. My cousin is Lord Marlborough's secretary. And I'm a good fisherman, like my dad and his dad before him. You know fishermen are good men, Mary B. And you know what? Among all the convicts here, I'm the only fisherman! Aye, the only fisherman!" He glanced toward Tench and lowered his voice. "I heard the officers talking about what we'll being doing in Botany Bay. You know, different kinds of work. Well, Jim's a good man, true enough, but all he knows is grunt work like the sawpit, and that's what he's going to do! Did you ever eat sawdust?" He asked this last with a coaxing smile, using a bent forefinger to lift Mary B's chin. "Well, it wouldn't taste so good, would it? No, if I had to choose between fish and sawdust, I'd choose fish. Wouldn't you? That would be the wise thing, wouldn't it? And I'm the fisherman! Aye, I'm the fisherman who's going to bring home the fish! Aye, bring home the fish to my wife. To my Mary B. Aye, to my wife, Mary B."

Albion
1788

When Lieutenant James Cook headed home from Botany Bay, he took note of a gap between towering headlands a dozen miles north that indicated another bay. Cook marked this find on his charts as Port Jackson, unaware that a sharp bend concealed an impressive waterway. Nosing into Port Jackson some twenty years later in his hurried search for a new settlement site, Arthur Phillip thus became the first European of record to discover the surprising reach of this magnificent riverain harbor, its 12-mile length uniquely scalloped with coves. Between the coves, forested peninsulas showed evidence of abundant fresh water. Oh, did Arthur Phillip's heart sing! Following the disappointments of Botany Bay he could scarcely believe his good fortune. The only sour note was his unsuccessful effort to engage the natives he spied along shore. Some hid in fright, others stonily watched the intruders sail by, while still others ignored his exploration party - no doubt hoping the strangers would move on and disappear like clouds.

After working several miles up the harbor, noting landmarks and making soundings, Arthur Phillip decided to found his colony in the seventh cove, one of the deepest. Situated on the south side of the harbor and reaching in a north-south direction for more than a half-mile, the cove narrowed to a marshy end fed by a vigorous stream. For a Navy man the anchorage alone was persuasive; in some places his ships could snug right next to the rock-shelf banks to offload. High ground promised healthy air and the grassy forest floor promised good planting. Why look farther? Arthur Phillip named his choice "Sydney Cove," after Lord Thomas Sydney, Home Secretary.

In this cove the remainder of the First Fleet dropped anchor at sunset on Saturday, January 26. The sight of forested heights silhouetted against a roseate evening sky, the comfortable embrace of the cove, the placid water, the quiet - the scene held much better promise than Botany Bay for the long-suffering voyagers. Drooping from a hastily erected flagpole at the end of the cove, an oversize Union Jack proclaimed the paternity of England's new colony.

Arthur Phillip had announced the name months earlier as his ships plowed toward their fabulous destination. Fixing his gaze over the heads of a dozen officers and officials at table in the Great Cabin of the *Sirius*, he proclaimed, "Gentlemen, I'm going to call our settlement 'Albion.' No one must mistake our object. New South Wales is English."

His table company exchanged glances and murmurs of approval. What could be more fitting? Young David Collins, designated as Judge Advocate for the new colony, raised his glass. "Then I propose a toast, Sir, to Albion - and to its esteemed Governor." Affirmations were enthusiastic as the men raised their glasses. Except for Major Ross. Peevish from what he believed was a succession of slights by Arthur Phillip, he rolled his eyes and waved his wine in a vaguely contemptuous joining.

Now finally arrived at the heart of Albion, Arthur Phillip's functionaries set about their duties with the same enthusiastic affirmation. The light of a new day was barely visible in the east when hundreds of convicts and Marines were fed a hurried breakfast and shuttled ashore. So much needed doing - find grazing for the animals, bed the plants and seedlings, clear land for tents, lay out streets, set up the hospital, dig saw pits, erect Arthur Phillip's large canvas house, build a boat landing, begin building storehouses. In the bustle that first Sunday few took time to attend Albion's inaugural Divine Service.

Major Ross wanted to build a palisade. Arthur Phillip overruled him: "If we have friendly relations with the natives we won't need a palisade." He ordered instead that Marines mark a perimeter and patrol it around-the-clock. "Make clear to the natives that they must not enter the perimeter, and make clear to the convicts and your Marines that anyone who straggles outside will be punished." Above all, Arthur Phillip wanted to avoid conflict with the natives. Again and again he cautioned, "The natives are not to be bothered."

That first overcast morning saw a flurry of activity as inferior axes rang against towering trees and grub hoes chopped at stubborn, rock-bound roots. Convict muscles long idle soon began to quiver. Soft hands blistered. As the sun advanced, sweat-drenched laborers were bedeviled in the warm, humid air by swarming insects. At mid-day some convicts dragged themselves back to the transports so drained by the heat and unaccustomed effort they could scarcely climb aboard to be fed. In the afternoon matters got worse. For most of Albion's labor force a felling axe or two-man saw were tools as unfamiliar as a compass or quadrant, and their use just as mysterious. Not knowing how to use their tools, convicts struggled with them, and produced little. They wore themselves out. They grew discouraged. Efforts flagged.

But some honed their skills in avoiding work. They mislaid axes and broke shovel handles. With time their only wealth, they frittered it freely. A call of nature might stretch into a long absence. To fetch a drink or clarify an instruction might consume an hour. The flight of a bird, a camel-humped cloud, a mouse in the grass, a bloat-belly fart, Marines passing by - all were reason enough for some to rest on their tools and perhaps begin a story, or simply wander off. Here was an unforeseen problem that would bedevil Arthur Phillip's Albion: it was no-one's job to make the convicts work!

Having carried the convicts to this new land, the transport contractors cared less if the convicts worked; they simply wanted to be rid of both convicts and cargoes so they could return home. Major Ross insisted his mission was to protect the young colony and keep the convicts from escaping, not for his Marines to serve as overseers of public works! And his officers agreed with him. The other civilian officials at Arthur Phillip's disposal - a commissary agent, a surveyor, a chaplain, the surgeons - all had more than enough to do. To make matters worse, on the final leg of the voyage dozens of convicts were laid low by an outbreak of scurvy and dysentery that now rendered them unfit for work. And dozens more were discovered to be too old, too feeble, or too crippled to do much of anything. As for the rest, ignorant and inexperienced, most were completely indifferent to the task at hand. Besides, government had to feed them. That was the law! And besides that, the sooner they got tents up, the sooner they had to sleep in them - leaky hovels stitched from threadbare sailcloth. And on the ground!

Work? Work hard? King and Country? Poppycock!

Thus, after the novelty of being ashore became hard, sweaty drudgery, convict efforts suffered a great falling off, and Arthur Phillip's assault on the forest primeval bogged down.

And thus, ten days later, Arthur Phillip sat in dying sunlight on a jut of reddish sandstone on the east point of Sydney Cove, numb with exhaustion, his thoughts on the chaos that was Albion. He absently scratched behind the ears of one of his greyhounds. His booted foot rubbed the supine back of another. He noted a new build-up of clouds and wondered if the rain would return. For most of the day the colony had sought shelter from explosive lightning and violent rain. Indeed, since their arrival almost every day had seen storms come beating through the woods, scattering his workforce and causing disorder and delay. And the convicts were already reverting to their old ways of petty thievery, quarreling, and fighting. Nothing was going easy! Unlike Tahiti, Albion was no land of friendly natives and bread on trees.

Nearby, his musket across his knees, a convict huntsman puffed on his pipe, a perquisite of his office. With Arthur Phillip's two other greyhounds at his feet, the huntsman sent up little clouds of sweet, contented smoke as he dreamily regarded the tiny blooms of native fires across the wide waters of the estuary.

Arthur Phillip squirmed from a recent, nagging ache in his right side, an ache that sometimes crept around to his lower back. He lifted his right arm and rotated his shoulder in search of relief. What other adversities would be meted out? Fifty times a day he had to make a decision, countermand an order, change his plans, look a fool. Just the trees were wearing him down.

- Excellency, we can't move the trees we cut down. They're too big!

- Excellency, these trees won't do! They're too hard!

- Excellency, the planks we saw are junk. They're splintery!

- Excellency, nobody cleans up the slashings.

- Excellency, Major Ross took twelve planks to floor his tent. He says your orders don't apply to him because he's the Lieutenant Governor.

But it wasn't just the trees. He was pecked at from every side.

- Excellency, it takes too long to bring the convicts back and forth for meals, and then we get malingerers.

Well, bring their meals ashore.

Of course, Excellency. But, er, who's supposed to do that, the contractors or the Marines?

- Excellency, the convicts aren't working hard enough. It takes a dozen men all day to fell a tree.

Then find some good convicts to keep them working.

But, Excellency, convicts don't want to be overseers of other convicts.

Then make them be overseers or cut their rations.

But, Excellency, convicts won't take orders from convict overseers.

Then punish them.

But, Excellency, then they punish the overseers.

- Excellency, the natives were very threatening this morning about our catch of fish.

Well, then, let them take some.

Yes, Excellency, er, how much? What if they want our whole catch?

- Excellency, who should tend the livestock?

- Excellency, will you be allotting garden land to the officers?

- Excellency, who should muster the convicts in the morning?

- Excellency, the Navy men want garden land, too.

- Excellency, who punishes the convicts? My officers say....

- Excellency, sailors are bringing rum ashore.

- Excellency, what if a sailor wants to marry a convict?

- Excellency, some of the officers are wondering, can they get convict women as housekeepers?

- Excellency, some convicts say their sentences are almost expired. Can they go back to England with the transports?

The greyhound at his foot twitched an ear at Arthur Phillip's long, soft sigh of exasperation. In the dusky light he saw the fresh glow of a lantern hoisted on the *Charlotte*, half a cable distant. The little transport lay at the mouth of Sydney Cove, a solitary prison. Earlier he'd ordered the women's ships positioned well out in the cove, safe from Marines who'd taken to swimming out at night to continue their trysts. Major Ross was unconcerned, "They're off duty, aren't they? Besides, they're just convict women."

Once again Arthur Phillip overruled. "Major, I want you assign a skeleton force to the women's ships. Shoot anyone trying to climb aboard."

Ross stiffened. His ash beat an irritated tattoo against his boot.

Arthur Phillip stirred at the memory of his constant quarrels with Major Ross. That fat-necked, rule-bound, dull-wit! The blustery nit! Nothing but trouble since the first day!

His greyhound's soft whine brought Arthur Phillip back. In stewing about Ross he'd pulled too hard at the creature's ears. He bent over and kissed her snout in apology and whispered in her ear, "So, Livia, what would you do, eh? Tell me, what would you do?" He chuckled when the greyhound raised her head and licked his nose and mouth in forgiveness. He put his head close and hugged her sinewy neck. The hound at his feet stood then, stretched her forelegs first, yawning wide, then her back legs. She crowded close for Arthur Phillip's attention. "Aye, you too, Lady Di, my lovely hunter, you too!" Aye, Arthur Phillip thought, if only he had more Navy men - good, solid, capable men who went beyond the bare minimum of what duty or rules required. Or brainy men like Tench. Of all the Marines, Arthur Phillip was most taken with Tench. As for the rest - nothing but quarrelsome drunks, strutting peacocks, half-grown men!

Arthur Phillip bent low, hugging his dogs, scratching their necks, thumping their flanks, confiding in a murmur. Well, it wasn't all bad! At least the Marines and most of the convicts were tented, the parade ground cleared, and his canvas house ready for the big day tomorrow - the official reading of his letters of commission. The baptism of Albion. Then a celebration.

But Arthur Phillip wasn't inviting the French, those two mystery ships that so incongruously appeared then disappeared off Botany Bay. Time had confirmed their identity. They were French exploration ships sent out more than two years earlier to probe the Pacific, and they posed no threat. Indeed, they carried Lieutenant James Cook's timekeeper and compass, even copies of his charts and observations, graciously loaned to the French in the spirit of international scientific cooperation by the Admiralty and the Royal Society.

Once safely anchored in Botany Bay, Captain Jean-François de Galaup de La Pérouse had welcomed Arthur Phillip's emissaries aboard *La Boussole* like a neighbor eager to share a sad tale. He explained that he sought refuge in Botany Bay after suffering an attack by natives in the Navigator Isles (the Samoas). In addition to killing a dozen of his men and La

Pérouse's fellow captain, the savages had destroyed two of his longboats. La Pérouse made for Botany Bay to repair his ships and build new longboats.

But La Pérouse didn't reveal all. Months before at the Russian port of Kamshatka he'd opened secret instructions from home: see what the English are up to in Botany Bay. He was on his way to spy on Arthur Phillip when the Samoans bloodied his expedition.

No, Arthur Phillip decided, not the French. The irony of conducting his official inaugural before a backdrop of brush piles, sagging tents, and several hundred bug-slapping reprobates would only produce French snickers. Moreover, convicts were already sneaking over to Botany Bay to beg for passage home, and Arthur Phillip would do nothing to encourage intercourse with the French explorers.

He sighed once again. What a circus! How was he supposed to create a self-sufficient colony with this miserable collection of lazy thieves and ignorant rascals? On any given day a hundred or more protested they were too sick to work, while fifty others were too decrepit to do anything but lean on their tools. Of the rest, one in three was a slobber-mouth dolt and the other two connived at avoiding work!

And the women! He'd kept them on the transports as long as he could, allowing only a few of the best behaved to live near his civil officials as housekeepers. Now another hundred had been put ashore, and tomorrow the *Charlotte* would be emptied of the last of them. After almost a year of strict separation, the sexes would mingle. What sin and debauchery would darkness bring tonight?

Arthur Phillip stood with effort and bent at his waist to ease the ache in his side. Governor! Governor of what? What did he govern? Bumpkin thieves and spread-leg whores? Tupping rams and laying hens? Rocks and streams and savages? He faced the fading roseate glow of the sunset. His companion hounds, attendant, eager, stiffened and glanced nervously back and forth from his face to the western sky, whining softly. He had the words by heart that conferred his powers and made him master of this chaos. He began to recite his commission in a quiet, semi-mocking voice: *"To our trusty and well-beloved Arthur Phillip Esquire: We, reposing especial trust and confidence in the prudence, courage and loyalty of you, the said Arthur, of our especial grace, certain knowledge, and mere motion*, *have thought fit to constitute and appoint and by these presents do constitute and appoint you, the said Arthur Phillip, to be our Captain-General and Governor in Chief in and over our territory called New South Wales....*" He paused. "With all its thieves and whores and rapscallion shirkers, its ne'er-do-wells and malingerers, its degenerates, dealers and foul-mouthed liars, its pickpockets, smugglers, and idiot ginpots, its...." His words trailed off. Tomorrow will be the real beginning, he thought. He'd announce his new rule: "If you don't work, you don't eat!" Thus far, all was prologue.

As he turned to go he suddenly realized he hadn't thought of his wife for several days, and he wondered if he still had a wife.

He called to the shadowy figure of his convict huntsman. "Come, McIntyre, let's close for grappling. As the man said, 'I've not yet begun to fight.'"

~~~~~~~~~~

While cast iron bridges were rising in England and steam-driven pistons drove belt-lapping wheels, the men they called sawyers cut trees into planks with the tedious craft of a century past. The convict labor of James Martin carried a long tradition.

His workplace was the sawpit; a hole in the ground about seven feet deep, surmounted by a stout wooden frame lugged from England. After the convicts cut down a tree they sawed the trunk into more manageable lengths and manhandled one end onto a brutish, two-wheeled axle. Like beasts of burden, they pulled this device to a sawpit, where they wrestled the log onto the frame and dogged it in place. With one sawyer teetering atop and another in the pit, using a long, two-handled saw, they sawed the log from end to end with an up and down motion. Finished, they slid the saw back to the beginning and sawed the length again. And then again. And then again. Depending on the size of the tree, the hardness of the wood, the sharpness of the saw, the thickness of the knots, it might take an hour to saw the length of a log. In this manner they reduced the log to slabs. In a second operation, they aligned several slabs and again sawed from end to end to square the sides. Thus did they tediously turn logs into rough planks.

Aside from being repetitious, boring, and exhausting, few aspects of James Martin's convict craft were notable. But any joint endeavor offered opportunities to quibble. Was it better to be atop the log or down in the pit? Atop was cleaner. And atop was better if the day was hot. Also, sawdust didn't come showering down with each stroke. But on a cold day the pit was warmer, and if a day began cool and turned hot, the pit stayed cooler. But on windy days the top was better because the pit swirled with sawdust that found its way into the eyes, nose and mouth. However, on a cold, windy day, whether to be atop and chilled or in the pit and inhaling sawdust was a subject of much debate. Sawyers usually traded places several times a day.

But put a convict at each end of a two-handled saw and argument rode the middle.

Saw, saw, saw, saw.

"You're not pulling!"

"I am!"

"You are not. I can feel it!"

"I am!"

"You are not! You're not pulling and I can feel it!"

"Well, you don't pull when I push!"

"I do so!"

"No you don't! I can tell! You don't pull!"

Saw, saw, saw, saw.

"What did you do that for?"

"What?"

"You pulled so I almost fell!"

"I did not!"

"You did! I almost fell!"

"I did not!"

"Well, don't do it again!"

Saw, saw, saw, saw.

"Damn your eyes! I'm going to give you one!"

"Now what?"

"You kicked that chunk of bark down here and hit me in the eye!"

"I didn't kick it! My foot slipped! You think it's easy up here?"

"You kicked it!"

"I did not!"

"You did so!"

"Oh, go to hell!"

"You go to hell!"

Saw, saw, saw, saw.

Thus, spat by spat, curse by curse, piles of green lumber grew for Arthur Phillip's Albion. In theory.

In fact, there was scarcely a usable tree in Albion - no English elm, American oak, Finnish birch, or Baltic pine. The strange trees that towered over Sydney Cove were so big, stone-hard, sappy, or splintery as to be worthless. After a few days of dulling their tools in futile efforts to produce usable lumber, the sawyers switched to palm trees, whose wood was pulpy and of little strength, but was at least easily worked. Structures built from these junky trees would last only a few years.

The same evening that Arthur Phillip mused on the rocky point of Sydney Cove, James Martin lay in his tent, utterly spent after a long day at the sawpits. He rested gratefully on a makeshift pallet of marsh grass. He heard the wind come up and hoped it would rain at the right time. If it rained too early his tent would leak, but he'd still have to work next day. The pit would be puddled, soggy, muddy, and reeking with the sour, acrid smell of wet sawdust James had come to hate. If it rained during the day, but not enough to quit work, the air would be heavy as molasses, and fine sawdust would cling to clothes and skin like stiffening glue. But if it rained just before dawn or early in the morning, they'd trudge to the pits and make a show of maintaining their tools. Under a sailcloth shelter they'd loaf, gossip, trade stories, and perhaps take naps. At ease in their tents, the officers wouldn't care, so long as no one wandered off.

James was ready for an easy day. After months of confinement he'd been eager to work hard, to feel the satisfaction of transforming the things of the earth into human things. He'd attacked his pitwork with almost demonic, mindless energy, like a dog released from tether that runs and romps. But he'd worked too hard and worn himself down. Now he dragged off to work.

He raised both arms straight up to flex his hands. They still hurt when he made fists. The day they dug the pits the new shovel handle filled his tender hands with oozing blisters. Now his newly callused hands were at an awkward stage. While they no longer blistered, the skin cracked with the first work of each day, revealing through its fissures the tender, pink flesh beneath that still complained at pressure. In a few days the fissures would fill in, his calluses would be whole, and his hands as hard as horn.

His muscles were no longer sore. Almost too tired to move, yes, but not sore. The first week he could have wept at the wonderful pain in his shoulders, neck, and back, his buttocks and the backs of his knees. To lift a cup with his wounded hands hurt so much he could have cried and gone thirsty. After a year of almost no labor, he'd forgotten how painfully a laborer is born, how the body protests.

This evening he was glad to be alone. Will Bryant was absent, finding out about Mary B. For days Will had been irritable, and when he learned the women on the *Charlotte* would be the last put ashore he went into a rage and pitched a chunk of firewood as hard as he could into the woods. He seemed afraid someone else - perhaps Tench - would lay claim to Mary B. He even accused James.

"Me? You know I can't do that, Will! That would be a mortal sin!"

Will rewarded him with a mistrustful smirk.

But James was jealous. Why could Will live free of the rules that bound James? Why should I have to live the rest of my life like a monk? he wondered. Even if I stayed honest and got back to England to my wife, what if she didn't want me? James groaned. Why did he ever get married? Things were so simple before. Just pick up and wander off. Now everything was a tangle. His wife's baby. Mary B's baby. God! if he could be as free as Cox, hiding over by the French ships, waiting to sneak aboard. "If I can just get to Djakarta!" Cox confided, "I'll be a free man." James nodded, wishing it were that simple.

James grew sleepy, his eyes heavy. He let them close and lay half-listening to the wind in the trees, the flutter of canvas, muffled voices, the hurried, heavy footfalls of a passer-by, an occasional laugh, the pop and hiss of a dying fire. He heard a faint groan of someone in pain, then a slow hawk, a spit, a weak cough, and another groan. James drifted. A trill of female laughter slipped easily through the tangle of sounds, and Mary B's face floated up, framed in the hood of her cloak, the face he saw after Rio, after the baby, fresh and blooming. He remembered how she held up her baby and her happy, expectant smile.

He drifted more deeply, and in that dream-place on the porch of sleep an image came to him of Mary B sitting on a bench before a sunbright wall. Next to her was the gaping darkness of an open doorway. Climbing flowers mottled the wall in breeze-blown shadow. She wore a light green dress, a starchy white apron, and rested against the wall with her eyes closed, perhaps asleep, her auburn hair magnificent in the sun. He came closer. She heard him, opened her eyes, smiled, and stretched forth her arms in welcome. He went to her and took her hands, cool and soothing against the skin of his own bruised hands. They embraced. She smelled of dew-rich morning air, new-mown hay, fresh laundry. He held her close, breathing in her smells, enjoying her firm softness, the luxuriant warm press of her arms around his back, her cool hands on his neck. Then he became aware of being watched and opened his eyes to see the pale face of a little boy looking out from the dark doorway...or perhaps it was a girl. He couldn't tell. Dressed in black, the child's hair was rich with blond ringlets. The child regarded them silently, eyes large and unblinking, face expressionless. The child looked at them for so long James began to wonder if what he saw was real. Suddenly the child slid back into the darkness, and James was alone. Standing in front of the door, he knew Mary B was also inside in the darkness and knew he was afraid to enter.

"Mary B," he called, "I'm here, Mary B, I'm here!" He stood waiting, feeling lost, helpless, lonely, sad, and suddenly overwhelmed with a great weariness, his body so heavy he could barely stagger to the bench to lie down. He lay on the hard bench and felt the sun disappear and a sudden chill in the air. He heard rain strike the leaves, the flowers, and begin to drip from the thatch to the hard ground. Why did it all change? He began to cry.

"Jesus! It's raining!" Tartop exclaimed, turning to brush the tent flaps closed. Startled, James thought for an instant it was Mary B. His throat was constricted. "Jesus!" Tartop said again, "Raining like cow-piss!" He brushed rain from his sleeves, ruffled rain from his hair, the fine drops a benediction on James. "You know that, Jim? It's raining! Jesus! I got caught in the rain! I'm glad I ain't out in the rain, ain't you, Jim, ain't you glad?" Tartop sprawled to begin pulling off his shoes. "Hi, Jim! you know what? You know what, Jim? Listen, those mountains? those blue ones? That's China, Jim! Ain't that something? China! Some of the boys told me. You get across those mountains and you're in China, Jim! Ain't that something? China! Just over those mountains!"

Tartop lay back and was silent for a couple of minutes. He began to explore his recent discovery, "Chiii-naaa! Gonna go to Chiii-naaa!" Silence. "Ch, ch, ch, ch, Chiii-naaa, Chiii-naaa. China! China! China! Yes, sir, gonna go to Chiii-naaa." Tartop quieted. James heard him move. "Hi, Jim, is China close to England. You know, close to home?"

James kept silent, pretending to sleep. In a little while through the sound of the rain he heard Tartop continue to vocalize more softly, "Chiii-naaa, Chiii-naaa."

~~~~~~~~~~

A sentry hunching past in his oilskin heard Malvina's muffled words even through the thrum of rain on Dobber and Malvina Thornapple's tent. "Well, what's she doing out there then?" she whined.

Inside their tent Malvina lay under the protection of a sailcloth coverlet near the edge of their makeshift bed, a pallet made of marsh grass and resting on the bare ground. This was a typical bed in Arthur Phillip's nascent Albion. Earlier in the evening a few thumping drops had announced the beginning of another night of rain, and shortly after that the first fat splat on Malvina's forehead announced the rain had once again penetrated their threadbare shelter.

"Oh!" she cried, bolting upright. Her sudden move rolled little Nell clean off their pallet. "Dobber! Do something! It's raining again! We're going to drown!" Soon rainwater seeped through a dozen seams and valleys to fall in irritatingly unsyncopated rhythms on their canvas coverlet.

Dobber retrieved Nell, who slept on through the commotion oblivious, her eyes glued shut with a gummy, yellow substance, her right thumb rooted solidly between protruding lips, her left hand anchored on her left ear. Dobber tenderly covered Nell with her worn little blanket.

But other than tuck the canvas coverlet closer around Malvina's chin, Dobber could do nothing more to ease his wife's discomfort. The battalion's tents had deteriorated in storage during the long voyage, and only a fortunate few didn't suffer from leaky canvas during the frequent rainstorms.

"I'll bet she's not getting wet!" Malvina declared.

"She," Dobber knew, was Hester Thistlethwaite, who'd become his wife's favorite subject, after food. Following the precipitous death of Willard Thistlethwaite just weeks earlier, Malvina had become obsessed with Hester's fate. Now she accused Dobber, "Well, you must have some inkling, Dobber, you must know something! Why does she get to stay on that boat while we suffer out here in the wild? Is it Captain Tench? Is it because she's going back to England? Is it because she can't get her pension till she goes back to England? My stars! she can't be on the parish here...we have no parish! So who will take care of her?" Scrunched as small as he could make himself, Dobber sighed and remained silent.

Poor Dobber! At least Malvina's nagging was a welcome familiarity in this unfamiliar land. After her abduction from Madame Labonne's house, Malvina had retreated for a time into an impenetrable silence that Dobber found more disconcerting than her nagging. She refused to talk, even to answer a simple question. She wouldn't even meet his eyes. Over the years Dobber had learned to block out her incessant nagging, and in the safety of his imaginary rebuttals sometimes gave as good as he got. But Malvina's sullen withdrawal left him defenseless. How was he to deal with this stranger?

After his wife's abduction from Madame Labonne's, Dobber became jumpy and even more indecisive. He'd hover near Malvina, seeming on the verge of speaking untried words of apology and sympathy, his words poised just behind his lips. But when she refused him the least sign of encouragement, his resolve to make amends would deflate with a sigh, his shoulders sag, and he'd retreat, unable to close the gap.

In truth Dobber wasn't the object of Malvina's spite. No, Dobber was just Dobber. It was Tench whose very presence set her teeth on edge, whose phony voice made her flush and tremble. That snake in the grass! That viperous dissembler! That rat! It was he who made her prisoner again on that hateful boat!

But even in the security of Madame Labonne's sanctuary, Malvina knew deep down that her fate lay in wait aboard the *Charlotte*. For a wife that was simply the way of the world, and when Captain Tench came to call she'd felt a sad foreboding that the beginning of the end was near. Despite her defiance, she sensed defeat; despite her bright plans, she sensed no change in her dependency. Oh, she was a victim, yes, a prisoner, yes, a woman who like all women counted for less, yes. But when she waked stuporous to find herself back aboard the *Charlotte,* what upset her most was that she'd been treated like...a thing! That was it! The humiliation of being hauled back to the *Charlotte* like a sack of corn! She might have been talked to! Reasoned with! Persuaded! She might have been allowed the dignity of returning under her own power, even if under protest! But no, she was treated like a dumb animal, fed sugar cakes and tea, and then dispatched with a blow to the head, so to speak, with no more ceremony than a dried up cow!

Even her son seemed a traitor, every day becoming more and more a midget version of Judas Tench, aping the awful man with his little wood sword and scrap of blanket after the manner of Tench's shawl. The little monkey!

But at least Nell provided some comfort. The darling seemed to intuit her mother's wounds, and she sought to offer succor by pressing her reassuring little body against a knee or sitting in quiet companionship at her feet.

Hester Thistlethwaite had also tried to comfort her. "Oh, how can you help?" Malvina rejoined, "You're as much a prisoner as me! What can you do? What can any of us do?" But Hester persisted, and as the *Charlotte* plowed towards Botany Bay the women talked for the first time in a serious manner, not of children and in-laws and cooking and husbands, but of their own lives; not as the wives of privates, but of their hopes and disappointments. They found much in common - what it meant to be born a girl and grow up with expectations heaped upon them, to learn what they could and couldn't do, could and couldn't think, could and couldn't be. Hester kept recounting the strange ceremony she witnessed in Rio of the old woman in the barrel. "It seems that whatever we do it's never good enough, or very important, or just silly. That old woman was a victim, Malvina, a sacrifice for the entertainment of men! They laughed at all the ways she could fail. They're always mocking us! Belittling us! Where's the justice in that? I see none! None at all!"

With their explorations of womanhood Hester and Malvina grew more excited, and more apprehensive. Their thoughts seemed dangerous. Were they entering a garden of forbidden fruit! "Why do we have to live this way?" they asked one another. "We bear the pains of birth, the burden of caring for the old and the sick, the day in and day out drudgery of cooking and washing, the tedium of spinning and gleaning and picking and sorting till our fingers are raw. Is that all we're good for? Why can't we do more? We're not dumb animals!"

Sotto voce, their intense, sibilant discussions were imbued with a mutinous hue. But other wives found their talk threatening. Not every woman was willing to look life in the eye and challenge its stern prescriptions.

So, although hobbled by ignorance and misinformation, the two schooled themselves, and bit by bit began to scrub away the encrusted layers of custom and culture, to sweep away the debris that buried their souls. Slowly, painfully, often fearfully, they began uncovering the source of their only hope for real happiness - the seeds of self-sufficiency. And Hester persuaded Malvina to try eating again.

Of course no one enjoyed the daily diet of salt meat and hard bread, but everyone choked it down. Perhaps Malvina's digestive malady resulted less from seasickness than from a malady of the spirit. Perhaps her vomiting and diarrhea were simply unconscious ways of saying, "I don't want to go to Botany Bay." After all, she'd committed no crime and deserved no sentence. But through the circumstance of being married to Dobber she found herself aboard the *Charlotte,* living under conditions not much different than the convicts, and facing three years of exile in an alien land!

In any event, with Hester Thistlethwaite's urging and after a few tentative experiments Malvina found she could once again keep food in her stomach, and was soon back to eating her full ration, which government had determined should be two-thirds of what Dobber received.

Tench later wrote in his journal, "*Much is diminished in the face of death, and death works wondrous changes.*" Perhaps he was thinking of Hester Thistlethwaite, for less than two weeks before the *Charlotte* anchored in Botany Bay, Willard Thistlethwaite's horrible, dysenteric death caught Hester up short. In fact, he was still in the process of dying when she suddenly realized that if he died she'd be stranded halfway around the world from everything familiar and important to her. And with absolutely no reason for being there! Without Willard as her husband, she was nothing. Was she entitled to a ration? Would she have a place to live? Would she have to stay until the battalion completed its three-year tour? What would she do for money? She kept hearing an inner voice, "What am I doing here? What will I do?"

Fortunately, Tench proved a great solace. Over tea in his cabin he comforted and consoled her, reassuring her again and again that everything would be fine. With a kind of somber charm he did his best to lay her concerns to rest. And Malvina, sobered by Willard Thistlethwaite's ugly death, regarded Dobber with renewed interest and began to speak to him again.

Absently brushing rainwater away, Malvina broached an idea. "Dobber, if you was to die I could go back to England, couldn't I?"

Dobber wondered what was coming next.

"Well, isn't that right?" When he remained unresponsive, Malvina answered herself, "Aye, I could." She was silent, picking her words. Then she said, "Dobber, suppose you don't die, but I got some money. I could go back then too, couldn't I? I mean, you're on that side of this bed and I'm over here. We're not one person, we're two. You're you and I'm me. We're married, yes, but I don't see why you should have to die before I can go back. Not if I want to. I don't see that at all. I should be able to go back when I want, simple as that! All I need is money to get there."

Their tent was silent, then Malvina said, "I could take Nell back with me. She's not very big yet, and you could keep Ned here. Heavens! do you even know where he is? Is he staying over with that nigger cabin boy again?"

She fell silent again for a little while, then said, "But where can I get any money? How can I get any money here? Nobody has any." The rain suddenly ceased and the tent was quiet except for the ragged plops on their canvas coverlet from the leaky seams. "Well, somebody has to have money," she said finally, "Not everyone can be poor as us."

In a little while the rain began again. Dobber rose and pushed out a tent flap to piss into the night. After he lay down again, Malvina emitted a little groan. "Ohhh, I just thought of a plum! A big, ripe, juicy, purple plum! I can feel one in my mouth right now, so plump, so firm. I can just feel myself biting down. Ohhh!" She began to weep quietly, her shoulders shaking. Dobber reached over with a comforting hand. She shrugged him off. "You can't do anything, Dobber! What can you do? All you can do is be a private. Privates get no money. Privates live on rations." She sat up and wiped her eyes and

191

embraced her knees under the canvas. "Rations! Do you know how long we've been on rations, Dobber? Do you? For a year! A year, Dobber! Salt beef, salt pork, salt fish! Salt, salt, salt! It's all we ever eat! Good Heavens! Nell asked today if I ever ate chicken meat! Imagine! our little girl doesn't even remember what chicken tastes like! Salt, salt, salt! We might just as well walk into Sydney Cove and drown ourselves!"

After another long silence, she said, "But that's salt too! I won't drown in salt water!" As she lay back she whimpered, "And no relief in sight. No relief." Weeping, she pulled the canvas over her head and turned away. After a few minutes her plaintive, defeated voice reached out to her long-suffering husband once again, "Oh, Dobber, the wet's coming up from the bottom too!"

~~~~~~~~~~

Again the brief, ambient lucidity of lightning showed Tench the pale stack of his manuscript. As the thunder rumbled he savored in his mind's eye the pallid, lingering image. Done! His days of guarding the women on the transports had worked out wonderfully, for he'd been able to spend the time fleshing out his narrative and polishing his prose. He smiled in the darkness, enjoying a sense of accomplishment, a feeling of completion. He had only to write out two clean copies, send them off by different ships returning to England, and his account of their voyage to Botany Bay would be the first to appear! With a little luck his book would be in Debrctt's hands in five or six months, and soon after in his subscribers' hands and in the shops!

Lightning revealed the manuscript again, the fading whitewash of his shipboard cell, the blanket-smoothed human forms in his narrow bed. Tench pressed his leg against the warm flesh of his bedmate and felt a responsive pressure. "But we'll have to be discreet," he said softly. "Perhaps later...." His words trailed off in the moaning wind, his thoughts lassitudinous in the aftermath of love. For several minutes he listened to the hiss of rain, his thoughts dissolving in the familiar swirl of archetypal faces, ragged words, the sliding off into sleep. Another bolt of lighting, so near the bang waked him with a jerk. "But you can stay aboard," he said. "I'll make arrangements. I want you here." He felt the warm pressure against his leg again and an appreciative hand move under the blanket to touch his arm. He reached over and patted the hand through the blanket. "You've made me very happy," he said, "I'm so glad we're here together."

"Oh, Captain," Hester Thistlethwaite said sleepily, languorous with well-being, "Me too."

~~~~~~~~~~

Sergeant Motherwell bent to peer past the light of his lantern. Rainwater ran from his hat in a rivulet, spattering Quint's blanket. Feigning sleep-fog, Quint shielded his eyes and half smiled. "Porter?" Quint said. "Gone to the woods, Sergeant."

"The woods?" Motherwell sounded skeptical. His wet oilskin caught the lantern light on a hundred planes.

"That's what he said, Sergeant. Isn't that what he said, Stork? Stork?"

Stork stirred and raised up on his elbows, blinked, swallowed, and peered squinty-eyed at the gleaming oilskin. "Whaaat?" he asked.

"Didn't Porter say he was going to the woods?" Quint prompted.

Stork's shadowed Adam's apple slid violently. He studied his stockinged feet protruding from his blanket with puzzlement, as if not sure they were his own. His hair sprouted a silly-looking rooster-tail.

"Well?" Motherwell demanded.

Stork blinked, swallowed again, then answered slowly, "Aye, Sergeant. He said he was going to the woods. He's got a bellyache."

Motherwell stared hard at Stork, then Quint, and regarded the two other snoring sleepers. He drew in a big breath and exhaled it slowly. "All right," he said at last, fixing each in turn with a serious look, "If that's what he said." After a moment he asked, "Baker in the woods too?"

"Yes, Sergeant," they answered in chorus. The other two soldiers began to rouse from sleep. Motherwell studied the empty bed spaces for several moments, then turned to raise the tent flap, admitting the sound of heavy drops splashing in the puddles of the footpath. Half over his shoulder he said, "Remember, Lieutenant's got us on detail first thing tomorrow. Then parade." He let the flap fall and turned back, "You men ready for parade?"

"Yes, Sergeant," Quint and Stork answered like obedient schoolboys.

"All right then." Motherwell regarded each of them at length once more, then raised the flap and exited with a slow headshake. Over the sound of the rain and the raucous noise from the women's camp his voice was barely audible, "I swear, you boys are going to put me in an early grave."

~~~~~~~~~~

Through the sounds of the storm, Arthur Phillip listened to the tumult from the women's camp - wind-blown laughter and shrieks and snatches of song and shouted words, the long-forbidden excitement of human congress. With a soft groan Arthur Phillip turned in his narrow camp bed to ease the pain in his side. For the guard to go crashing through the wet undergrowth in darkness and rain to chase down a few revelers...no, this wasn't the time. Tonight he'd let matters run their course. Getting here was done. Tonight was the end of getting here.

But tomorrow the revelers would wake in these unforgiving wilds once again, and if not tomorrow then one of these days they'd wake with the realization that this rock-bound forest was where they'd live for a long, long time. Perhaps then, when they realized the finality of things, they might begin to see...but see what? Most saw no farther than the next day's dinner, the next week's ration, and Arthur Phillip had spent many a troubled night wondering, how can I turn these reprobates into useful citizens? How can I begin to build a community? What can I say to them that will make a difference?

His side throbbing, Arthur Phillip mulled over these and a hundred other concerns until the storm diminished at last and the eastern sky lightened. Then he fell into fitful sleep, lulled by early morning birdsong and the poppety-pop-pop on his canvas roof of rain dripping from the trees.

~~~~~~~~~~

After that memorable night when one storm after another marched across Albion, the last of the women were shuttled ashore from the *Charlotte*. Mary B and Charlotte rode in the very last boat, as did Katie with feverish John Matthew and Liz Bason with Little Jacob. In the last boat too were Seedy and Hannah and a good-hearted Bristol woman named Jane Fitzgerald. Dumpy Ann Carey and another were so debilitated by illness they were passed down into the boat like sacks of grain. Leaving the two sick women at the hospital tents, the boat made for a marshy landing place where a handful of Sergeant Motherwell's squad jostled and flirted with those already landed. The Marines looked as spiffy as when the women first glimpsed them aboard the *Charlotte* back in Plymouth.

Ever in the right place at the right time, lucky Quint lifted Seedy ashore and was rewarded with a big kiss on his mouth, a reward that Liz and Jane also bestowed on the cheek of the brawny Marine.

For Mary B the moment possessed a more symbolic significance. Dressed in new clothes given out a couple days earlier, she giggled nervously as she swayed and wobbled, unaccustomed to the earth. For the first time in more than a year and a half she stood on solid ground! Unbundling Charlotte, she pranced her baby's bare feet on the ground. "There, sweet Charlotte, that's how it feels!"

Sergeant Motherwell led his procession up a steep, muddy path to high ground. Burdened with his musket, Seedy's bundle and Little Bill's box, a cooking pot and tea kettle and tin bowls and cups, Quint wore an irrepressible grin as he bumped alongside Seedy, who carefully protected a withered geranium in a cracked clay pot. All were content to listen to Porter's excited words tumbling nonstop. "Aye, I already saw a kangaroo, two in fact, and more strange birds than you can shake a stick at, and the alligator too that lives over there by the steam. Aye, ten foot at least, with teeth like spikes, aye, like spikes. And savages too, with bones in their noses, aye, and women naked as newborns, aye, naked. You have to be careful, aye, and all the time keep an eye out...."

By turns horrified and fascinated, the women nodded and gasped and punctuated his prattle with little exclamations of alarm.

As the uneven, puddled path leveled off, the breathless women skirted tangles of felled trees, brush piles, and countless stumps half-rooted in the earth. Despite their escort and the sight of convicts at work, many trembled with trepidation. The rough, muddy ground and hanging haze of smoke and the clutter of woods-work gave the scene an eerie, disorganized, depressing appearance. Most of them urban dwellers, the women had never set foot in a primeval wilderness, and here were snakes and wild beasts and savages and unseen demons that prowled at night! They screamed when a flying squirrel swooped across their path, spooked from its tree.

Arthur Phillip's plan for Albion divided the convicts into two groups: the favored few, and the rest. For tending government gardens and the colony's precious livestock, for hunting and cooking and laundering and otherwise serving Albion's elite, the favored few lived on one side of the stream that provided the settlement's water. That's where Arthur Phillip's own spacious tent-house stood, where his official residence was under construction, and where the other civil officials lived. On the other side of the stream, near the hospital and the hospital gardens, almost 1,000 convicts, Marines, wives, and children were tented.

Passing through the women's camp on the way to their own quarters, the women from the *Charlotte* stared wonderingly at the hangdog inhabitants. Some of the women sat forlorn, their eyes vacant or introspective. One retched in the bushes. Others were desultory as they wrung out clothes, draped laundry to dry over bushes and low-hanging limbs, poured water, stirred smoky fires, and nursed babies.

Quint winked, "Some of these ladies had visitors last night." They sidestepped a tiny, gray-haired woman slowly turning in a stiff, private dance. Her clothing in disarray, oblivious to the passing procession, she quavered a single refrain again and again of Mary B's song, "*Botany Bay-ay, Botany Bay-ay.*" A gleam of recognition came into her eye when Porter passed near, and she seized his coat. "Soldier boy! My soldier boy's come back!" She hauled herself up Porter's arm while beaming a foolish, toothless grin up at him, "Aye, my lovely soldier boy's come back to see old Annie!"

Red-faced, Porter sent the old woman reeling as he pulled himself away. At Quint's questioning look, Porter raised a burdened arm to spin a finger near his head. Quint regarded Porter with a small, knowing smile and slowly shook his head, then grinned. So she was the one? Porter shrugged and stuck out his lower lip.

As the women were dropped off here and there to their assigned tents, Quint led them to the very end of the encampment. Through the trees they glimpsed the hospital tents. He paused before a tent marked with black char, "18th Cpy." "This means us," Quint said, pointing to the writing,

"Captain Tench's Company. This is where we guard." Seedy and Katie exchanged glances. They were practically in the woods! Quint winked at Seedy and pointed with his chin, "See that sentinel over there?"

Seedy stepped close to follow his gaze. Through the trees she caught a glimpse of red. Quint nudged her hip and grinned. "That's my post too!"

Smiling broadly, Seedy said, "Oh, Quint!"

From the direction of the Marine camp came an insistent tattoo. Sergeant Motherwell hurried up to admonish the women, "All right now, ladies, get yourselves ready. Fix yourselves up for the Governor, ladies, fix yourselves up!" He studied a cloudy sky. "We'll have the ceremony in a couple hours. Fix yourselves up now!"

Mock kisses and playful boo-hooing followed the grinning Marines as Motherwell took his men back to prepare for parade, and for Arthur Phillip's investiture.

~~~~~~~~~~

As Mary B and her friends settled into their tents, Arthur Phillip came ashore from the *Supply* to join a ceremonial group waiting at the flagpole. Freshly wigged, powdered, and perfumed, the waiting group was small - Major Ross and a color guard; David Collins, the young Judge Advocate; Augustus Alt, Surveyor of Lands; Andrew Miller, Commissary of Stores and Provisions, and his assistant, Zachariah Clarke; and Arthur Phillip's old shipmate and newly appointed Provost Marshall, Henry Brewer. A little apart, as if unsure of his place, the Reverend Richard Johnson stood with his mousy bride of less than a year. The welcoming party gave Arthur Phillip a polite round of applause, to which he bowed acknowledgement and smiled graciously.

Although bewigged, beplumed, beribboned, and bemedaled, Arthur Phillip looked tired. Major Ross would later say that Arthur Phillip was in his cups when he stepped ashore, that he'd spent the morning drinking with his Navy cronies on the *Supply* and almost fell on his face getting out of his cutter. In fact, Arthur Phillip did stumble as he stepped over the side, but not because he was drunk. He was exhausted. After pushing himself for two weeks and not sleeping well because of the pain in his side, he'd spent several hours that morning on the *Supply* going over plans to take possession of Norfolk Island. A thousand miles distant, that uninhabited dot in the ocean

197

reportedly possessed wild flax and virgin pine that England coveted, and Arthur Phillip had orders to insure England's claim. He'd soon dispatch the *Supply* with a handful of the best-behaved convicts to establish an English presence on the tiny island.

As the procession of dignitaries approached the Marine parade ground, ensigns dipped and fifes and drums struck up a spirited rendition of "The British Grenadiers." Four companies of Marines snapped to attention. They looked good. Their uniforms bright from sponging and brushing, their crossbelts and leggings gleaming with fresh pipeclay, their gorgets polished, their muskets and bayonets shining with new oil, their hair carefully powdered and pulled back in tight pursetails, they wore their ceremonial tall, black, visored hats with the battalion's perky, white plummets. They looked incongruously elegant in the rough surroundings. But their appearance was deceiving.

Weak-stomached, sleep-short, weaving with horrible, sour-breath hangovers, the rat-tat-tatting of the welcoming drums punished many a pounding head that had sneaked off to spend the previous night in riotous carousing with the women. Of course such intercourse was forbidden! Of course the orders were strict! Of course those caught would be punished! But the prospects of so many women after so long, and so close, and no longer under guard! Oh, the prospects had been too good! So when the wind came up and the heavy rains began to fall, Marines and convicts alike slipped from their tents to stumble off into the stormy darkness, many with bottles, for enterprising sailors were ever ready to sell their private stocks to those who could pay, although that too was forbidden. And thus did garish lighting on that novel night illume these happy acquaintances come to call, and thus did doughty drafts of liquor gladden many lonely hearts. And thus did some, like Porter, welcomed as a stranger, share those sweet, lubricious pleasures of the flesh, those long-forbidden fruits of female love, even with the agèd Annies.

Nonetheless, as Arthur Phillip approached the rough parade ground, officers posted stiffly, the ranks straight, the sergeants muttering, "Stand tall! Hands still! Quiet in the ranks!" all eyes attendant on their colonial leader, the Marines looked good.

At the front of their formation stood the stump of a tree that had taken a full day to cut down. More than six feet through, the bulk of the massive trunk now rested in a slough a short distance away, too big to saw, and too big to bother with further. Let it rot. Two more days of sweaty, cursing, futile effort to dig out the stump had finally led to a practical solution. Let's make the stump the reviewing stand! So it was this makeshift podium that Arthur Phillip mounted when - oh, happy omen - the sun broke through!

Glancing at the sky, Arthur Phillip grinned through his fatigue with a stage-motion shrug, as if to say, "Are you surprised? I do it all the time!" An appreciative chuckle ran through the ranks. Officials nudged one another with knowing grins.

But for the convicts Arthur Phillip would have a different message.

~~~~~~~~~

After the parade ground ceremony with the Marines (in which Arthur Phillip thanked the Marines for their loyalty and support), a clanging iron triangle called the convicts to assembly in a rough clearing. Here the entire convict colony came together for the first time. In large it was an assembly of strangers. Bleary-eyed from the previous night's revelry, they huddled in mess groups or gathered with shipmates, rumors rampant.

Searching anxiously, Will Bryant at last spotted Mary B and sidled next to her. Charlotte on her hip, Mary B looked into his eyes with friendly candor and smiled back, then nodded and leaned to whisper something in his ear. He grinned back and squeezed her arm. They'd exchanged only a few words when their attention was drawn to the fifes and drums of approaching Marines playing a tune some recognized as "The Drunken Sailor." They watched a living fence of red coats encircle them.

In the shade of a large gum tree Albion's officials took their seats on chairs from the *Sirius* behind a camp-table draped in red. Clustered here and there stood Navy and Marine officers, civilians from the transports, the wives of civil officials and Marines, a few children. Murmuring and nervous laughter died as the Marines tightened their circle, herding the convicts into a more compact mass. Striding up to place two red-leather cases on the table, the youthful Judge Advocate spoke his first official words to the convicts, "Sit down!"

They sat on the ground. After they more or less settled, Collins unclasped the first of the cases containing the Crown's official commissions. Despite his fatigue and outward air of calm, Arthur Phillip's eyes flicked nervously from the middle distance to the Judge Advocate as he watched him remove the first of the parchment rolls, hold it before the congregation like an offertory chalice, and slip the heavy seal with a thumb.

Hokum! Of course David Collins had opened the commissions before, several times in fact, as had Arthur Phillip. During the voyage both men reviewed the documents and discussed them at length. Why, just the night before Collins had rehearsed his speech once again, his muffled words a recondite incantation to the ears of the miserable sentry splashing back and forth in the rain. But now, in this crude theater growing warm under a midday sun, Collins played his part for real, enouncing in a resonant baritone the words that bestowed on Arthur Phillip his official powers and made manifest this unrepleviable act of possession, appropriating for a tiny nation some 14,000 miles distant this vast, unexplored land, this homeland of thousands of years for thousands of invisible natives.

How odd, the juxtaposition of ideas in the cluttered clearing that day. Did Arthur Phillip, Collins, or Tench see the irony? Not a single white person alive knew the language of the people whose lands they were taking, or knew their religion or rituals, their customs or taboos. The natives were shadows, insubstantial. Arthur Phillip actually hoped these shadows would go on living as they had, moving with the seasons along this small corner of seacoast to fish and scrape up oysters and shellfish and gather birds' eggs or whatever came to hand, occasionally bringing down larger prey, and occasionally a tribal enemy. Thus, Arthur Phillip's policy was not to threaten or harm them, not to molest their women or disturb their crude, bark shelters. Not to steal their dugout canoes or filch their belongings as souvenirs - no, none of that - but nonetheless, by some inherent, presumptive, pale-skinned right, to dispossess them of their homeland.

Arthur Phillip was not ignorant of the destructive effect Albion would have on native land and culture. But to his thinking, Albion was progress. Albion was God's Plan for Man's dominion, and if God's Plan dovetailed with England's plan, well, so much the better. No, what bothered Arthur Phillip more was that he couldn't explain to these shadowy natives that he meant no harm, an increasingly vexatious conundrum as Albion spread its tentacles and cut down their forests and polluted their waters and killed them with its European diseases. Not knowing their language, Arthur Phillip was unable to reassure them that their interests would be protected and that, if things really had to change, well, of course they would be better off for the change.

And everything was changing now. Albion was England's. Every pebble, every drop of water, every beating heart now belonged to Arthur Phillip's England. From this time forward, whether convict or keeper, settler or native, to live here was by leave of His Majesty's government.

Thus, in that humid clearing on February 7, 1788, David Collins played out his role before an audience mostly illiterate, and mostly indifferent. The sounds of the language were familiar, but what did the words mean? Most had no idea. Nevertheless, Collins intoned the articles from a distant world that described Arthur Phillip's powers and established in this strange wilderness the English system of property, of governance, of law.

And much law it was. A criminal court for felonious crimes, a civil court for disputes and misdemeanors, a military court for the Marines and Navy men, a vice-admiralty court for crimes on the high seas. And how complicated! Why, within a year Albion would be so entangled in its laws it would ask England for more paper to document the litigious colony's complaints and indictments, notices and subpoenas, judgments and appeals. Were these white-skinned creatures really a superior society of reason, of order, of the rule of law? Was their complex legal system a measure of their enlightenment, or of their mistrust, greed, bickering, and depravity?

To Arthur Phillip, looking out over the several hundred seated convicts - lawbreakers all - depravity had the decided edge. He ran through the formalities of introducing the various officials and briefly described their functions. Then he stepped around the table, strode back and forth once, and faced the convicts with a serious mien. "So far you've had it easy," he began, "but it's clear that most of you are not working."

Of course Arthur Phillip knew that many were sick with scurvy and other ailments, and many more were too old and too infirm to do the hard work at hand. Almost 200 were women and children whose roles could hardly include clearing the

201

land, tending the livestock, or putting in crops. Nevertheless, he railed, "You number more than 700, but only 200 of you are working, and that isn't enough!" He paused to let his words sink in. The faces were impassive. "You all have to work, every one of you, every day! Our food will not last forever, and we don't know when more will come." Were they listening? Did they hear what he was telling them? That this was serious? "You don't have to work as slaves, with no hope for your futures. Your futures are here, with land to build lives as useful citizens!"

Behind him the heads of the officials nodded approval. "But we must work together. We have to clear land and plant crops and grow food. We have to husband our livestock and make it multiply."

By this time his audience was yawning and scratching, looking around and muttering among themselves. Arthur Phillip masked his resentment by rubbing the back of his neck. "In England," he began with a little smile, "if you steal a chicken, which I am sure a few of you did...." He paused for a rise of laughter, the convicts' first response of the day, perhaps a relief for them as much as for Arthur Phillip. "If you steal a chicken in England...," again chuckles and nudging, "the punishment isn't very bad, because we have many chickens in England. But do you know how many chickens we have here?" He gave them a serious face again and then a Heaven-pointing finger. "We have 143 chickens! 143! We could all eat chicken for just one day, and be left with none!" Official heads nodded agreement. A brilliant stroke! "So if you steal a chicken here, if you steal just one egg, believe me, you will wish you hadn't! Believe me, you will wish it mightily!"

For nearly an hour Arthur Phillip harangued, humored, instructed, and cajoled. Sometimes he got a laugh, but more often he spoke to indifferent yawns. One of the best responses came when he talked of marriage. Marriage was good for both men and women, he said, because it settled them, brought comfort and companionship, and produced children who would care for them when they grew sick or old. "His Majesty has sent a chaplain," Arthur Phillip said, gesturing to the Reverend Richard Johnson (who beamed and bowed his thanks for the acknowledgement), "not only to lead you in worship, but to marry you and baptize your children. Life will be better here for those who marry. If you wish to marry, see Mr. Johnson."

That night, through the glow of an extra ration of grog provided in honor of the day, Stork looked across at Quint staring dreamily into the rose-yellow coals of a dying fire. Still moved by Arthur Phillip's sun-brightened speech of appreciation to the Marines, young Stork asked, "Are you proud to be a Marine, Quint?"

His thoughts on Seedy, on home and his mother, on his book, Quint leaned back with his hands between the thighs of his outstretched legs, unable to tear his eyes from the fire. "Proud?" he asked at last, "Did you say proud?"

Stork nodded, eager to share his own pride in their great enterprise, of having a place in an experience that few were privileged to share, just as the Governor said. But Stork was reticent to say this aloud.

"Proud?" Quint asked again, his gaze still locked on the hypnotic wavering glow of the coals. He remembered Seedy's long fingers on his chest and how she called him her big, strong Quint, her handsome Marine. At last dragging his eyes from the fire, he said, "Aye, I'm proud, I'm proud." But his mood and his voice caused Stork to study his face in the long silence that followed.

Aye, Arthur Phillip had charmed the Marines, telling them how important they were to Albion, how well they'd acquitted themselves during the voyage, how much he appreciated their loyalty and courage. And when he doffed his hat with a flourish and bowed his thanks to the smiling soldiers, they broke into a spontaneous cheer, followed by three enthusiastic hip-hip-hooray's.

~~~~~~~~~

Mary B and Will Bryant were married after Divine Service three days later, on a perfect day for a wedding. For the five couples who stood before the Reverend Richard Johnson that morning, the sun shone from a clear blue sky and a gentle breeze cooled the air. Mary B's shipmate, Fanny Anderson, whose craft was the robbing of dirty old men, would marry a weaver she'd consorted with on the hulk, Simon Burn, sentenced to seven years for accosting and robbing one Thomas Scrivenor. The other three couples were from the *Friendship*, the youngest pair being Susannah Holmes, 23, serving a 14-year sentence for stealing clothes and silver from the house of one Jebez Taylor, and Henry Cable, 21, a laborer convicted with his father of burgling the house of one Abigail Hambling.

A Liverpool dealer in stolen goods, 25-year-old Mary McCormack, would marry William Parr, swindler, also from Liverpool. And Hannah Green, a London shoplifter, would become the wife of William Haynes, cabinetmaker and now convict overseer, who'd drunk himself broke one evening with an old friend and, wanting more to drink, threatened and robbed one Robert Becket of one shilling, nine, and so was exiled from England for seven years.

Beneath the tree that would be his principal chapel for almost three years, the Reverend Richard Johnson began the ceremony with relish. Thus far his lugubrious ministry had been to bury the dead, comfort the sick, and baptize far too many bastard children. For him to administer the sacrament of matrimony was to lay the foundations of progress! He sent a satisfied smile to his young, pale, mousy wife and turned to the participants, his voice a tad oleaginous, "Are you all ready?"

Wearing her new dress and most colorful handkerchief, her knees trembling, Mary B's hand perspired heavily in Will's as her thoughts fluttered like a panicked bird in a cage about Will, her mother, Charlotte, Dolly, Ann, Little Bill, her grandmother, her father, Captain Tench, Seedy, Katie, Charlotte again, James. Did James skulk nearby? When the preacher began the sonorous words of the marriage ceremony, "*Dearly beloved, we are gathered here,*" her vision blurred and she barely heard his words. With sudden mild panic she looked over where Seedy, Katie, and Liz Bason held the babies. They smiled reassuringly and Seedy winked. Mary B winked back and felt better. She watched the minister's moving mouth for a few moments, now hearing his words, then regarded Will's intent, listening profile. "*...but reverently, discreetly, advisedly, soberly, and in the fear of God. Into this holy estate these persons present come now to be joined.*"

Conscious of Mary B's eyes, Will turned, smiled, and squeezed, and they both faced the minister together again. "*...as ye will answer at the dreadful day of judgment when the secrets of all hearts shall be disclosed, that if either of you know any impediment, why ye may not be lawfully joined together in Matrimony....*" Another flutter of panic! What if Will really did have a wife? After all, what did she really know about him? Oh, James! she thought, how could you? She was angry with James, yes, but she liked James! This was foolish! She didn't even know Will! She should run away! "*...persons are joined* *together otherwise than as God's Word doth allow, their marriage is not lawful."*

The minister closed the book on his finger. Will smiled at Mary B, puckered his lips in a little kiss, winked, and smiled again. Mary B smiled back, but could not hold his eyes. "...ask each of you in turn to repeat these words after me. Now then, Fanny and Simon, let's begin with you...."

Katie's baby began to squall before Mary B and Will's turn came, which set off Charlotte and Little Jacob, and after two benign pauses and tolerant smiles, the minister looked over at the third interruption without humor and the women stepped to some shade at the edge of the clearing, jiggling and cooing their charges. As she watched them depart, Mary B suddenly realized that she and Will were next. She listened to Will repeat the words and then it was her turn and she stumbled over "ordinance" and had to repeat it twice with the preacher's prompting. And then Will surprised her with a gold ring hammered from a guinea coin and she didn't mind that it was a just a little too tight. She heard his words *"for better for worse, for richer for poorer, in sickness and in health, to love and to cherish, till death do us part..."* and then it was her turn and she repeated the words with sudden certainty that she was doing the right thing. She would try to be a good wife and would surely come to love him.

The Reverend Johnson intoned more words and prayers that Mary B scarcely heard, for her mind filled with confused thoughts about the rest of that day. What would they do? Where would they go? What would they say? How would they act? After almost three years of living with other women, Mary B suddenly realized everything would be different. She'd no longer be with these women with whom she'd shared so much laughter, terror, misery, sickness, and hardship. She wouldn't return to the women's tents, but would move to an encampment for the married - with a man she scarcely knew and who now possessed rights to her person she'd never given up before, even as a prisoner.

And what of her baby? Would he really accept Charlotte? And James! She remembered Will's announcement after the inaugural ceremony, flung towards James like a challenge, "She says she's going to marry *me*!" And as Will bustled her away by the elbow, she'd time only to look back and say, "James, I'm sorry."

In the months ahead, as she grew familiar with Will's boasting, his showy acts of independence, his foolish risk-taking, his outbursts of anger, she'd wonder whether she should have confronted James with more persistence. And especially later, when Will grew distant, brooding, churlish, and still later, when in a black, fearsome mood he struck her, she'd begin to wish she'd taken more charge of her life. But now, as the marriage ceremony drew to an end and she heard the minister's cheerful invocation that they all enjoy the gift and heritage of children, that their homes be havens of blessing and peace, that they were joined together by God, that they were all pronounced man and wife, her spirits rose with the smiles, handshakes, and kisses of mutual congratulations that came from every side. Will kissed her on the mouth for the first time, hungrily.

The minister's wife held the parish register in its stiff leather binders and William Bryant, William Parr, and William Hayes signed their names with the minister's quill. Illiterate, the other seven celebrants made X's. Mary B's X resembled a child's uncertain sign of addition.

On succeeding Sundays dozens more would enter into the sacrament of marriage with people they scarcely knew.

~~~~~~~~~~

The first of Mary B's shipmates was hanged less than three weeks later. Originally sentenced to 14 years for stealing, Thomas Barrett had mutinied on the *Mercury*, then was recaptured and sentenced a second time. This time his sentence was a life term, which he completed at the hanging tree before the assembled convicts near day's end. Many felt he got his just deserts. A bully and provocateur on the *Charlotte*, he always seemed able to edge safely away just as authority came down the ladderway. Tench suspected he was one of the counterfeiters in Rio, but his final crime was to steal food from a storehouse tent.

The hanging did not go well. Thomas Barrett seemed unremorseful until he spoke his last words to an acquaintance and mounted the ladder. Then, visibly shaken by the indifferent noose dangling beside his head, he paled and began trembling so violently he nearly fell. And the convict hangman couldn't bring himself to take a life. Weeping, shaking, the hangman refused to tie Barrett's legs, refused to put on the hood and set the noose, and finally fell to his knees to beg that someone else be chosen to hang the prisoner. Only when Major Ross drew his pistol and threatened to shoot him did the unfortunate wretch fumble through his thankless office, both he and Barrett sobbing all the while. After the Reverend Johnson, sweat-drenched and shaking, briefly mounted the ladder to embrace Barrett's knees and pray fervently, the hangman kicked the support away with surprising alacrity, and Thomas Barrett was left swinging. The foot-thick branch of the hanging tree bobbed savagely, leaves rattling, as the victim writhed for almost three minutes with diminishing effect, and near the end of his fight for life produced a noticeable erection. The crowd tittered, glad he was gone, and no one came forward to hang on his legs to speed his death. Inevitably he hung limp, twisting casually first one way, then back the other, as if the white hood surveyed its audience with all the time in the world. Although he was unconscious, Barrett's last spark of life would not leave his brain for several more minutes.

Four more convicts would be hanged before the end of the year, and four others be murdered by natives. Fifty-six would die from sickness or the infirmities of old age. Fourteen would be unaccounted for and presumed dead, perhaps by perishing in the wilderness, perhaps by suicide or by escape. But escape to where?

Among the dead would be another of Mary B's shipmates, dumpy Ann Carey, one of the twenty women originally embarked in Plymouth. Ann was a young retarded woman convicted of stealing one linen apron, one linen handkerchief, and five linen caps. She died of scurvy.

Those deaths did not include the death of Katie's baby boy, who succumbed only five weeks after his baptism, which took place immediately following Mary B's wedding. At her wedding, seeing Katie trying to cope with a fever-fussy John Matthew, Mary B felt a rush of pity for this lame, single woman with a bastard child. But it was pity mixed with a sense of relief. At least now Mary B had a man!

Katie, however, wasn't the same mean, petty woman who limped aboard the *Charlotte*. After almost dying in childbirth, she'd begun to examine her life and had seen the light of a revelation: Blackpool's seed was God's punishment! Not for the sin of fornication, which was only a small part. And not for a life of petty thievery or for abusing Agnes Lakeman. No, you get caught, you pay. Rather, Katie finally began to see that the sin God caught her at, grabbing her like an errant puppy by the scruff of the neck, so to speak, and shaking her good to get her attention, was, quite simply, her way of life. God had grown weary of her jeering mockery, her meanness and little cruelties, her cynicism and resentful service. God knew she was bad and punished her by making her suffer with Blackpool's seed. In a further revelation she'd later see that John Matthew's death was also a sign from God, that He'd accepted her blood and pain and suffering in bringing her baby into this world as atonement, and that by calling her baby to Heaven, He absolved Katie of her sins. In fact, after she buried John Matthew, Katie's life did improve.

But for many, life did not.

Albion's convicts - mostly thieves, mostly male, and mostly young. Some were sent off to be rid of them, some to mend their ways. Some were being punished for many crimes, some for only one. Seasoned London pickpockets, thieving maids, clever counterfeiters, dumb sheep-stealers, strumpets, inept burglars, country bumpkins, wild, young, urban animals, Fagin's future apprentices. The youngest was nine, the oldest 82 - one Dorothy Handland - who after two years in Albion hanged herself from a gum tree. From this population sprang Arthur Phillip's Albion. When they stepped ashore they carried their pasts with them, and promptly got into trouble again.

Thomas Barrett remained the only convict put to death in the new colony until early May, when 17-year-old John Bennett was caught stealing biscuits and sugar from one of the *Charlotte*'s shoreside tents. Young John admitted in the noose that he'd led a bad life and certainly deserved hanging, which would teach him a good lesson. By this time Albion had a new hangman, a convict named James Freeman, age 20, whose startled eyes gave him a look of constant surprise. Freeman had also stolen a few pounds of flour and been sentenced to hang, but was given a choice: he could either hang others, or be hanged himself. Freeman chose to hang others. But Freeman didn't find his first commission easy, especially as Bennett reminded him of his own younger brother, and more especially when he heard the youth whisper up a prayer asking God's blessing for his mother and father, for Governor Arthur Phillip, and for Freeman himself.

But hanging was only one form of punishment in Albion. Two of Tom Barrett's accomplices had also stood hooded, bound, and noosed at the hanging tree when a messenger from the Judge Advocate rushed up with a reprieve. Not to be hanged, they were instead sent into exile. As Albion already was exile, this latest sentence meant they'd be dropped off at the mouth of Port Jackson harbor to shift for themselves. If they survived, fine. If not....

The most common punishment, however, was flogging. After the flagpole, one of the first structures erected in Albion was a tripod of stout poles about eight feet high, a familiar device called "the triangle," to which a prisoner was tied hand and foot for flogging. The hangman also served as the flogger. For idleness, fighting, drunkenness, minor theft, or mouthing off, convicts received sentences of 50 to 500 lashes. Stiffer sentences were usually meted out in installments so the transgressor could recover enough to continue to be punished. If a convict expressed remorse and promised to be good, Arthur Phillip sometimes reduced punishments. Several women were also flogged, usually for theft. Their bare backs exposed, their hands tied to the tail of a cart, the women were pulled around the camp while Freeman followed, administering the punishment of the lash. Some women had their hair cut off or were sentenced to wear a canvas doublet daubed with large letters - RSG for Receiver of Stolen Goods or T for Thief. When two seamen and the black cabin boy were caught in the women's camp, the boy was dressed in petticoats and all three drummed into the wilderness with their hands tied behind. Some malefactors were confined without food on a barren, rocky islet at the mouth of Sydney Cove known as "Pinch Gut."

The Marines complained their punishments were harsher than those of the convicts. Arthur Phillip might forgive a convict and forego a flogging, but a Marine's flogging was never forgiven, and the Marines were constantly getting in trouble over women, mostly because of the unequal proportions of the sexes. Quint's tent mate, Corporal Baker, was charged with the murder of another Marine over possession for the night of Mary Phillips, a soft, affable thief from the *Charlotte*. Abed with the pliant Mary Phillips, Baker rose to answer the challenge of Private Thomas Bulmore, and bested him in a fistfight. But at dawn Bulmore confronted Baker again, this time with companions, and insisted on continuing the matter according to form, with rules and seconds and rounds. They stripped, fought a dozen rounds, shook hands, and separated. However, Bulmore died four days later of head injuries suffered in the fight. Cleared of murder but convicted of manslaughter, Baker and the other participants each received 200 lashes, 50 a week for four weeks.

But trouble in Albion came from more than just the women and squabbles over food. On the voyage to Botany Bay life was contained. Although the transports were crowded, people lived with at least some sense of personal space in a community of shipmates. Despite flare-ups and petty grievances that were part of ship-life, there was at least a vague sense of community. If conflict arose, all could be settled down and put back in order. A shipboard intimacy prevailed, and one made do.

But Albion was a wilderness. The people lacked bearings. They might wander into the woods, get turned around, and be lost for days, sometimes forever. Surrounded by savages, strange animals and birds, unlikely trees, a dangerous unknown awaited on every side. The stress of these anxieties, the hardships, disconnection, loneliness, and isolation - all contributed to a level of irritability that soon saw snapping and quarreling everywhere. Even John White wasn't immune. On a night that began with a celebration of the Prince of Wales' birthday, he and another surgeon banged away at each other with pistols at twenty paces. Five times they loaded and fired in the moonlight, without seconds, and without effect, until well-liquored friends finally realized the two medical men really were trying to kill each other, and intervened. Later that month White also wanted to have it out with Creswell. The gossip was that White's problem was really with Tench, some private matter that went back to the *Charlotte*. Whatever the issue, White was very edgy.

~~~~~~~~~~

Poor John Bennett! Hanged at age 17 for stealing biscuits and sugar, he died in possession of little more than his name, and some doubted the veracity of that. If he possessed a family, they didn't possess him, for they were ignorant of his whereabouts, and how can you possess something that cannot be found?

Young John Bennett was a Portsmouth rascal, no doubt about that. Looked for by the law, he'd sneaked himself among the convicts in all the to and fro when White was fumigating the *Alexander*. Had sneaked among them for a price of course. A clever lad, John sold himself for three pounds and his benefactor's name on the *Friendship*. Once the fleet sailed it would be months and perhaps years before authorities might discover that John Bennett of the First Fleet wasn't the man he said he was. Perhaps they'd never discover the truth.

But such ambiguity and confusion were commonplace. Take Mary B. Baptized as Mary Broad, she was sentenced at Exeter as Mary Braund, a common error in hearing and spelling by a court clerk. She sailed on the *Charlotte* under the name of Mary Bond, a name on the list of transportees that John White brought down from London. But for some reason, perhaps a bribe, Hugo Blackpool held back the real Mary Bond and substituted Mary B. Now married to Will Bryant, Mary B's legal name was Mary Bryant. However, Albion's commissary agent continued to carry her on his rolls as Mary Bond, because he said it was too confusing to change his records with so many women getting new names through marriage and so many babies getting born out of wedlock. Thus from Broad to Braund to Bond to Bryant in just three years. Who was she, really? Her woman friends still called her "Mary B." So let it be.

On a quiet summer night at year's end, she sat on a stump near the women's camp and watched Will's boat slip out of sight around the big fallen tree on the point. Charlotte snoozed in her lap. Will would be out for most of the night again, for Albion had come to rely on his fish. If Will said he had to fish at night, he was free to fish at night.

Already a darkening of the spirit had crept into Albion. Almost a year ago seeds that sprouted and briefly flourished had shriveled and died. Plants that survived the poor soil and intense heat were ravaged by voracious ants and rodents. The first harvest was a joke. Except for a few carefully tended vegetables, the colony reaped less than it planted. Most of the sheep they brought with them were gone - some killed by lightning, some by natives, most by slow starvation on nutrient-poor grass. The last cow was gone too, shot by Major Ross before she too died of starvation. The rest had vanished without a trace the night they celebrated the Prince of Wales' birthday, a disappearance that still baffled Arthur Phillip. Many in Albion thought natives had spirited away the cattle, although others believed it was convicts. A few suspected forest demons. Little Ned was forlorn over the disappearance of Lily, the brown-eyed Jersey cow, and Lieutenant Creswell much missed her rich, yellow cream.

Neither did the people fare well. With few native plants to augment a salty, starchy diet, scurvy was prevalent. Some of

211

the salt meat was going bad. Much of the flour was weevily, and some had suffered water-damage during the voyage. The palm-log huts were leaky, drafty, and already rotting. Clothes and shoes were wearing out. Even the colony's stock of seed was low, a situation made worse when gray-faced Nat Mitchell, who could eat no cheese, ate the beans he was supposed to plant for John White. For his troublesome appetite he suffered 50 lashes. Meanwhile, storehouse foods that once seemed endless, barrel after barrel after barrel, steadily disappeared.

Except for the little *Supply*, all of the ships were gone. The French explorers sailed away in early March, leaving reports and letters for Arthur Phillip to send back to Europe. They also left one dead priest, a crude palisade, and the enmity of the natives. With them they took the longboats they built and one of Arthur Phillip's convicts, a Frenchman named Peter Paris, who was smuggled aboard by crew. His freedom would be short-lived, however, for both French vessels would sink in a storm a few weeks later. The scattered, rusty evidence of their mysterious disappearance would lay undiscovered for almost forty years.

In early May the *Charlotte*, the *Scarborough*, and the *Lady Penrhyn* sailed for China to take on tea for the folks back home, the *Scarborough* by way of Tahiti on a secret mission. Convicts who said they'd completed their sentences wanted to return on these transports, but of course Arthur Phillip held them back. With no court records, how could he let them go?

Several women went down to wave good-bye to the *Charlotte*, for more than a few of its sailors were sweethearts sailing away forever. Never love a sailor. Thomas Gilbert would later write a book about his voyage to Botany Bay, then to China, then back to England by way of Cape Horn, a plucky circumnavigation of the globe by this laconic sailor. On his voyage he would discover a group of islands that came to be called the Gilbert and the Marshall Islands, the first name being his own, the latter being the master of the *Scarborough*. After its return to England the *Charlotte* would be sold to a West Indies company and would shuttle molasses and rum between the Caribbean and north Atlantic seaports for almost thirty years before she sank in a storm off Newfoundland, taking to her watery grave the captain and three seamen.

In the midwinter month of July, a raw, rainy, gloomy day, the little colony suddenly seemed much lonelier with the sailing of the *Alexander*, the *Prince of Wales*, the *Borrowdale*, and the

Friendship, all bound directly for England. After lingering for five months in Albion, they carried with them the longing of many to return home. Convict schemes were desperate. One convict ended up starving to death; rather than eat his food, he sold it to get passage money home. Another put out a story of finding gold, hoping to be rewarded with his freedom. His reward was a flogging. Hester Thistlethwaite, free to return to England, now served Tench and decided not to return. The returning ships did carry skins of birds and animals, dried plants, seeds, samples of the wood and soil and flax from Norfolk Island, and probably two smuggled convicts. Although the homebound ships were thoroughly searched, after the ships sailed the two convicts were never again seen in Albion. Spread for safety among the vessels were dispatches, reports, lists, requests, Tench's manuscript, and dozens of letters, including several petulant communications from Major Ross, one of them being a plea for Tench's immediate recall. The two were feuding.

The *Golden Grove* and the *Fishburn* were last to depart, their vast stock of wine and spirits safely cellared in storehouses guarded around the clock by Marines. The *Sirius* was also gone, off to fetch seed and flour and more livestock from Cape Town.

Mary B sat with her baby and watched the little *Supply* lose its identity in the gathering gloom. The ship looked absurdly small at the mouth of Sydney Cove, like a lapdog crouched in the doorway ready to yap at intruders. After almost a year of isolation, the colony listened more eagerly than ever for the boom of a cannon that would signal the arrival of a ship and news from home, a ship bringing fresh faces, fresh food, and new topics for table talk. Every week a party of Marines hiked over to Botany Bay to see if a puzzled ship lay at anchor wondering what had become of the convict colony.

If Governor Arthur Phillip had confided in those he trusted - his Navy cronies and perhaps even Tench - had told them Albion would wait another year and a half before it heard the boom of the signal cannon announcing relief for the colony, they wouldn't have believed him. The world was too small, too well organized, and the intricate links of maritime communications too well established. Being cut off for so long was unthinkable! Of course Arthur Phillip didn't know how long the colony would wait, but he was a prudent man, and realist enough to know that despite his hope and faith and determination, Albion was not a land of milk and honey. Indeed, there was no honey at all, and after their cows disappeared, no milk either. Among the keepers and the kept he alone wondered, what if....

The late December air grew chill, and Mary B drew Dolly's cloak more closely, settling the baby more snugly beneath its warmth. Almost sixteen months old, Charlotte was taking steps, babbling and becoming more and more a person. Her appetite at Mary B's breasts was vigorous, but Mary B had begun feeding her porridge and giving her crusts of bread to chew. She enjoyed her child, and as her life settled into a comfortable routine, for much of the year her thoughts had been serene,. She awakened early to cook a good dinner for Will's return from a night of fishing, and lay with him a while before he slept. Their new hut was livable, although scarcely large enough for more than a pallet mattress on a rope frame, a cradle, a table, two chairs and a stool, all crudely notched and tied from saplings and sinewy strips of bark. Situated on the "good" side of the cove, their home was specially built for Arthur Phillip's "official fisherman." Mary B enjoyed the first real privacy she'd known in more than three years. She was faintly amused by their upside-down world, where winter was summer and summer was winter; where Will slept by day and fished by night; where tree-dwelling creatures slept head to the ground, and the stars were as strange as the savage tongue.

But that morning she and Will had quarreled, and Mary B was reliving their quarrel, recalling again and again Will's threats and his cutting remark about Charlotte.

Although it was their first serious quarrel, Mary B had seen it coming for some time, as things had not been good between them. This latest business started the day after Christmas, when Will was arrested.

He'd fished as usual, but returned home that morning with rum on his breath. At Mary B's questioning sniff, he said he'd taken only a nip or two in honor of Christmas from a Marine he knew from the *Charlotte*. Later Mary B learned that he'd lied to her and had actually bartered some of his catch for rum. After sleeping until late afternoon, Will rose surly and silent and made his way down to his boat. There he plunked down on a rock and didn't budge except to extricate a near-empty flask from beneath a bush. When his two dim-witted helpers showed up for work, Will announced he was unwell and wouldn't fish that night. Marlowe and Mariner didn't care. They still got their rations.

Will drained the last of his rum in one long pull, tossed the flask, and begun to harangue Marlowe and Mariner about how dumb the Governor was not to let him fish outside the headlands. "What does he think, that I'll sail off to Tahiti and take up with a *wahine*?" He continued in this vein until dusk began to fall, by which time his blustering had attracted several idle convicts, which caused a passing Stork to see what was going on, and after an increasingly heated argument between Will and Stork, the Marine ended up arresting him for drunkenness and Marlowe and Mariner for idleness. The three spent that night in the hut used as a jail.

At the hearing before Judge Advocate Collins next morning, his sliding Adam's apple a give-away to his nervousness, Stork admitted he hadn't actually seen Will with liquor, but he'd smelled it on his breath and wanted to do his duty. Will called a Marine in his defense, a Private Brown who admitted he'd given Will some rum in the spirit of Christmas and apologized to the Court for his offense. Collins put Brown on report, praised Stork for doing his duty, and released Marlowe, Mariner, and Will with an admonition to diligence, but without further punishment. After all, it was the Christmas season.

The incident frightened Mary B. When Seedy, her belly bulging with Quint's baby, had come through the woods to tell her Will was in jail, Mary B was disbelieving. He'd been crabby and railed against going to work and gone down to his boat surly, but...in jail?

Despite an occasional outburst of temper, Will usually treated her well and had accepted Charlotte. Their lovemaking had been all right too, as they found ways to pleasure each other with little rituals. But a couple months earlier something began to slip. Will grew moody. He slept fitfully. He started finding fault. He complained of Charlotte's occasional fussing, of the monotonous food, his inept helpers, and Mary B's failure to make things better. Sometimes smelling rum on his breath, Mary B sensed trouble coming but kept her apprehensions to herself. She tried doing little things to please him. She'd warm water for a wash, give him a back-rub or a kiss on the cheek as he ate. But he seemed indifferent. Sometimes he even seemed to resent her. Their first serious fight began when he trudged home after spending the night in jail.

He'd taken to bed, said he was hungry, and told her to fix something to eat. Putting Charlotte in her cradle, Mary B took pot and teakettle out to the cooking place, heated water, and began warming up a beef pudding she'd made early that morning. She mulled over questions she wanted to ask about his arrest. Then she heard Charlotte yowling and Will yelling and she rushed back into their hut to find the baby on the floor half under the overturned cradle and Will on the bed, his eyes squeezed shut, his hands pressed over his ears and yelling, "Shut up, Goddamn you, shut up!"

Mary B snatched up Charlotte and carried her outside. She was calming her with a breast when Will called, "Where's my food, damn you!" She called back that it was at the fire.

He called out, "Well, dish it up proper, damn you," and she called back that she was busy. He yelled, "Do it, woman!" and she yelled back, "I'm not your slave!" He swore and told her in a sinister voice that she'd better dish up his dinner proper and do it now or.... And she said quietly, "Do it yourself. You don't own me." He said then anyone could own her and called Charlotte "a whore's brat." She said he was "a disgusting pig." He called her "a useless whore." She called him "a liar" and told him to mind his words or he'd be sorry. He said he'd mind Jim Martin if he ever came around again. Mary B didn't respond to this threat, but stood and began to walk away.

"You hear me?" Will yelled as she made her way to the path that would take her around the cove. "That man comes to see this bastard brat again and he'll be sorry. You tell him, you hear? No more! No more!"

At her continued silence he suddenly came boiling out of the hut.

Sydney Cove
1789

*I*n his journal Tench mused: "*Who knows what fruits an enterprise will bear? We sow to reap progress, but perhaps we sow destruction.*" Gloomy thoughts for the captain, but first was the hanging of his three Marines, and then the smallpox.

~~~~~~~~~~

Not since David Collins read Arthur Phillip's commissions had so many gathered. Called by the rat-tat-tat of battalion drums, ragged processions of male convicts shouldered their hoes and shovels to trudge in from work. They joined the women, who brought their babies, something to eat, and something to sit on. But no Marine wives appeared. Although soldiers were rarely put to death for their crimes, six Marines would "haul the hemp" this overcast day in March, 1789, three of them from Tench's company.

Tench was dazed by this terrible turn. He thought he knew his company, and he took pride in his concern for its welfare, in his fairness and awareness of life in the ranks. And now this! How could his men be criminals? How could they be so stupid? What had they gained?

It all went back to the convicts of course, and the women. At first Arthur Phillip tried to reason with malefactors. Their early thefts were punished with floggings that were often forgiven. "Arthur Phillip's velvet glove" Major Ross called the policy. Whatever the name, it had little effect. Thieving grew bolder. Convicts returned from work to find their tents rifled, their food and belongings gone. More severe floggings were ordered for those apprehended, and were not forgiven. But still the thieving continued, even from the storehouses and hospital.

If the thieves set out to test Arthur Phillip's limits, they succeeded. He clamped down. With no deliberative body to temper his actions, he decided or decreed as he wished. His word was law. To protect Albion's food he declared that stealing from the storehouses was a hanging offense. Thus Barrett and Bennett were hanged, proof that the Governor meant business.

"But why did my Marines steal?" Tench wondered. Why would Marines conspire to carry out a systematic looting of the storehouses, hauling away untold liquor, flour, and salt meat, when the battalion always received its full ration? He hadn't wanted to believe Stork when the lanky young private drew in the dirt with the toe of his boot, hemmed and hawed, and at last revealed his suspicion that Marines were robbing the storehouses. His Adam's apple jumping, his long, narrow face wet with nervous sweat, he told of seeing Private Bloodworth pass a ring of keys to Porter.

Tench actually shuddered. Thieves? In his company? Only two buildings required keys, the main storehouse and the spirits cellar. Upset and disbelieving, Tench wanted to dispute Stork's reasoning. Where was evidence of their loot? A crime was not yet proved! And when Stork expressed concern about his own safety for revealing his suspicions, Tench exploded, "Damn it, man, why didn't you think of that before!"

"But it was my duty to come to you, Sir!"

"Your duty?"

"My duty, Sir!"

"Your duty?"

"Aye, Sir."

"Aye." Tench said in a resigned voice. He would have done the same.

Tench fretted before he confided in Creswell. They decided the watch officer would conduct random checks whenever Bloodworth or Porter was involved with the nighttime guard.

A few days later the Commissary Agent came to work in the morning to discover a counterfeit key broken off in a lock. Summoned to examine the evidence, a convict locksmith confessed he'd made such a key for Bloodworth from an impression the Marine brought him. Bloodworth denied everything, but Porter spilled the story. In exchange for immunity, he told how on the night in question he was letting himself into the storehouse when he heard a patrol approach. Panicking, he forced the key the wrong way. It broke it off. Then Porter surprised everyone by implicating not only Bloodworth, but also five others.

Even after the trial, during which the Marines admitted their guilt and expressed remorse, Major Ross couldn't believe the Governor would actually require the punishment to be carried out. Hang six of his Marines? For filching a few pounds of provisions? And the guilty informer goes free? The idea was monstrous! He was livid when he confronted the Governor. "You can't hang six Marines! You can't treat my men like damn convicts!"

Arthur Phillip was obdurate. "We can't have two sets of laws. There can be only one law, and the law is clear!"

"Aye, your law! Your law is going to hang my men, your convict law!"

Surprisingly calm, Arthur Phillip answered, "No, not my law, Major, the colony's law. And the King's patent says I am Governor, and you, Sir, are my subordinate."

That three of the wrongdoers were from Tench's company gave Major Ross some satisfaction. Affecting commiseration, he twisted the knife in Tench. "Pity," he said. "These things never look good on one's record."

Tench shrugged, unwilling to give his sneering superior the satisfaction of a response. He wondered if their growing enmity might lead to a duel.

But Tench was also bothered that Arthur Phillip couldn't be budged. The evening before his Marines were to be hanged Tench made one last appeal. Placing an avuncular hand on Tench's shoulder, Arthur Phillip steered him out of his capacious tent towards the Governor's house rising nearby. The gloaming of the day was lovely, the air mild and perfumed with the faint smoke of cooking-fires. The light was roseate, gossamer on the brick walls of the house, on huts and tents and rocks and tree trunks. In such tranquility how could six lives hang in the balance? It was an evening for contentment and well-being, for taking pleasure in the joy of life!

The two-story walls of Arthur Phillip's official residence were up and the roof-work in place. In a few weeks Arthur Phillip would be moving in. Pausing to regard the unfinished structure, he asked Tench, "Do you know what they call this house?" He didn't wait for a reply, but answered himself. "They call it 'Phillip's folly.' Aye, and I understand that. Most of you still live in miserable little huts. Your men have lived under canvas for more than a year. Is it any wonder this incongruous brick building where I will live is called a 'folly?'"

The men resumed their perambulation of the first permanent building in Albion, its lower walls laid with bricks carried as ballast from England. "But it's not folly," Arthur Phillip said, "It's the Governor's official residence, a symbol of government, of reason, of law, of justice. Long after I'm gone this building will remain, and from under its roof others will govern this colony. That's what we must remember, the long view. Aye, the long view."

Arthur Phillip paused to regard the skeleton roof against the darkening sky. Nearby James Martin tended a trash fire. An apprentice mason now, he'd asked to learn brick laying as a way of escaping the sawpit. His stirring caused fine ash to swirl briefly into the air, then settle on the raw earth like snow.

Arthur Phillip resumed his stroll. "The long view requires a leader to act with authority but be able to explain his acts so they make sense later, even years later. And sometimes that's hard, Captain. It would be much easier to let the law slide, to compromise, to grant this man's request for an exception, to bargain for that man's support. But after those men are gone and are no longer an influence, how do you explain your decision? That he was a friend? That he was a good man?" He kicked a small stone from his path. "No, sir, that is neither good leadership nor good government." He gestured vaguely at the rough scene with a flap of his hand. "Especially in a situation like this, Tench, where we all depend on each other. Here, leadership and government must be consistent. Yes, consistent and authoritative. That's the only way we'll survive in the long view, Tench. And that is why your Marines, my Marines too, I remind you, must hang tomorrow."

The two had completed their route around the Governor's house and now stood where they could look down the slope to the dark, smooth waters of Sydney Cove. Tench swallowed. Here he was, 31 years old, and he felt like a schoolboy. Of course the Governor was right! There must be just one kind of justice, or there was no justice at all. Impulsively he turned to grasp Arthur Phillip's hand, his tears welling. Arthur Phillip returned his grasp and again put a hand on Tench's shoulder. "I'm sorry, Tench," Arthur Phillip said, "I truly am."

Tench nodded, took a deep breath as if he would say something, but instead turned and strode rapidly into the dusk.

~~~~~~~~~~

As the crowd assembled, the six condemned Marines were marched down from the jail, hands manacled behind their backs. Stripped of coats and leggings and boots, they stepped tenderly on their bare feet. Dressed only in undershirts and breeches, a couple of them had combed their hair, but none had shaved since the trial. "Quiet in the ranks!" was heard again and again as their comrades voiced encouragement. No single branch of the hanging tree was long enough to hold six, so planks on trestles waited on either side. Divided three and three, the condemned were helped to mount the planks to stand by their nooses.

James Freeman moved a short ladder down the line, mounted it, asked if the man wanted the hood, pulled a white hood from beneath his blouse and drew it over the victim's head, set the noose, climbed down, moved his ladder, and climbed up again to repeat the ritual. To the waiting assembly, strangely quiet, he seemed to move so slowly and deliberately that he would never finish. To those about to die he seemed to rush through his office.

"Blimey!" Quint breathed as Freeman pulled the hood over his head, "I can't see a thing." Freeman said nothing, but patted Quint's shoulder and moved on to Dobber. Last winter Freeman had asked forgiveness of his sentence on the occasion of the King's birthday, when Arthur Phillip forgave several pending punishments as a sign of His Majesty's mercy. On a day free from work, with extra grog for the Marines and a half-pint for convicts, June 4, 1788 had been memorable. Before a roaring bonfire that engulfed an entire tree, convicts and Marines alike raised their cups and voices in the colony's first communal toasts to the King's health, to the Governor, and to Albion. Meanwhile, a few absent convicts took the opportunity to pilfer tents.

But Arthur Phillip turned down Freeman's request.

That day had proved fateful for Quint too, for that was the day he told Seedy he wanted to marry her, and the day their child was conceived.

The six condemned men perched precariously on the trestled planks, like hooded hawks tethered to the tree. On one side were Bloodworth and two men Quint barely knew, Jenks and Catchpole. Dobber and Brown flanked Quint. Immune from punishment, Porter guarded the Judge Advocate's tent, more for his own safety than from any threat to Collins' belongings.

"This is not how I wanted it to end," Quint thought once again. In a recurring scene in his mind's eye he saw himself trying to explain his fate to the grief-stricken faces of his mother and Seedy. He could think of no more to say, so kept repeating, "This is not how I wanted it to end," as if those few words were enough. The sad, averted eyes of his mother and Seedy asked for more.

Quint felt a lump in his throat, an awful sense of loss, and thought he might begin to cry, so he forced himself to go outside of his head as the Judge Advocate finished reading the charges and the verdict. Then Quint heard the preacher begin and on his right heard Dobber sobbing, "Oh, God, I'm sorry, oh, God, I'm sorry!" On Quint's left Brown prayed earnestly with the Reverend Johnson, a make-do priest for the Catholic Marine. Brown's muffled voice was high and tight, "...and forgive us our trespasses, as we forgive those who trespass against us, and lead us not into temptation...." Quint heard a few voices in the crowd joining the prayer against a background of coughing, throat-clearing, murmuring, and here and there a nervous laugh. He was glad Seedy wouldn't see him. Suddenly he realized he didn't care if the crowd laughed at him forever if only he could remain on this narrow, cruel scaffold, alive! He would gladly suffer storms and night and hunger and pain forever if only he could continue to hear the indifferent voices of these unseen people, could continue to live, if only as a mute or a bird or even an ant!

He straightened, pulled mightily at his manacles, and tensed his legs. He bent his knees to feel the pressure of the rope against his neck. He couldn't still his hammering heart. It pounded like a fist trying to break out of his chest. He put his tongue in the familiar gap of his missing tooth. He thought, I don't have to stand here waiting. I can jump. I can control my own end. He tensed his legs again, wondering if he could bring himself to jump.

"Quint!" he heard in the distance. "Quint! Quint!" Quint's heart leapt. Word from the Governor! A reprieve? Was he saved, he alone? He barely heard the excited buzzing of the crowd as he strained to make out the words coming from a distance. The voice grew closer and at last he could hear Porter calling, "Quint! I'm sorry. Quint, I'm sorry. I'm sorry, Quint. Forgive me, Quint, please, I'm sorry!"

Quint heard Porter's pounding footsteps halted by the Provost guards and heard Porter beseech them as if Quint couldn't hear his words. "But I've got to tell him I'm sorry, I've got to tell him! Please, I've got to tell him."

In his disappointment Quint wondered if his heaving stomach might betray him. "It's all right, Porter," he heard himself calling, surprised at the closeness of his voice in the hood. "It's all right! It doesn't matter! It's all right! Do you hear? All of you? It's all right!"

"Is it all right, Quint, really, is it all right?"

"Aye," Quint said, this time quietly, nodding his hooded head. He felt more in control now, and felt that of all the ways to die this might be an all right way too, and he repeated the words of forgiveness Porter needed to hear.

Moments later, moments both infinitely long and infinitely short, the trestles were knocked away and the stout branches of the hanging tree thrashed violently. Leaves rattled, limbs clattered, vegetative debris rained. With tears streaming down his cheeks, Porter cried, "Permission to aid the prisoner!"

"Permission!" Motherwell bawled, his chest heaving.

"Permission!" others cried, "Permission!" At Motherwell's violent nods they broke ranks to embrace the kicking legs of their comrades to speed their strangulation, to deny brain-blood, to end the desperate pumping of diaphragms. They bore down for quicker death.

Dobber was the first to cease struggling, then Brown, then Jenks, then Catchpole. Finally just Bloodworth and Quint fought death, both brawny young men who would not succumb easily. A few convicts wagered who would die last. Blood from Quint's bitten tongue soaked his hood and his shirt and dripped onto Porter's head. His own breath rasping, his voice desperate, Porter renewed his grip and pulled harder, "Die, Quint, die, please, die! Oh, God, please die!"

Poor Porter! His efforts to help Quint unwittingly pulled his friend's breeches down enough to reveal a last wonderful erection. Hundreds saw a vigorous jet of milky fluid arc from Quint's throbbing penis and fall fruitlessly beneath the hanging tree. His penis drooped. Tartop laughed, then some other convicts, and then a few clapped and whistled. The Marines would gladly have murdered them all.

At last all six soldiers hung quietly, twisting rather gracefully back and forth. The hanging done, the punishment completed, the law preserved, equal justice served, the people turned away from the largest execution yet witnessed in Albion, Tartop still chuckling.

~~~~~~~~~~

"Oooh, it hurts," Porter said. He was slumped on his knees at Quint's grave. "I loved him, Stork. I loved him. Is it a sin to say I loved him? I did. I loved him, and now he's dead. I'm sorry, Quint. I'm sorry for what I did! You were going to be something, Quint. You were!"

Stork stood by Porter's bowed back and patted him absently on the shoulder and looked through the trees to the placid waters of Sydney Cove. Stork knew he'd never forget the sight of Quint's swollen, blackened face with his tongue protruding, bitten half through, his shirtfront crimson. Aye, Quint was dead. Stork's mind went back to the fateful confrontation almost three months earlier, when he marched Quint to the perimeter of the encampment.

"Goddamn, Quint, it isn't right!"

"Well, Goddamn, Stork, it isn't right the other way either! Why should Seedy get a two-thirds ration? She's going to have my baby! She's hungry all the time. You should hear her stomach! I won't let it happen!"

"But what about Dobber? His wife isn't having a baby! Why should he steal?"

"Aw, Stork, what would you do? He's just trying to raise money to get her out of here!"

"But not by stealing, Quint. That's wrong!"

"Wrong? Wrong, Stork? Sometimes you don't make sense! You think it's not wrong to keep us living here without a proper barracks? You think it's not wrong to stop a man's rum if he doesn't go to Sunday service? You think it's not wrong to keep a man here who's served his time and then flog him if he argues? Or send young boys here, or old men who can hardly hobble? Jesus, Stork, we're surrounded by wrongs, and Dobber's only doing what he has to! She never should have come. She's not cut out for this!"

"Well, she did! And she decided! And what about Porter? Who's he stealing for, huh? Some come-hither whore?"

Quint glared at Stork, then shrugged and looked away. "Well, Porter," he said, as if the name alone were explanation enough.

"Aye, Porter! He better stop too, Quint. You all better stop. You're headed for trouble. That Bloodworth's no good. You know that, Quint. Believe me! You can't steal from the stores! That's everybody's food! It's not right!"

Quint gave Stork his foolish, missing-tooth, boyish grin. "Ah, Stork, if you had a woman who's having your baby...." He raised a hand to Stork's shoulder, but Stork shrugged it off. Quint shrugged in turn. "Jesus, Stork, you don't have to take it all on yourself! There's misery enough for all."

"Just remember, I've got my duty," Stork said, and walked away.

A gentle rain began to fall and the last of the Marines began drifting down from the graveyard, but still Porter and Stork remained. Aye, Pottle thought, misery enough for all! He saw the reddish earth splotching dark from the raindrops. As the rain fell harder, clods began to crumble and water trickled in muddy little rivulets. How quick, he thought, Already Quint's melting. What's left of Quint now? Where is he?

Stork wiped rain and tears from his face, put on his hat, and helped Porter to his feet. Only in my head, he thought. Quint's gone except for in my head, and when I'm dead, he'll die there too.

~~~~~~~~~

Seedy gasped at the picture. Quint must have spent hours on it! Ten panels on the first page of Quint's book told a story in fine-lined black ink. The top two panels showed the *Charlotte* sliding past the Eddystone lighthouse in Plymouth and at anchor against the steeples of Portsmouth. Down the right side of the page the little ship sailed against the backdrops of El Pico, the Sugar Loaf, and Table Rock. At the bottom the *Charlotte* lay at anchor in Sydney Cove, with native canoes gathered near, and another scene showed Marines facing spear-bearing natives. On the left side, the lowest panel depicted a tent with three Marines in front, one holding aloft a pennant with the number 18, the number of Quint's company. The next contained a kangaroo, an emu, and three flying birds with black-tipped wings. In the last the Governor stood atop a stump addressing three Marines.

The page was entwined with a sweet-tea vine sprouting leaves and flowers that wended serpentine around the panels and gathered in the center, where it doubled to form a subtle, flowered figure of the letter "Q." Within the "Q" Quint had drawn another scene. Smoke curled from the crooked chimney of a palm-log hut. On one side was the suggestion of a garden, and on the other a chicken and a pig sitting on its haunches wearing a grin. On a pathway stood a smiling Marine and a smiling woman holding a baby. In each corner of the page he'd written a word: "Peter, Quint, His, Book." Seedy, of course, could not read.

"Did he do this?" Seedy asked, caressing the paper, "Did he really do this? It's so beautiful! I didn't know!" Tears slid on cheeks shadowed with defeat.

"Aye." Stork swallowed nervously, his Adam's apple jumping. "He wanted you to have it."

"It's so beautiful, so beautiful," Seedy repeated. Then she said "Thank you" so softly Stork scarcely heard.

He edged towards the door, wanting to be gone. The sight of Seedy's ravaged, lopsided face made him wonder once again what strange power this woman possessed to have made Quint risk nighttime visits, to dig and hammer and thatch and chop and carry for these bedraggled women in their leaky hut, to court death by stealing a few pounds of flour.

Stork nodded a farewell to Mary B, who rose with Charlotte on her hip. "Thank you," she said at the door. She smiled reassuringly. "You were very nice to bring Quint's book." She again offered him a cup of sweet-tea, the slightly bitter, licorice make-do brew derived from the leaves of the local vine depicted in Quint's book. Stork mumbled "No," nodded again, and ducked out, painfully bumping his head on the low doorway. "Ow," they heard from outside. From his stool by the fire, Will snorted his amusement.

"Will?" Mary B threatened.

Will shrugged and hunched closer to the small, smoky fire. Made of latticed sticks plastered with crumbling clay, the sad state of the hearth and chimney mirrored the sad state of Albion. Clay also chinked the hut's walls and showed the ravages of sun, wind, and rain. The colony still had no limestone, so convict women scrounged for clam and oyster shells and other ocean debris that might be pounded fine for the lime of precious mortar. But pickings were slim, and much of Albion was held together with mud and clay.

Life in Albion mirrored life for Mary B and Will, who'd suffered a humiliating and painful tumble from grace.

Poor Mary B! What had she got herself into? She thought their marriage would begin like a slate wiped clean, but she feared now that something festered in Will, something ominous from his past. His withdrawals were a mystery. When she saw him sitting on a half-rotted chunk of cabbage tree, dabbing at the dirt with a stick, brooding and unapproachable, she felt as if she'd married a stranger. She wondered where his thoughts were and whether she was anything more to him than a bedmate and housekeeper. She worried whether she'd done something to be the cause of his gloom, and could not shake the premonition that someday Will was going to fail her, that he'd take up with another woman or announce he was going back to another wife or simply walk away.

But Mary B carried her own secrets, perhaps the most troublesome being her deliquescent feelings for James, shapeless longings that rose to feed her doubts about Will. Afraid for her future with Will, perhaps her thoughts of James were as simple as this: if Will failed her, would James still be there?

Poor Will! Another man might have tempered his deeds and avoided his troubles. But Will was not temperate. Roused by anger or bedeviled by a black mood, he was as likely to do something foolish as not. And that little cherub Mary B tended so lovingly worked at him, a constant reminder of her past with a man much too conveniently present. All Albion knew that when Mary B fled from Will after his first arrest, James Martin camped on her doorstep like a faithful watchdog, and kept a club handy in case Will came by to bully her again. For Mary B to have sought the protection of her former lover pained Will more than his recent flogging.

When Mary B fled after Will's first arrest she hadn't returned until the Sunday after Epiphany. On that particular Sunday it came to pass that the Reverend Johnson preached to the assembly how Jesus taught for a time in the temple but returned to Nazareth with Joseph and Mary and was subject to them. "We are all subject," he instructed. "We are subject to God, to His Divine Son, to the King, to the Governor. Soldiers are subject to officers, servants to masters, wives to husbands, and children to parents." He admonished everyone to remember they were subject until they reached the Kingdom of Heaven, when all would be equal. Taking the preacher's words to heart, Mary B decided she had to try again to live with Will, and be subject to him.

Will's welcome was surly, as if Mary B had done him an injustice, but they resumed a life together of sorts, although they shared little more than space and food. Then a few weeks later Will was arrested a second time. Convicted of selling his catch for coin, he was stripped of his post as the colony's fisherman. And flogged. And evicted from his hut. And that caused their second big fight.

They'd been gathering up their belongings. His back still seeping pinkish fluid, Will was growly, and his only words were snarling orders. Nor was Mary B happy, and she let Will know they were paying a mighty high price for his foolishness. Then at some comparison she made to James Martin, Will snatched a stick of firewood and clunked Mary B alongside the head with a roundhouse swing that sent her staggering. She fell over Charlotte's cradle, which upended and spilled the baby. Livid and swearing, Mary B grabbed Charlotte and lurched past Will with a good stiff-arm that sent him backward over the stool, a fall she learned later dislocated his shoulder. Once again she sought refuge with Katie and Seedy. She'd been living with them for a couple weeks when Will, free of his sling, came to make amends. It was at this time that Stork brought Quint's book to Seedy.

But Mary B was unaware that behind her back Will still belittled her shortcomings, that he thought her to be an ignorant woman with too much a mind of her own. He complained to his cronies that she failed to keep a good hut, was a bad cook, and spent too much time with her baby and her women friends. He said she was too tall, was letting herself go, and was no good in bed. Indeed, Mary B's shortcomings were a familiar litany whenever Will felt a need to bolster his ego. Did he like her at all? Had he ever liked her?

Perhaps, but in the same way a man might like a good dog. From the day Will awakened aboard the *Charlotte* with the idea that he wanted to possess Mary B in Botany Bay, he'd pursued his objective with determination and guile. But once he possessed her, why! she lost value! Oh, he still enjoyed having a regular bedmate and having his meals cooked and his clothes mended and washed. And he liked that she listened to his opinions and was impressed by his learning and laughed at his entertainments. He liked being master of the house. But he didn't feel connected to her. And for sure he didn't regard her as a real wife. Their marriage was a convenience, not a commitment. She was a convict wife in a penal colony, and after she became a bother - after she began standing up to him and talking back, and especially after she stalked off the second time - well, good riddance.

Mary B's intuition was right. In truth Will did have a secret he brooded over, but it had its beginnings earlier, after Mary B stalked off the first time. He'd awakened one morning in his solitary bed with a terrible headache and the glimmer of an idea. As the day advanced the idea distilled, at last becoming so simple he wondered why he hadn't thought of it earlier. He'd return to England alone! After all, Mary B had left him, and he was devil-sure not chasing after the nag. James Martin was welcome to her. As his plan took form Will realized about midmorning that his headache had disappeared.

His plan was simple. He'd sell part of his catch for coin, and use the money to bribe a place for himself as crew on a ship returning to England. With money to sweeten the deal, he'd have a big advantage over convicts who had only their labor to barter.

His problem was that as Albion's "official" fisherman every fish he caught belonged to the colony, except for two pounds a day he could keep for personal use. But who was to say who caught a particular fish, or when? He could use go-betweens to sell part of his catch, and no one would be the wiser. Thus, when Mary B returned on Epiphany Sunday of her own accord, Will's plan suddenly seemed complete. He resumed his duties as husband and provider and seemed as happy as a man who had everything in the world he wanted.

That was Will's plan, but his foolish ego did him in.

The affairs of life dictate that the higher one's office, the more one's authority. And the more one's authority, the more one needs a little moral elbowroom. A governor cannot be held to the same moral strictures as a goat boy, because a governor deals with more complex issues. As Albion's "official" fisherman, Will deserved at least a little moral elbowroom, and if he'd filched only an extra fish or two Arthur Phillip might have overlooked his transgressions. But Will was flagrant in the abuse of his office. He sold too much of his catch - sometimes several pounds a day - and the Governor could not tolerate such behavior. Word got out, and after a careless transaction with one of Captain Meredith's convict servants, Will was arrested a second time.

Will's flogging was his first, and he was unprepared for the awful pain. But except for involuntary grunts and an occasional groan, he bore each of the 100 strokes in stoic silence, unwilling to let the lash defeat his plan. Indeed, the authorities never knew that behind Will's thievery lay a plan. In their minds Will was no different than other convicts who committed foolish crimes, who seemed to steal for the stealing itself, whether a plank, a pot, a tattered shirt, a cup, a comb, or an empty box. If no one is looking, take it! Amid such mindless thievery, who would have thought that Will had a plan? Most officialdom thought Will Bryant to be little more than a puffed-up bother, a moody braggart, a brutish wife-beater. Judge Advocate Collins thought him simply unlikable. But because they saw Will as less than he was, the authorities discounted his intelligence, or at least his cunning. In short, they underestimated Will.

Nonetheless, to be whipped at the triangle, demoted from his job, kicked out of his hut, and left by his wife took some stuffing out of Will. He moped for days, finding fault with everything and everybody. But he found few sympathetic ears. Some were so bold as to tell him he'd squandered an enviable situation. Finally, feeling such overwhelming sadness and loneliness that he cried one night in bed, Will decided to seek reconciliation with Mary B.

"Oh, look," Seedy said, "There's writing too!" She'd turned the page and now held the book up for others to see a page nearly filled with ragged lines of writing. "What do you think its says?" Seedy asked, caressing the words. She turned another page, then several others, but the rest were empty.

"Give it to Will. He can read it," Mary B said. "Maybe it's a letter. Read it for her, Will, please."

Careful of his sore back, Will rose, took the book, and turned to stand near the doorway where the light was better. He noticed where Quint had carefully excised several pages at the beginning and guessed they were false starts for his picture, which he studied for several moments, drawn to its symmetry, its painstaking detail, its completeness. Even the smiling people and the grinning pig didn't look silly. Will nodded grudging admiration as he turned the page, then almost laughed aloud.

Quint's writing was tortured, and Will stood puzzling out the words so long Mary B finally asked, "Will, what's wrong?" Will said he was just trying to get the sense of it.

"Thank you," Seedy said, and slid down a little on her pallet. Since Fate conspired to deprive her of both her lover and her baby in the same afternoon, she'd become indifferent, and she was unsure if she wanted to continue living. Nothing made any difference anymore. Although she drank the colony's sweet-tea, she'd eaten nothing for a week. She felt guilty about food.

Earlier she'd laughed when she told Quint of being hungry all the time, but Quint went into a rant. "Why should you get only two-thirds of a ration? You're eating for two! Our baby in your belly is more important than anyone!" He'd begun sharing his own food, and when she protested his sacrifice he made a muscle. "Does this feel like I'm getting weak?" So they shared his ration until one day someone broke into the women's hut and stole their entire week's food, more than twenty pounds of flour and a dozen pounds of salt meat. It was after that calamity that, unbeknownst to Seedy, Quint got involved with the Bloodworth gang.

Will Bryant read aloud.

"*My nam is Peter Quint an ths is my book. I sald from Plymuth with 18th Cmpy marins (capt Tench) for Botany bay in the yr of ar Lord 1787. On th voiaj I cam to luv a convic nam Zede Haydn with al my hart.*

"*I was a gud marin but lif was hard in port Jackson wehr we wer. We luvd to much for such a tim. I men Zede Haydn an me. I stol food for ar chiuld in her woom. Now I wil hang and my yers ar onli 23 but I wod do th sam agen for ar chiuld an th wumun I luv but I wod not hang.*

"*Now I kis my mothr and my fathr and my bruthrs and my litl sistr in my hart an Zede and ar chiuld I wil nevr see so gud by my lif an God bles th grat wuruld. Peter Quint pvt Hs Majstys Royl marins*"

"Oh," Seedy said, "did you hear that? He loved me with all his heart! And he even wrote it in a book. I'm in his book! It so beautiful! Doesn't it sound just like Quint, Mary B? Read it again, Will, please, will you read it again?"

~~~~~~~~~~

When presiding over John Matthew's simple little burial, the Reverend Johnson had run through a formulary of condolence and a brief reconciliation of God's way to Man, but even as he turned from the tiny grave more pressing matters demanded his attention, and in her grief Katie had to look elsewhere for support.

But Katie's world had changed. Newly married and living across the cove, Mary B devoted her life to Charlotte and Will.

Seedy was wondrously distracted by Quint. Liz Bason and Jane Fitzgerald made gestures of friendship to Katie, but for some reason she'd never been attracted to either. So she was left with Hannah.

From the days of her mocking rejection of Hannah on the *Charlotte*, Katie had come to appreciate the warmth and comfort of Hannah's sisterhood and the riches of her soft-covered Bible, carefully protected in oilcloth. Hannah's good-humored spirituality embodied an understanding and acceptance that gave Katie genuine comfort, and the two women were a familiar sight as they sat side by side, Hannah's Bible opened on her knees, her finger tracing sacred text across the narrow columns as she declaimed in earnest voice the words that seemed to provide a wise precept, a thoughtful admonition, a vigorous exhortation for every vicissitude, every temptation, every despair, every doubt.

It was after Quint's death and after Katie had endured almost three years of imprisonment that she first made a connection between the Word of God and the works of man. Rather, of woman. Who knows whether the months she spent in the filthy confinements of Plymouth and Exeter and the *Dunkirk* had anything to do with her insight? Who knows whether her ordeal on the *Charlotte* began some internal transformation? So far as Arthur Phillip could tell, there was no dependable connection between punishment and reform, between an apparent stimulus and an apparent response. So far as Arthur Phillip could tell, to generalize such a connection was to tempt the gods. The more he flogged, the more he had to flog. But the opposite seemed true also. The better he treated the convicts, the better they behaved. At least sometimes.

Perhaps her imprisonment and suffering did prepare Katie for the afternoon of her epiphany. Perhaps being exiled in a heathen wilderness and losing her baby and seeing her friends move on with their lives prompted Katie to look for more meaning in her life. Or perhaps it was the native sweet-tea. Or perhaps it was that she ate more greens. Or perhaps it was the influence of an age-activated hormone, the chemistry of Hannah's friendship, or the owl Katie saw fly across a full moon. But one particular afternoon Katie suddenly did see that if she performed good works she'd achieve salvation, and that salvation would ease the burdens of her soul, and perhaps even bring happiness. She came to this belief on hearing the exhortation of St. James as she and Hannah sat in afternoon shade, several weeks after Quint was hanged, Hannah read:

"What doth it profit, my brethren, though a man say he hath faith, and have not works? Can faith save him? If a brother or sister be naked, or destitute of daily food, and one of you say unto them, depart in peace, be ye warmed and filled; notwithstanding ye give them not those things which are needful to the body; what doth it profit? Even so faith, if it hath not works, is dead, being alone."

Of course the words were familiar. Katie had heard them many times before. But on this special afternoon they went straight to her heart, and the simplicity of the message and the sad recognition of her failures powerfully moved her. Aye, she had faith. But where were her works?

Her legs outstretched, her eyes on the dilapidated remnants of her shoes, her big toe peeping through the hole worn by her draggy left foot, Katie asked, "But, Hannah, what can I do? We hardly get enough to eat as it is. My clothes are rags. I have nothing to give, nothing to share. Nothing! How can I do good works? How can I feed the hungry and clothe the naked? I don't have anything?"

Sure of the moment's significance, Hannah thought carefully before replying. She searched Katie's face. "You can give yourself, Katie. After your soul, the best thing you have to give is yourself."

Katie looked shocked. "What? I have to do that? Jesus wants me to do that?"

"No, no, no," Hannah laughed, "Not that! I mean, good works. You can do things for others besides pray. You can comfort them, you can tend them when they're sick, you can wash them. You don't need things to do that...."

And that was how Katie began her work at the hospital, where three years later a larger-than-life convict hero would bring her news of Mary B's escape and sweep Katie off her feet.

At first John White was suspicious of Katie's motivations. That a convict wanted to work was remarkable, but that the convict was the woman who'd suffered so greatly at his own hands was even more remarkable. Would she pilfer from his patients? Was she after the medicinal wine? As a consequence of White's suspicions, Katie found herself assigned to the most distasteful tasks. She emptied slop buckets, cleaned up vomit, scrubbed the rough plank floors, and washed bandages caked with blood and pus. She didn't enjoy her drudgery, but it made sense that the more demeaning and hard the task the more her service must count as "good works." She came to regard her most odious duties as expressions of love, and for the first time since she was a little girl she went to bed at night feeling good about herself.

~~~~~~~~~~

As if the hanging of the Marines were a harbinger of death, ravaged corpses of natives began turning up in the harbor and on the beaches, on the natives' fishing rocks, and in the shallow caves they sheltered in. Sometimes torn asunder by birds and animals, sometimes corrupted beyond recognition, they lay where they died, abandoned by their people, as if their deaths were the wrath of an evil spirit that terrorized the survivors and sent them fleeing for their lives. The corpses bore the unmistakable signs of smallpox.

The news could hardly have disheartened Arthur Phillip more. Every attempt to develop good relations with the natives had been frustrated. Despite his efforts to control the convicts, their continued thievery and abuse provoked retribution from the natives, who murdered several convicts, wounded several others, and chased and frightened dozens. But who knew how many natives had been robbed of their belongings or secretly dispatched by souvenir hunters? Long before the smallpox began to devastate Albion's natives, virtually all intercourse between the cultures had ceased. On seeing the red uniforms of the Marines, most natives fled. Somehow, Arthur Phillip reasoned, the two peoples had to learn to communicate.

A native named Arabanoo was a chance victim of Arthur Phillip's first effort to improve communication. Enticed with an offer of freshly caught fish, Arabanoo was seized by Marines and dragged away while his shocked comrades stood cowered by muskets. Within a few days the English had transformed their captive into a dusky representative of western culture. They cut his hair, shaved his beard, washed his body, and attired him in hat, shirt, trousers, stockings, and shoes. They also fastened a manacle on his wrist, to the other end of which they fastened a convict keeper.

At first Arabanoo suffered from terrible diarrhea, but he amazed his captors with his appetite for half-cooked fish and meat, gulping down several pounds of each at a sitting. Hosted at a sidetable during the Governor's own meals, Arabanoo showed abhorrence for Arthur Phillip's alcoholic spirits and only slowly developed a taste for bread. He also seemed to develop some attachment to the Governor, who was clearly the headman but nonetheless treated Arabanoo with kind attention, showing by gesture and deed that he wanted Arabanoo not to feel put upon by his captivity. Arthur Phillip even joined some of Arabanoo's language lessons.

During the five months he lived in a little hut near the Governor's house, Arabanoo's teachings laid the beginnings of the first English glossary of the area's dialect. Unfortunately, his efforts had only a short-term effect. Several years and several thousand convicts later, Judge Advocate David Collins would lament that it was still impossible for white men and natives to carry on a natural conversation. At best they could communicate in a crude pidgin. Later, with the demise of the last of the natives from smallpox and other causes, the Port Jackson dialect disappeared entirely.

Arabanoo might have become the kind of emissary Arthur Phillip needed, but after several natives were brought in for treatment, Arabanoo helped nurse them, contracted the dread disease, and died within a week. Of the other native patients only an adolescent girl and a nine-year-old boy survived. The girl stayed on with the Reverend Johnson and his mousy wife under the name of Abaroo, and John White incorporated the boy into his own growing menagerie of animals and humans. For the next two years, White and his native boy Nanbaree and Tench and his orphan boy Ned Thornapple comprised parallel families of bachelor father and foster son.

~~~~~~~~~~

From its beginnings the Botany Bay project was controversial, but in Prime Minister Pitt's inventory of concerns the new penal colony did not loom so large as the impeachment trial of Warren Hastings or the increasing turmoil in France. Success was better than failure, however, and speculation served no good end, so promoters of the Botany Bay project waited nervously for word from the new colony. After Arthur Phillip's brief reports from Cape Town, England heard nothing from the Governor for more than a year. Meanwhile, critics complained of the cost and characterized the distant colony as a folly that coddled criminals in a South Seas paradise. Others complained the project was a heartless separation of families.

Thus, when the van of the First Fleet returned to England in March, 1789, word zipped from Portsmouth to London in a day, and hundreds flocked to Portsmouth and government offices in London to learn the fates of their spouses, relatives, comrades, and friends.

Even as the first ships returned, the *Lady Juliana* loaded convicts in the Thames, the forerunner of the next fleet, a fleet that would sail under very different circumstances. Home Office functionaries who sputtered over Arthur Phillip's insistence on high-quality provisions, on putting Marine guards on the transports, on keeping the ships in convoy, were having their way. True, the First Fleet was well superintended and loss of life low, but it was all too much bother. "Billhooks and petticoats!" they carped, "We spent more time buttering toast for that damned First Fleet than for the rest of the colonies combined!"

For the next batch of convicts a single contractor would receive a flat fee of £17 for transporting and feeding each convict loaded. Seventeen pounds a head for misfits and felons? At a time when a skilled worker might earn two shillings a day, the government was squandering the equivalent of a year's wages simply to carry a thief to Botany Bay? Preposterous! But of course there had to be more to Botany Bay than just providing an out-of-the-way dump for England's criminals!

Of course. England's domestic well-being depended on global mercantilism, on buying and selling, on getting and spending, and the Botany Bay project was seen as a way to secure the lucrative spice and silk trade of the orient. Not only that, but the penal colony would provide a base for Pacific whale fisheries and for the fur trade with the northwest coast of Spanish North America. It would provide metals and minerals yet to be discovered. Indeed, England coveted even tiny Norfolk Island for its forest wealth of maritime masts and spars. But such distant resources required a permanent presence in the Pacific, an implicit objective of the plan for Botany Bay. With a temperate climate, a strategic location, a sparse native population, and a vast territory free for the taking, Botany Bay seemed a perfect opportunity. All Botany Bay needed was people.

Did convict men know their government intended to keep them in Botany Bay simply by not assisting their return to England? Probably not. What became known as the Second

Fleet would carry instructions for settling emancipated males, with the assumption that every sneakthief and ne'er-do-well could become a self-sufficient farmer. But a convict wanting to return to England had two choices: help crew a ship, or pay for passage home. Too bad there were far more convicts than could possibly find work as crew. Too bad that getting money enough for passage back to England was a very dim prospect.

Did convict women know their main purpose for being shipped off to Botany Bay was to help settle the men and serve as breeders in this new colonial outpost? Probably. The great majority of them knew only menial labor and the familiar tasks of cooking, washing, sewing, mending, tending the sick, and birthing babies. For most, staying alive and managing their reproductive machinery were their most essential functions.

Ah, the glorious age of empire and enterprise! To put tea in England's pots required a little social engineering!

Thus, as the first news from Botany Bay reached England, the *Lady Juliana* took on women from jails all over the kingdom. This transport would be six months in the loading, first on the Thames and then on the Hamoaze, which seemed a curiously protracted affair. But as most of the women lived as prostitutes before imprisonment, the *Lady Juliana* soon became a floating brothel for mariners under the benign eye of its master.

Perhaps James Boswell, whose *Journal of a Tour to the Hebrides* had crushed Tench's flower aboard the *Charlotte*, knew some of the women. For years Boswell had been making the weeklong journey from Edinburgh to London to dabble in politics, hobnob with the famous, and pursue his friendship with the eminent Samuel Johnson. And to pursue whores aplenty too, sometimes two or three a day in his obsessive quest for carnal relief. But Samuel Johnson was dead now, and Boswell labored in London to get his eagerly awaited biography of the great lexicographer ready for the printer. He might instead have tended to his dying wife, but theirs was a curious relationship of love and separation. During Margaret Boswell's long invalidism, he tended a fruitless legal practice in London as a pretext for escape from dreary Scotland.

So, on a mild day in mid-May, James Boswell, Laird of Auchinleck, let slip a sour belch as he fingered the light blue paper cover of a new book in Tom Dilly's bookshop. He'd stopped by to see if Dilly had Jeremy Bentham's new treatise on morals and legislation, but not finding Bentham's work, he'd taken time to browse. Now he riffled the thick, uncut pages, and held the slim book up to the clerk's view.

"Two and six, Mr. Boswell. We've nearly sold out this second printing in two days."

Boswell nodded and wiped the drippy evidence of a spring cold from his nose. Not surprising, the little book's success. Since the return of the first ships from Botany Bay, newspapers had labored mightily to feed the public's hunger for information, but making sense of the confused reportage was like trying to understand the purposes of a mob. However, a favorable notice of this account in the *Monthly Review* had prompted Boswell to make a mental note to seek the book out.

Boswell unclasped a little knife, then for almost a half hour slit uncut pages, dabbed at his drippy nose, and read from *A Narrative of the Expedition to Botany Bay* by Captain Watkin Tench.

Tench showed ability, Boswell thought - clear prose, balanced but not ponderous. Although they were a comfortable convention, his classical quotations were apt, and unlike so many travel books, Tench seemed judicious in what he wrote, with hardly a word of compass-bearings and sail-settings. Indeed, Tench was spare to a fault. "Where are the people?" Boswell wondered. Tench seemed disinterested in the convicts as people. He wrote with an odd detachment. What a waste! That rich mine of stories reduced to generalities! Boswell's own life of Johnson was teeming with personalities, anecdotes, conversations, laughter, exclamations. His friend and editor Malone wanted Boswell to cut, cut, cut, and Boswell wanted to add, add, add.

The clerk heard a loud sigh and looked at the paunchy barrister snuffling by the window. Laird of Auchinleck, indeed! Buffoon! Boozer! Whoremonger! All London knew that while Boswell whored in the side streets his wife died in Scotland. Oh, he'd burn!

Boswell had read almost fifty pages and Tench was describing the safe arrival of the fleet in Botany Bay: "*For it must be remembered that the people thus sent out were not a ship's company starting with every advantage of health and good living, which a state of freedom produces....*" Suddenly the printed words disappeared and Boswell found himself staring at the pale, emaciated face of his dying wife. He staggered, instantly bathed in sweat, and when he reached for support he sent a noisy cascade of books to the floor. Good God! he thought, please, no, Margaret, not now! Don't die now!

"Good journey!" were her parting words! Now his premonition was overwhelming. No, she can't be dead! Oh, God! I'll never find peace if I don't see her again and tell her what a good friend she's been and...and...and how much I regret my damnable debaucheries! Good journey! And she'd tried to brighten her farewell with a smile.

Hands on his knees, panting for breath, Boswell groaned aloud.

"Mr. Boswell, are you all right?"

Boswell slowly straightened. The clerk saw a pale, shaken man with wild, panicky eyes, as if Boswell were looking right through him at some fearsome apparition. Boswell was gasping and glistening with sweat. Suddenly he rushed out the door.

After his own heart slowed and he regained his composure, the clerk said half aloud, "Of course, Mr. Boswell," and bent to make a note in Boswell's account for the book he carried away.

~~~~~~~~~

Breadfruit Bligh could barely stand. Nevertheless, he helped each of his seventeen starving men out of the longboat and shook his hand. Safely landed at the brief jetty of Kupang on the island of Timor, they'd sailed more than 3,800 miles from where Fletcher Christian and his mutineers forced them off the *Bounty* some 46 days earlier. Since then Bligh and his stalwarts had battled storms, monstrous seas, a blazing tropical sun, fatigue, starvation, thirst, and despair. But except for one man's violent death by natives when they put in for water, Bligh had brought his entire company to safety. Now they knelt to recite the prayer for deliverance they'd offered every day during their hazardous voyage, although this time it was also a prayer of thanksgiving.

To the authorities of this Dutch East India Company outpost, the story of Bligh and his starving survivors was scarcely to be believed. But the shaggy, skeletal figures who staggered to the fortress were proof of their feat: men suffering from exposure, open sores, their remnant clothing hanging from stick-thin limbs, their eyes haunted, their voices hoarse. Aye, those skeletal men, and Bligh's log, which detailed day by day and hour by hour the progress of their ordeal.

"Old Beak 'n Claws" some of his men fondly called him now behind his back, a reference to the lucky day they snagged a noddy and Bligh divided its pigeon-size carcass. By then almost a month at sea, their twenty-three foot boat lightened by throwing out everything not essential to daily survival, their salt pork long exhausted, they'd been living on an ounce of bread given out each morning and evening, and a daily teaspoon of rum. The hapless seabird represented their first meat in days. Fortunately, frequent storms that had them bailing for their lives had also kept the tropic sun away and slaked their thirst and provided water enough most days to share out a half pint for each man. But this day, too parched to salivate, they watch their captain carefully apportion the plucked bird atop an empty water cask, then summon Robbie Tinkler to the stern. Thirteen years old, Robbie was a "young gentleman" sailing under the sponsorship of his brother-in-law, the *Bounty*'s sailing master.

"Do you know your job, Mr. Tinkler?"

Robbie couldn't tear his eyes from the tiny portions on the barrelhead. His mouth worked but his swollen tongue was cemented in his mouth. The rich, bloody aroma of the meat scraps was overpowering. Interpreting his silence as ignorance, Bligh instructed, "You will turn your back and close your eyes, Mr. Tinkler, and I will point to a portion and ask, 'Who gets this one?' and you will call out a name and that man will receive that portion. I will not point in any order, so you won't know where I am pointing. It's fair for every man. Do you understand now?"

Robbie nodded, reluctant to turn, but he scooted around to kneel facing the bow. Eyes closed, he folded his hands as if in prayer and hoped he'd remember all the names. Bligh pointed to a bloody scrap of wing. "Who gets this one, Mr. Tinkler?"

Thus was the bird divided, and Bligh's own portion, which caused the men to laugh for the first time in days, was the bloody beak and clawed feet.

Perhaps his portion was fateful irony. Much later, one school of gossip in England would insist it was Bligh's obsession with his private supply of fruits and vegetables that finally drove young Fletcher Christian to mutiny. Loaded with more than a thousand breadfruit plants, the *Bounty* was headed for the West Indies and home when Bligh apparently pushed Christian to the edge. A mutineer who kept a journal recorded Bligh's tirade: "*Goddamn, you scoundrels, you are all thieves alike, and combine with the men to rob me. I suppose you'll steal my yams next; but I'll sweat you for it, you rascals. I'll make half of you jump overboard before you get through Endeavor Straits.*"

Perhaps their capture of the noddy was providential. Later that same day the castaways feasted on a booby, a larger seabird usually found near land, and during the next couple days were fortunate to catch several more, including some with fish in their stomachs. These birds they carefully preserved to divide later, but let the three weakest men suck their blood.

The island of Tofoa, where Fletcher Christian and his mutineers seized the *Bounty*, lay more than 2,000 miles from the coast of New South Wales, but Bligh knew if he and his men sailed due west he would ultimately reach the great reef Cook encountered two decades earlier, a reef later called the Great Barrier Reef. Once through this hazard, Bligh figured he could work his way north to the tip of the new continent, where he'd again head west. After another 1,000 miles of open water he'd reach the island of Timor and one of the few permanent outposts of Western civilization in the South Pacific, a sleepy port called Kupang.

Now safe-come to Kupang, the determined lieutenant had proved the possibilities of open-boat navigation over great distances on the high seas.

His return to England with his men several months later put a feather in the cap of the English Navy. But lest other disgruntled seamen get the wrong idea and decide to mutiny in the exotic South Pacific, the Navy prepared a warship to seek out the *Bounty* mutineers and return them to England for punishment. And thus would an impressed Irish sailor named Mackesey be fated to enter the life of Mary B.

~~~~~~~~~~.

But news of Bligh's remarkable voyage to Kupang was still a year away in Albion, and on a cold, dark July night Charlotte slept soundly near the fireplace of Will's hut, dreaming that in the glow of the dying embers she felt the warm glow of a sunset. Will and Mary B lay uncharacteristically naked beneath their thin blankets and Dolly's cloak as Will traced a fingertip over Mary B's middle. "Now this is the coast between Fowey and Plymouth. You see how it goes in and out along here till it rounds the head near the Eddystone...." He traced his finger around Mary B's left breast, then bent under the blanket to kiss its nipple. "Thank God for the Eddystone!" he whispered. "Then it snakes back here along the Hamoaze where the hulk was and where I first saw you and wanted you, and then on the Dock side it comes along like this till it runs into Stonehouse Creek and then over to Sutton Pool, and then around the Pool...." Here he traced his finger around her right breast and bent to kiss its nipple. "Thank God for Sutton Pool! And then it runs along like this over to the Catwater...." His finger moved lightly down her left side to her hip, and then farther down the inside of her thigh. "And then it goes into the Catwater and back out again...." Eyes closed, smiling, sometimes shivering with pleasure, Mary B followed the delicious path of his finger.

How far she and Will had come since their knockdown fight! Their physical wounds healed, their emotional wounds still tingled, and would perhaps never heal entirely. But how persistent Will had become! Coming to see Mary B almost every day, sometimes with fish, sometimes with flowers, sometimes with just an amusing story. Once when James happened to be there the two men had exchanged curt nods and stony names.

"Jim?"

"Will?"

The hut was silent for an interminable minute as James by the hearth and Will by the door studiously avoided each other's eyes. Mary B, Seedy, and Katie tried the kind of small talk that invites participation, but it ultimately expired like a fish dying out of water, several efforts of futile flopping followed by gulping silences. Mary B was left looking from Seedy to Katie to James to Will for help not forthcoming. Then with a sigh and a small smile and nod, she indicated it was all right for James to depart, which he did with his own pained smile, again exchanging curt nods with Will. Then Seedy and Katie found excuses to go, and Mary B and Will were left alone to deal with their future.

During those fumbling visits Will tried to regain a small measure of Mary B's trust. Another man might have sought recourse from the law, for a wife who left her husband could be flogged and sent back where she belonged, which had happened more than once in the brief history of Albion. But Will didn't stand very tall before the law, and if Mary B were the

242

complainant Will might have been the one flogged. So Will apologized repeatedly to Mary B for his drinking, his harsh words, and his bad behavior, and over the next several weeks courted her proper until she at last consented to return to their marriage bed. But not before he met several conditions.

1. He had to apologize to her friends.
2. He had to promise never strike her again.
3. He had to accept Charlotte like a good father would.
4. He had to apologize to the Governor.
5. He had to get his job back.

"Otherwise," she said, hitching Charlotte a little higher on her hip, "I'm going to stay right here with Seedy and Katie. I get my own ration and can make do by myself."

Subsequently, in celebration of the King's birthday, when the Governor forgave sentences and granted pardons, he reinstated Will as the colony's "official" fisherman.

But all was not the same. Each morning Will had to report to the Provost Marshall and tell him exactly what he intended to do, and each evening had tell him exactly what he'd done. And there was still James, who lived across the cove, but lurked just beyond the pale. Will was certain if matters got out of hand again Mary B would seek out James and not come back again. In short, Will Bryant realized his situation was rather precarious; he had to appear stable and trustworthy and live as if his soul were at peace - until the time was more propitious for him to escape from Albion.

And neither was Mary B the same. She'd grown up learning that things were meant to be a certain way. The world had a natural order made manifest a hundred times a day, a natural order preached in the gospel, written in law, and hallowed by custom. But since that fateful day in Plymouth four years earlier she'd endured one shock after another to her faith in the natural order of things and to her sense of what was fair. And not just in the general sense, but fair to her. She now had her own ideas.

Oh, she'd sinned, she knew. As a little girl she'd snitched from the poor box. She knew she'd done wrong plenty of times. But in the matter of Agnes Lakeman she was paying out of all proportion. And Will's treatment of her was wrong. She'd taken her marriage vows seriously, but after he knocked down with a stick of firewood and she hobbled off to seek protection from her marriage for the second time, she said, "Enough is enough!" Easy-going, fun-loving, trusting, passive Mary B realized then she must do a better job of looking out for herself.

What a fool I've been! she thought. If I was old or ugly or crippled he would've ignored me. I was just something he wanted to own, like a dog. She knew she wasn't beautiful. Oh, her features were regular, and her gray eyes were frank and friendly and could light up her face with a playful gleam. They were eyes one wanted to trust because they looked so honest. And her auburn hair looked good in the sun. But she was taller than average and had broad shoulders and her clothes were just convict issue and she could neither read nor write.

But she saw now it was the same for all the women. They'd no sooner set foot on Albion's soil than the officers were after them. Servants, they said. Housekeepers, they said. But from laundry tub and cooking pot to camp bed was but a short step, and plenty of convict wombs soon plumped with little officers and little officials. Mary B found herself complaining to the other women, "Does anybody really care, or is it just a matter of owning us?"

Liz Bason had disappointed her. "Lord," she said, "why stir that pot! That's just the way it is, and there's nothing we can do about it! You just try enjoy what you can!" Seedy was no help either. Thin to a fault, her eyes so deep-shadowed as to be almost unseen, her hair lank and lusterless, Seedy after Quint drifted with the tide, seeming not to care. Only Katie offered hope. Her hospital work was bringing fulfillment long absent in Katie's life. She had purpose and was happy. "Come work with me," she told Mary B. "It's so good to feel worthwhile, like I'm making a difference!"

But Mary B wasn't yet ready to abandon the idea of marriage and family. Even after James skittered away and Will's abuse, she wanted nothing more than to live in peace with a man - someone attentive and considerate, someone who was a good provider and would treat her with respect. And would be a good father. Too bad she couldn't roll James and Will into one! Will was a good provider, but didn't respect her. James on the other hand respected her and even cared for her a great deal. But how dependable was he? Oh, he took great pleasure in Charlotte. And he was always ready to help. And more than once Mary B caught James looking at her with such a confused expression of affection and longing and torment she felt like cradling his head in her bosom. He carried such guilt! And was in such a pickle!

"What a fool I've been!" he exclaimed one day when he stopped by to see how she was getting along. Mary B looked up in surprise and wondered what brought on his outburst.

"A fool, James? Why do you say that?"

His eyes roamed the fisherman's planked floor, a rarity in convict quarters. After a long moment he looked at her sorrowfully. "I don't know, but when I look at you and our baby, I just know I'm a fool!"

Poor James! Some men grow up simply wanting to fit in. They seek not greatness, or fame, or honors, or preferments - just a useful life of fitting in, a life that puts cream in the pitcher and children in church. That was all James really wanted, a simple life of cows and crops, of children and church. But born to a tenant farmer threatened by rack-rents, having to leave home at an early age to shift for himself, James had drifted, years of footloose existence Mary B knew nothing about. When he finally did risk commitment he found the responsibilities of marriage overwhelming. Perhaps that foolish act of picking up a sack of old lead and iron was a kind of wandering away again. Like Mary B, James began his convict life still unsure of who or what he was.

Aye, Mary B thought sadly as she regarded her friend's forlorn, downcast face. You are a fool, but at least now you see that. A fool, and a disappointment. But I was too.

But now it was Will she had to deal with, and whatever Will might think went on between her and James, she'd never given him grounds. And although too often it was for worse, she'd married Will for better or worse. And she was determined to keep her marriage vows. She and James were friends and former lovers. They shared Charlotte. And that was all.

What finally brought Mary B back to Will for the second time? Nothing special. No marvelous transformation that changed her from easy-going, fun-loving, passive Mary B to a new woman. Yes, she'd changed. She'd become more self-aware and would continue to change, but life alone with an infant was hard, and Will's persistence probably made a difference. One thing about Will, once fixed on a course he didn't veer. Perhaps her loneliness mattered too, and her desire to be done with a hard thing. Perhaps she was tired of struggling, of having to think through everything by herself. Then too, wounds become

245

scars, and strong feelings become memories. Time blurs. In any event, after all was said and done, Mary B's decision to return wasn't so difficult. It simply wasn't in her nature to remain critical and unbending, and she sensed that Will was genuinely happy to have her back.

"...and then it comes back out around the Sound and out around the East Head...." Here Will circled her left breast again.

"I thought that was the Eddystone side," Mary B giggled softly.

"Oh, no, was it? I must've got lost in the fog. Now I'll have to be very careful and go along the coast very slowly, into every little bend and bay so I don't get lost again." He continued to trace his fingertip and whisper and kiss and nip until Mary B drew him to her with a soft moan.

Her moan was not entirely the result of Will's foreplay, although perhaps that helped. For several days an inchoate, powerful urge had been building in Mary B, an urge that seemed to rise from deep within and demand fulfillment. Whatever Will thought, he knew that never before had Mary B entered so completely into union.

In the supine aftermath of their lovemaking, Mary B was stirred by a sudden spasm of the earth, gentler than had visited the colony soon after disembarkation, when a nighttime tremor frightened the jittery new settlers. She nudged Will. "What was that?" she asked.

"What?" Will asked sleepily. He'd felt nothing.

Next day when Mary B mentioned the spasm of the earth to others she found no one else who'd noticed. Perhaps the earthquake's epicenter was too distant to register much effect in sleeping Albion. Mary B shrugged off her experience as another curiosity. But several weeks later, when she recognized the first signs of her second pregnancy, she remembered exactly the time when it must have occurred, that night in mid-July, 1789, when she alone in Albion felt some distant, earth-shaking event.

~~~~~~~~~

Mary B was unaware that Captain Tench still dreamed of her, almost always after he happened to see her at the commissary or at Sunday service or on the rough path that rose and fell as it followed the shore of Sydney Cove between the fisherman's hut on the east side and the hospital on the west. Tench usually took a minute to chat, to chuck Charlotte under the chin, and to inquire after the family's health. For a while he thought Mary B looked grim, which he attributed to gossip that all was not well between her and the fisherman. But her season of grimness passed, and by late summer (rather late winter, for it was mid-August) he thought Mary B seemed a picture of health and happiness. That was just before little Nell, then eight years old, was raped by a Marine, an incident that involved Tench in Malvina's life for the second time.

Something of a naturalist, John White once observed that no matter how fierce or powerful the predator, it prefers a victim least able to defend itself. So too did Private Henry Wright. After his arrest for the rape of little Nell, it came out that several wives had suspected him all along of an evil predisposition toward children, and the women spent hours reconstructing the history of the past two years to determine when and how the short, stolid Marine might have pursued his evil ways. Half-forgotten instances of prolonged bed-wetting and nightmares and uncharacteristic thumb-sucking and sudden cries in the night became horrible hints of the dark histories of their little girls. "And what horrors has that devil worked on his own children?" some asked, for Henry Wright was a married man.

Although she steadfastly denied any knowledge of her husband's activities and was at first an object of pity, Alma Wright became a pariah. She must have known of her husband's evil ways! Even Malvina was held culpable. After all, she should have seen that slow-talking Henry Wright was a little odd, and she should never have let little Nell go off with him! Poor Malvina! She already carried guilt enough. The hanging of Dobber had seen to that.

The Sunday afternoon of the rape, Henry and his daughter Amanda and Nell went into the woods to pick flowers. After a time Amada grew tired, and Henry Wright sent her back with the flowers they'd gathered. He and Nell continued their excursion. He held her hand as they walked, lifted her over windfalls, and boosted her up by her bottom so she might better look for pretty blossoms. "Nell, shall we play?" Henry Wright asked when they reached the further slope to Cockle Bay. They'd stopped to rest, and sat on a low, rocky ledge.

At Henry Wright's trial Nell testified that she tried to wriggle free when he slid his hand under her dress and put it over her pee place, but he was too strong. She ceased her struggles in momentary puzzlement as he unbuttoned his front flap. His engorged penis looked enormous. She writhed and whimpered when he began to position her little body, but he had her trapped, one forearm under her neck, a hand forcing her little body over his bulbous member. She screamed in pain and childish outrage, but he clamped a hand over her nose and mouth. Her panicky struggles seemed to arouse him still more. Possessed in the act of possessing, his arms and legs moved spider-like as her small, sinewy body squirmed in protest. He smelled fear in the sweat of her hair and sucked the back of her neck as he thrust and thrust and too soon came with such a powerful, pumping, ecstatic spasm he thought he'd swoon, but in a couple minutes, still hard, he forced his bloody, greasy penis into her anus and soon came again with glassy-eyed pleasure.

As they made their way back to the settlement, Henry Wright held Nell's hand tight and told her that what they'd done was a secret little girls had to learn so they could be good wives. Now Nell could be a good wife. When she wanted to practice being a wife again she could whisper to him, but mustn't breathe a word to anyone, even to her mother, because if she told anyone the Devil would come to her bed at night and carry her down to fiery Hell for telling the secrets of grown-ups. He squeezed her hand very hard then. But Malvina found out that night when Nell woke screaming and bleeding.

Found guilty and sentenced to hang for raping Nell, Henry Wright was instead shipped off to Norfolk Island for the rest of his life. Arthur Phillip's reasoning was somewhat murky as to why he reduced Henry Wright's sentence; something having to do with Wright's unnatural appetites being rare and about convict women being easily available for carnal relief. In any event, Arthur Phillip decided the rape of Nell did not merit extreme punishment.

Tench was aghast. "With all due respect," he opined to Arthur Phillip in a moment of candor, "I think the rape of a little girl deserves at least as severe a punishment as stealing a little food." Arthur Phillip fixed Tench with a baleful eye.

~~~~~~~~~

Food! Arthur Phillip had sailed from England with enough food for two years, along with the sanguine expectation that the colony would be self-sufficient before it was all eaten. But as the end of 1789 neared, almost two years after the First Fleet dropped anchor in Sydney Cove, the colony faced a more melancholy situation: with its cattle disappeared, the sheep dead from slow starvation, and the crops failed again, the colony's Commissary Agent reported he had enough rations for only a few months. Convict deaths and disappearances had proved salutary, for there were about ten percent fewer mouths to feed, but some of the food had been ruined in passage, some had spoiled in storage, and a healthy population of rats in Albion ate their daily share. The barreled brine-meats, now several years old, were dry as wood, shrinking, and beginning to weevil. The last firkin of butter had been scraped clean three months earlier, and with no cows to produce milk, butter would not reappear in Albion until brought from the outside world. A year earlier, acting on the assumption that garden produce could supplant some of the rations doled out every week, Arthur Phillip reduced the flour distribution by a pound a week for men and a proportional amount for women. He also dispatched the *Sirius* to Cape Town to purchase new seed, for Albion's seed was mildewed and impotent. In a hurried circumnavigation of the globe, the *Sirius* returned bringing seed, badly needed medical supplies, and sixty tons of flour, a four-month supply. But the story of how the *Sirius* nearly wrecked off Van Dieman's Land on its return was a grim reminder: supply ships seeking the colony faced hazards aplenty.

In the 2½ years since the First Fleet weighed anchor in Portsmouth, Arthur Phillip had received no news from home but the chit-chat of a few private letters brought back from Cape Town by the *Sirius* - the King struck spittle-chin mad, the Dutch Stadholder restored and in alliance with England, France in foment. God save the King and all very interesting, but where was the Second Fleet? Where was Albion's food?

Anxious ears kept half-cocked for the signal shot that would announce a Second Fleet ship heard only the guns of the *Sirius* occasionally signal its work parties shore-side or the little *Supply* saying Good-bye and Hello as it shuttled between Albion and Norfolk Island. Where was their food, their tools, their livestock, their overseers and artisans? Albion's entire food supply could now be inventoried in a few depressingly small piles in one storehouse. Arthur Phillip ordered the weekly food ration cut by one-third.

Food? Self-sufficient in two years? The idea of Albion being able to grow its own food was a bitter joke. Of the hundreds of inhabitants housed in leaky huts perched on the hardscrabble verges of Sydney Cove, only the Reverend Johnson and a few others could point to any botanical success. Amateur gardeners who saw their seeds sprout and sent enthusiastic letters back to England with the returning ships of the First Fleet saw cloudbursts, giant ants, rodents, heat, drought, and thieving convicts reduce their once-promising gardens to ruin. Although a small acreage on Garden Island bore well enough, the large public gardens on the mainland were a bust. The soil simply lacked proper nutrients. Fortunately, Norfolk Island's deep, loamy soil was wondrously productive, and even after a rare hurricane flattened the tiny settlement, prospects for self-sufficiency looked better there than in Albion. For certain, Albion needed to find better farmland, but finding it was hard.

The difficult terrain and the lack of navigable waterways made inland exploration arduous. Despite several attempts, no one had yet reached a range of western hills called the Blue Mountains, just thirty miles distant. An intervening washboard of steep-sided gorges exhausted both men and animals. Every expedition returned defeated by the relentless up and down. So far the colony had discovered only one major river in the area, the Hawkesbury, which also proved a disappointment. Coursing through swampland or steep, rocky gorges for much of its navigable length, the Hawkesbury showed evidence that its floodwaters reached as high as forty feet, high enough to discourage any thought of settlement along its banks. Thus, after two years and several efforts to do better, Albion was still confined to the thin-soil environs of Port Jackson's salty blue waters.

Thankfully, Arthur Phillip had discovered a patch of better land a dozen miles farther up the estuary, where he sent a detachment of Marines and convicts to settle a place they called Parramatta, a name taken from the native word for a prominent hill that overlooked the area. Crops promised better at Parramatta. At least the ground was more open and fertile, and the surrounding countryside showed little evidence of native habitation. When the next convicts arrived in their hundreds, Arthur Phillip planned to send them to Parramatta.

Poor Arthur Phillip! Not only was his food supply dwindling, but his energy and spirits were dwindling too. Now age 51, his strenuous explorations of the wilds to seek better farmland had aggravated an old kidney stone problem. Most mornings he rose stiff and achy from a sleepless bed, where during the long nights he stirred the myriad problems that beset him. They were ever the same: lazy, thieving, conniving convicts and quarreling soldiers and hostile natives and a harsh, unforgiving land and failing crops and no word from home and....

More and more in the dumps, he found himself thinking of his wife, or rather, of his wife's death. Although not sure whether she was alive or dead, she was dying when he left, and was probably dead, and he felt bad that he'd never be able to make amends to her. Never ever. That was the thing. He'd simply huffed off in a petulant snit after an exchange of words and never gone back. Now he realized he should have returned to make amends; he should have bid her a decent farewell. He should even have brought her a little present, told her she was a good woman and that she'd made him happy. He should have done *something* to ease her departure from this world! But as he lay abed in pain and remembered his good homecomings and their laughter and lovemaking and her crinkle-eyed smiles in the morning, he felt like a heel. "To my beloved Arthur" she'd inscribed on his little hourglass. How often when he scribbled at his table he thought of the obverse, *Omnia Tempus Revelat!*

Had it really made a difference that he failed go back to say good-bye? Did she die sooner because he hadn't? Does an act not performed make a difference? Or is that just an idea?

Consider James Freeman, still Albion's hangman after two years. Although sentenced to hang, he wasn't hanged, and for certain his not being hanged wasn't just an idea to James Freeman. With his neck in a noose James Freeman would argue with strong conviction that an act not performed is a reality. Not hanged, James Freeman has potential. He can vow to live an exemplary life. Not hanged, all things are possible for James Freeman. But once hanged, what's done is done and dead is dead. Depend on it.

Such speculations about such consequences might entertain Jeremy Bentham, a budding moral philosopher back in London, but to James Freeman the issue was more immediate. As the colony's reluctant hangman, how could he comfort Ann Davis, a convict woman sentenced to death for theft?

"Well, at least you got your full ration to the end," he finally said, for he put her to death on November 1, 1789 the day Arthur Phillip cut everyone's rations by a third.

251

Despite his nocturnal tossing and turning, Arthur Phillip continued to fight the good fight. His vision would not die easily of fruitful fields and neat cottages row by row.

Item. He ordered the *Supply*'s carpenter to construct a small boat to shuttle between Sydney Cove and Parramatta. The result was a launch badly built of dense, native timber. An ugly thing, it rode dangerously low, moved like a log, and almost immediately came to be called "The Lump." But it was progress!

Item. With Marines refusing to serve as police, Arthur Phillip established the "quarter-guard." (It was actually a convict who suggested that he divide the settlement into four quarters and appoint some of the best-behaved convicts as constables). With a cudgel as their staff of office (handy for threatening malefactors), Arthur Phillip empowered the convict constables to detain anyone who threatened the peace, acted suspiciously, or was caught in an illegal act. Within a few weeks Albion was experiencing fewer thefts, ewer disturbances, and fewer complaints. But of course Major Ross complained. Ye Gods! was the world turned upside down? Convict constables detaining Marines? But it was progress!

Item. Arthur Phillip tried once again to communicate with the natives. The sad truth was, relations had become so sour that anyone venturing unarmed outside Albion's perimeter was in danger of being attacked and murdered. Although frightened of the Marines and their guns, if the natives chanced upon anyone without a weapon they threw stones and spears with deadly accuracy, as if the only good white was a dead white. They never sought to parley, never took a captive, and sometimes mutilated their victims. This state of affairs vexed Arthur Phillip. After every fresh incident he struggled for a solution. Finally, he dispatched a young Navy officer to kidnap more natives.

"We've got to learn to talk with them," Arthur Phillip said, "If they always run away we can't learn to talk with them."

So in his blue Navy uniform (not an object of fear to the natives, like the red coats of the Marines) young Lieutenant Bradley rowed down the estuary with a small party until he spied a couple of stalwart-looking natives, enticed them with a proffered gift of fish, knocked them down, dragged them into his boat, fended off their friends with a show of arms, and took his captives home. Albion was becoming quite adept at such operations.

The captives were taken amid a hubbub of gawkers to the Governor's compound, where each was fitted with a manacle around his ankle and put in the care of a convict. After ropes were spliced to the manacles, the convicts were given their instructions: "Let go of that rope and you'll feel the lash." For his part in the kidnapping, Lieutenant Bradley wrote, "*It was by far the most unpleasant service I was ever ordered to execute.*"

But it was progress!

~~~~~~~~~~

"Songs? You sing songs?"

Mackesey knew his fate was sealed. The jowls of the fat magistrate quivered with indignation as he snorted his contempt at Mackesey's explanation of why he was in Somerset without kith or kin or means of support, other than for his little guitar.

"Well, sir!" the magistrate continued, "We need no songs here! We need no Irish vagrants to pick our pockets while they sing their songs! No, sir, we need you not!" Titters caused the magistrate to raise his eyes in warning before he bent with a hissing belch to paw through papers. "What an assembly of clod-pates, grog-blossoms, numbskulls, and buffleheads...."

Mackesey shifted his feet to ease the weight of the unfamiliar shackles. No matter that he'd never picked a pocket; to be Irish and in the wrong place at the wrong time was wrongdoing enough for this press-court, and enough for the other victims too, landsmen all, rounded up to serve His Majesty's ships.

"And you have no family, even there? Did they disown you?"

Mackesey knew nothing he said would make a difference. "No family," he answered. "They're all dead, Excellency." In his mind's eye he saw the stone he'd chiseled and set outside his father's little schoolroom, *Memento mori*.

"All dead? How all dead?" The magistrate still pawed his papers, half-paying attention.

"Ague...and other things, Excellency."

"Well, then," the magistrate said, having now found the paper he was looking for. "You shall have a new family, Mr. Mackesey, a family of shipmates. I'm sending you on board the *Pandora* in Portsmouth. It is a noble enterprise, sir, for you will bring the *Bounty* pirates back to justice, and you will serve your King and Country. You can sing songs on the *Pandora*, sir!" Fixing Mackesey with a stern look, the jowly magistrate pronounced, "I remand you to the immediate custody of Captain Edwards of the *Pandora* until discharge."

Instead of using a gavel, he slapped his palm on the table to signal the end of the case, and thus alerted the gods of chance to continue the fateful joining of Mackesey and Mary B.

~~~~~~~~~

Food? Self-sufficient? Arthur Phillip's cutbacks angered Mary B. She was hungry! Five months pregnant, Charlotte between breast and bowl, Mary B had seen her portion shrink to two-thirds of two-thirds, less than half of what a man normally got! Damn the Governor! She was of a size with many men, and worked just as hard too! Oh, she was angry! And angry with Will too, who was already angry on this Christmas Eve of 1789.

"Damn your eyes!" Will swore, "You didn't have to take it!"

"What do you mean, I didn't have to take it! I'm hungry! I'm hungry all the time! And Charlotte needs better food! I wanted it!"

"You didn't have to take it! Not from him!"

"Him! Him is not him, Will Bryant! Him is the father of that child sitting there, and he wanted to do *something*. After all, it's Christmas!"

"Well, you didn't have to take it!"

"All right, I didn't have to take it! I won't take it! Here," she said, thrusting the rush bag at Will's chest. He wouldn't raise his hands to receive the vegetables and flour. "Here!" Mary B challenged, "Take it back! If you're so cock-sure you'll find us a better Christmas dinner, take it back. Here, I don't want it! Take it back!" she said, bullying Will backwards. "Take the damn chicken too! Take it all back!"

"Back off, woman!" Will threatened, his hands twitching.

Tied by a leg to the table, a young rooster craned its neck to cast an inquiring eye at the scene above, then went back to scratching and pecking at the rough plank floor.

"All right! I'll back off. I'll back off right over here to the table. And I'm going to start making something to eat. And I'm going to eat it too!"

"You didn't have to take it! I could've done something! How was I to know you wanted something special?"

"This isn't special, Will, it's what we need!" Her voice dropped. "Between us we just don't get enough. I'm hungry, Will, all the time. That's the way it is when a woman is making a baby. We've got to have more. Remember how Quint stole for Seedy? It's not because it's Christmas, Will. It's what we need all the time, and one good dinner won't do either."

"Well, I can't do anything, so don't nag! The ration's the ration. I don't like it either, but what can I can do!"

Mary B gave him a fish-eye. "You could've taken care of the garden. Just because I wasn't here you didn't have to let it go. It was still our garden!"

"Our garden? It's your garden, damn it! I'm a fisherman, damn you! That's woman's work!"

"Oh, pardon me, m'lord, I forgot how mighty a fisherman you be! And are you the same fisherman who told me on the *Charlotte* how well off I'd be, how you'd take care of me? I don't remember you telling me I'd have to keep my own garden to get enough eat, or did you tell me and I just didn't hear?"

"Oh, shut up!"

"I won't shut up! Why should I shut up? This is your baby too, and just because I carry it doesn't mean you don't have a duty."

"I know my duty."

"What? What's your duty!"

"What I do."

"Hmpf! What do I say to that! What you do! Psshh!"

They stood faced off, Will near to striking her. Charlotte sat a plump lump on the bed, close to tears, her suckling lips a pout of grave concern, looking intently at her mother's flushed, angry face. But it wasn't Charlotte that restrained Will's quivering rage. And certainly not a promise never to hit her again, for after a few months of relative tranquility Mary B's announcement that she was pregnant triggered the familiar smoldering anger she feared might flare up again. Perhaps what held him back was the noticeable swell of Mary B's belly, his own forming child.

But Mary B was unsure sure how Will felt about the baby. She thought the baby might give them something to share, something to bind them as husband and wife, perhaps even heal their emotional wounds. "Don't you want the baby?" she asked when she first noticed his grim silence. Will's snappish response surprised and confused her.

"Of course I do!"

"Well, what is it then? Is it something I did? Something I said?" Will shrugged and turned away. What could he say? To say he wanted the baby was a lie. He desperately did *not* want the baby, or anything else that might complicate his plan to get away. After all, if it was just Charlotte he could slip away with a clear conscience. She was James's brat and James would be sniffing around before he was even clear of the headlands. But knowing Mary B carried his baby made things harder.

Will's shoulders fell. "Ah, let it go. Let's have a nice Christmas. Jim's all right. Besides, the Second Fleet will be here next month and we'll all get back on full rations." From habit he raised his hands to the flickering flames and flexed his fingers.

"Of course, sweetheart," a relieved Mary B said. She stepped behind him and put her arms around his chest. She kissed the back of his neck. "We'll have a nice Christmas dinner with the chicken, and before we know it the Second Fleet will be here with new food and new clothes and new faces."

~~~~~~~~~~

On that same Christmas Eve, several thousand miles distant, the last of the crew and convicts on the *Guardian* struggled for their lives. They'd already cut the throats of the horses, cattle, and other livestock and pushed the expiring animals overboard. They'd also cast the remaining belongings of the passengers into the icy water, along with tons of personal goods destined for the Marine officers and officialdom of Albion, for the *Guardian* was a special ship bringing relief to the colony.

Late in getting the Second Fleet away, England dispatched the *Guardian* to bring Arthur Phillip's supplies for the colony's third year. Its just-in-time arrival before the end of January would fill Albion's warehouses with more than a half-million pounds of flour, salt beef and pork, would bring bales of clothing and blankets, new shoes, medicines for the hospital, tools, wine, potted fruit trees - in all more than 1,000 tons of bounty. It would also bring two trained agriculturists, a second preacher, convict artisans, overseers to make sure the convicts worked, paper and writing supplies, books, letters from home, and news.

A converted 44-gun frigate, the *Guardian* left Cape Town on what thus far had been a swift and uneventful voyage. But the past 24 hours had turned ugly. Encountering icebergs, the ship soon found itself surrounded by dense fog. As night fell, huge swells lifted and dropped the ponderous ship with indifferent carelessness. The watch was tense. Eyes strained into the white darkness. Again and again an imaginary iceberg evoked sudden alarms and running footsteps, only to dissolve in swirls of fog. Every four minutes the bow gun fired, a fog-muffled bang whose echo might announce a mountain of ice that loomed dead ahead. Hopefully not.

But the iceberg the *Guardian* struck just before midnight had not loomed. It rode the swells as a surface-level shelf of ice that showed only intermittently, and the hole it stove in the bow quarter sent a torrent of frigid water into the hold.

Sydney

1790

*P*ease *porridge hot, pease porridge cold*
Pease porridge in the pot, nine days old
Some like it hot, some like it cold
Some like it in the pot, nine days old

Astride Mary B's knees, her hands guided through the motions of the rhyme, Charlotte chortled her pleasure. "Mo!" she demanded.

"Mo?" Mary B laughed, "You want mo?" She tickled Charlotte's stomach, "All right, little piggy, mo' porridge! Here we go! Ready?" Charlotte tensed with rapt anticipation. Again moving her daughter's hands, Mary B repeated the little rhyme. "Mo!" Charlotte said again, gripping Mary B's thumbs and trying to clap her mother's hands, bouncing her bottom on Mary B's knees, "Mo!"

"No more, you little piggy! You ate all the porridge up. All gone! Porridge all gone!"

"Mo!"

"No, baby, no more!" Mary B smiled. "Get down and play now. Mama wants to talk to Auntie Seedy." Mary B bent awkwardly over her pregnancy to help Charlotte slide down. Seedy's fond gaze followed Charlotte's determined toddle to the shade, where she plopped down and began filling an old clay cup with dirt while babbling nonsense. Seedy smiled at the scene.

"Will you have more tea?" Mary B asked. Seedy studied the remains at the bottom of her cup, shook her head no, and placed the cup on the ground with hands and fingers so thin her bones seemed visible through translucent skin. When she looked up, her wan, gap-toothed smile looked skeletal in the mottled sunlight. But the smile belonged to a childhood friend, and when Mary B saw those familiar, canted features wasted from months of self-neglect, haggard with Seedy's heavy sadness, her heart swelled with compassion. Seedy had come with a cutting of her geranium in a greenware pot as a memento, a token of friendship, a farewell gift, an acknowledgement that after several years of sharing the incongruous consequences of that carefree outing in Plymouth so long ago, their lives were moving in different directions.

The two women sat on crude chairs carried out from the fisherman's hut on a warm, pleasant afternoon in late February. Across the placid waters of Sydney Cove the diminished figures of a convict gang moved barrows of brick and material for the new dispensary, the strain of their toil lost in the distance. The pace had slowed on Arthur Phillip's public works. After living on two-thirds rations for almost five months, the convicts had little stamina to shape the raw stuff of Sydney Cove into buildings and roads.

The two women had consumed several cups of sweet-tea as they revisited old times and laughed at Charlotte's antics, but then a long silence descended, as if both knew the time was at hand for something important to be said. At last Seedy took an object wrapped in oiled paper from her rush bag. "Mary B, I want you to have this," she said, leaning to offer the gift. Mary B looked puzzled as she reached for Seedy's parcel. "It's Quint's book," Seedy explained.

Mary B froze. She began to shake her head no, but Seedy jiggled the book at her. "Mary B, listen to me...." She continued to jiggle the book insistently, saying, "Listen to me, listen to me," until Mary B reluctantly took the gift in her hand. Seedy folded her hands in her lap and sat looking at them. She sighed several times. At last she said in a quiet voice, "I know I'm never going home."

When she heard Mary B's gasp, Seedy raised a hand, "No, no, Mary B, it's all right. I've thought a lot about this and I know it's true. I'm never going to see England again." She paused, flustered at the sight of Mary B's suddenly brimming tears. "It's all right, Mary B," she pressed on, "There's nothing for me to go back to anyway, and once I'm sent to Norfolk Island I may never see you again, but I wanted to give you this one thing before I'm gone." She paused, then swallowed hard. "It's the only thing I have means anything."

Mary B caressed the book as Seedy continued with a self-absorbed, approving nod, "You're my best friend, Mary B. You've always been good to me and I know you'll get home some day. You're strong, and you'll make it through this misery. You'll get home with Quint's book. I know you will."

Her voice dropped. "I know you will." Her eyes fell. Her unspoken resignation to her own fate hung in the silence.

Mary B absently stroked the oiled paper with a finger as she tried to think of what to say. Seedy's had been a hard life. Born with a touch of dullness, marked by a crooked body and a crooked face, she'd lived with the consequences of her handicaps without complaint. From the cruel teasing and mean tricks Mary B so guiltily remembered, from the succession of boys who made easy use of her ripe, young body, Seedy had lived on the edge of acceptance. Too soon had she learned resignation, and too easily had she given herself up at the most cursory promise of friendship. Only later did she learn that by cooing and flattering and drawing back, by not being so easy, she could sustain a man's interest. But never for long. Although her distortion intrigued some men - her crooked trunk, her crooked face, the novelty of her delicate tongue in a broad, gap-toothed mouth - when the bother of getting her seemed greater than the reward, their attention wandered.

But by the miracle of Quint, Seedy had come to know a wondrous, brief joy, a genuine love, an extravagance of rapture that built through weeks of anticipation when they were kept separate on the *Charlotte*, a love that when finally consummated demanded consummation again and again, that sent the balance beam of Seedy's happiness plunging to the stop with an excess of good fortune. Few are blessed to know such love. Although they lived in hardship, hardship mattered not. Each found in the other all of consequence. But just as life with Quint brought joy and fulfillment to Seedy, so his death left her utterly bereft. Life no longer mattered. Without Quint the balance beam of her happiness plunged to the other extreme, and there was Seedy stuck.

Mary B wondered if Seedy was really ready to let everything go, if she was really certain her life would remain empty. After all, there was always tomorrow. And who knew what tomorrow might bring? "Are you sure you'll be going," she asked. "Are you sure you won't be coming back?"

Elbows on her widespread knees, chin in her hands, Seedy studied the remnants of her shoes, their soles bound on with twine the women fashioned from a tough, viney plant that grew in abundance. With their shoes disintegrated beyond repair, most convicts and Marines now went about barefoot. Seedy nodded. "Katie says she heard my name is on the list. And I know I'll never come back."

Although nothing official had come down from on high, everyone knew of the Governor's latest plan. Assured he would always have at least a year's food in reserve, Arthur Phillip had expected re-supply ships long ago. Now that relief was so late, who knew when (or if) ships might arrive? War, disaster, accident, or miscalculation might have cut the long lifeline between England and Albion. With the colony's food running dangerously low, lookouts at the head of the harbor scanned the empty horizon each day with hope they'd spy a supply ship in the distance. But each day ended with disappointment.

Earlier that month Arthur Phillip had summoned Major Ross. "Listen," he said, "It's only prudent to assume our relief ships won't arrive, so I'm transferring half your men and a couple of hundred convicts to Norfolk Island. The land's more fertile there. And if worse comes to worse the boobies are due back, so they won't lack for meat."

As Major Ross stood to go, Arthur Phillip said, "And I'm sending you too." Ross darkened. He opened his mouth to protest, but Arthur Phillip silenced him with an upraised hand. "Look," he said, all reason, "With half of our population there the Lieutenant Governor should be there too."

Ross was livid. To be uprooted from his house, second only to Arthur Phillip's in size and comfort, to be shipped off like a common felon to that God-forsaken dot in the ocean - this was too much!

"Prepare to leave by the end of the month," Arthur Phillip said, bending to shuffle papers, thus signaling an end to their discussion.

Ross made a sound in his throat, a choked growl. He glared for a long moment at the shine of Arthur Phillip's fringed pate, but the Governor purposely ignored him. Finally, his chin quivering with indignation and rage, Ross spun on his heel.

Charlotte did the splits, got her hands and knees under her, pushed to her feet, and stooped to pick up her half-filled cup. She toddled toward Mary B, bearing the cup. Midway she tripped and fell, spilling most of her dirt. She pushed herself up again, retrieved the cup, and completed her journey. "Tea," she said solemnly as she offered the cup to her mother. Mary B smiled her pleasure, and with a wink at Seedy pretended to drink.

"Mmmm," she said, "Good! Thank you! Will you give some to Auntie Seedy?" Charlotte looked in the cup, then at her mother, then again in the cup, and made her way to Seedy. Gravely she offered the cup, which Seedy took with a smile, pretended to drink, and repeated Mary B's praise and thanks. After Mary B drank once more, she gave Charlotte a good-natured pat on her bottom and sent her back to her play-place.

"She's such a fine little girl, and such a pleasure to watch," Seedy observed. "She's grown from a little nothing into a little person. Aye, talking now and all, like a flower opening up."

Mary B nodded, feeling proud, and they sat watching the little girl painstakingly scoop more dirt in the cup. Mary B thought: To her the cup must seem so big. Each handful she picks up is no more than a spoonful, and half of it dribbles away before she gets it to the cup, yet how patiently she works, as if her task were the most important thing in the world. Mary B wondered, does she find this little kind of work a pleasure? Is her intense little work a joy, or is it just life, a way of passing the time?"

As if their unspoken thoughts had been a conversation between them, Seedy mused, "Do you suppose she knows the times are bad?"

Mary B glanced at her friend. Coming from Seedy, who had no one but herself to care for, the question surprised her, because Mary B had pondered the question also. How aware was Charlotte of their hard life? Because of the shortages, Mary B regularly denied herself and Will to provide her little girl with something extra. Now teething, Charlotte constantly wanted something to chew on, and Mary B was always fishing slobbery stones and twigs from her mouth. But getting enough food was a daily worry, and as she swelled with her new baby Mary B's own appetite also made demands. Happily, James helped every week by sending over vegetables from his garden and a cup or two of flour. "It's all right," he said one day at the commissary, "I want you to have it for Charlotte. You don't have to tell Will."

Mary B hoped Charlotte had never known hunger for more than a few hours. Nonetheless, feeling defensive she asked Seedy, "What do you mean?"

"Oh, I don't know." Then Seedy gestured towards Sydney Cove and its surround of a struggling, half-starving colony,

"But if this is all your little girl has ever known, does she know life could be a lot better?"

At Seedy's hint that Mary B had somehow failed her daughter, Mary B felt a little peeved. Charlotte's life was as good as Mary B could make it! She started to protest, then held her tongue when she remembered how Seedy sometimes misspoke. As she watched Seedy watch Charlotte at play, Mary B realized Seedy was only asking the question of Charlotte's life she'd come to ask of her own. Having known only a hard life, Seedy had accepted her lot as the way life was meant to be. But Quint had brought her fulfillment and real happiness for the first time. Now reduced to her former state, Seedy felt the profound loss, and did not want to endure a return to the past.

Mary B had realized months earlier there was nothing more she could do for Seedy. Juggling her own baby and her marriage, Mary B had nursed Seedy, comforted her, counseled and encouraged her, had tried to interest her in projects, had even raised the prospects of another man - certainly not to replace Quint, but as a comfort. But Seedy wouldn't budge. She pined over her losses and blathered on about Quint until even her friends could hardly stand to listen.

"My God," Liz Bason complained, "He wasn't a saint! You'd think the way she carries on he was Jesus Christ himself!"

"Oh, Liz, we know he wasn't a saint," Mary B said, "but he was the only man who ever truly loved her, and that makes him a saint to her."

Now Mary B chose to change the subject. "I hear Norfolk Island is a healthy place, so much better than here. You might feel better there, Seedy. You might be better off there."

Seedy sighed. "Yes," she said noncommittally. She looked across Sydney Cove at sunlight gleaming on the thatch roofs of the convict huts. "But I suppose we'll be living in tents again. I never thought this place would look good to me, Mary B, but the thought of leaving you and Charlotte and Katie, and all of you...well, it makes this misery seem like home. I'll find it hard to leave Sydney."

Sydney? Did Seedy say Sydney?

Aye, for Arthur Phillip's Albion had not stuck. Albion simply didn't sound right. In their first letters back to England nearly everyone called the settlement "Port Jackson" or "Sydney Cove." No one used "Albion," and within a few months the conversational name for their new home had become "Sydney." Before long even officials were referring to "Sydney." Thus, from Port Jackson, to Albion, to Sydney Cove, to Sydney - a rush of linguistic change. Now the people used "Albion" only sarcastically, as in "Well, ain't things grand in Albion!" In queue at the commissary one day, Mary B had laughed at a topical verse declaimed by a convict.

How sweet is Phillip's Albion,
This land of gentle clime,
Where labor's lot is two-thirds fare,
And hunger fills the time;
Where naked neighbors bearing spears,
Put out their hands for bread,
Where Phillip gives them our small share,
And pats them on the head;
Where every dawn new hope doth bring,
To hungry convict huts,
Where every night we soothe with sleep
Our empty, growling guts.
Oh, let me home to England soon,
Though far across the sea,
For if I stay in Albion,
I'll ever hungry be.

But despite his claim to the name, Arthur Phillip wasn't disappointed that Albion failed to stick. He saw soon enough that Albion bore little resemblance to the Pacific Cockaigne of his daydreams in England. Where were his golden fields of ripening grain, his neat rows of happy-hearth cottages, his fatted sheep grazing beside sweetwater streams? Fatted sheep, indeed! He still ground his teeth when he remembered Albion's first case of sheep-theft, his own fatted animal! He'd planned to kill a young ram for the colony's first celebration of the Prince of Wales' birthday, but some bastard stole the animal, the only one they could spare! Argh! Convict thieves feasting on Arthur Phillip's own fatted ram! Infuriated by the theft, he offered emancipation to anyone who identified the culprit. But in vain. No one snitched, proving there really is honor among thieves.

Seedy sighed. "Well, I'd better be going. You'll be wanting to get Will's dinner." Charlotte looked up when she heard the change in conversational tone. "And yours too, you sweet, little piggy!" Seedy said. She stood and in two quick steps swooped Charlotte up with surprising strength. She kissed her several times. "Ummm. Sometimes I'd just like to eat you up, you're such a sweet, little piggy! Yes, I would! Will you give your Auntie Seedy a big kiss?" Charlotte ducked her head. "No? Won't you give your Auntie Seedy a kiss? No? Please? Just one?"

Despite Mary B's own entreaties, Charlotte would not be persuaded, and Seedy finally slid the child to the ground with a resigned chuckle. Seedy tugged her threadbare dress down on her hips, straightened the kerchief she wore as a shawl, and put a hand to her lank hair. Legs apart, she bent over and surveyed herself. "Look at me!" she exclaimed in self-mockery, "From head to foot, rags and bones, off to a new life!"

"Oh, Seedy, you aren't rags and bones! You're beautiful!" Mary B squeezed Seedy's hands, kissed her on the forehead, and looked into her friend's tear-filled eyes. Seedy swallowed several times with an averted face as she struggled with her emotions. Finally she raised her hands to Mary B's cheeks, nodded, and kissed her softly on the lips. She put her head on Mary B's shoulder and rocked. "Thank you," she said in a voice so small that Mary B scarcely heard the words. After a few moments Seedy pushed away and turned to go.

She'd taken several steps when Mary B called, "Wait, Seedy! I'll walk with you! I'm not going to let you go off like this!"

Mary B picked Charlotte up to ride on her hip, and the women exchanged smiles as their arms encircled each other's waist. Heads together, laughing softly with the pleasure of these last special moments, they strolled side by side down the path towards Sydney.

~~~~~~~~~

Poor Tench! Too bad he was such a loner. Sydney might have seen a new Watkin Tench step ashore from the *Charlotte* - more generous, more open, more honest with himself. Imagine! A new little Watkin born that afternoon on the bed above Mynheer van Jaarsveld's sitting room in Cape Town! For when John White kissed the tearful captain's forehead, when Tench in turn kissed the surgeon's hands, he might have let the moment develop. He might have plumbed those emotional depths that boiled up in his dreams. But, no, fearful of confronting himself, he wiped his tears and broke the communion with an embarrassed "Ahhh" and retreated to the good ship *Narcissus*.

Luckily, Private Willard Thistlethwaite's death delivered Tench a timely diversion in the person of Hester, a diversion that saw Tench safely to Botany Bay. And once he disembarked in the new land, the excitement of discovery and the work of settlement soon crowded out his disturbing dream memories. Before long he came to see his oceanic preoccupations as silly. To construct stage-drama metaphors on the *Charlotte*, to meditate on the origins of a piece of wood in his cabin, to fuss over dreams - why, that was the fancywork of idleness! Dreams were but vaporous visitations, febrile works of the imagination. "A dream isn't real," he told himself, "not at all!"

But despite his denials, sometimes his thoughts returned to those almost palpable memories of melting bodies and lingering lips during his illness on the *Charlotte*. Bird shadows of doubt fluttered near. Was it memory or dream, was it real or not?

But Sydney was real, and when Tench dreamed in Sydney his dreams fled on waking, and he often couldn't say whether he'd dreamed at all. Reality was daily life and demanded his complete attention. Once he was done with his book, his soldier duties, setting up his household, managing his place in the rivalries and enmities of battalion politics, and his explorations of the wilderness - such business filled his days.

Hacking out a place for civilization in this second New World proved restorative for John White also, and after a few weeks of petulant avoidance the two shipmates and roommates resumed a regular and comfortable society. Add to these satisfactions the domestic comforts he enjoyed through the efforts of Hester Thistlethwaite, and Tench saw his life replete with manly pleasures.

Hester lived in a tent pitched near his own, employed as his cook and laundress. That she continued as his periodic bed partner was a private matter, of course. Policy entitled each of the privileged class to two male convicts to tend gardens, look after the pigs and fowl, cut wood and fetch water, clean boots, catch fish, and do whatever might need doing. But a few like Tench chose not to keep male servants. A few kept convict women instead. And no one said a word.

Despite their difference in station, Tench and Hester enjoyed each other's company. He might playfully call her "wife" and she rejoin with "husband," but both knew they were only marking time. When she returned to England (which Hester thought she might do when the time felt right), she knew her chances were slim of finding much of a match the second time around. She was, after all, used goods. On the other hand, if life in Sydney was hard, the prospects were better. So far as being looked after through the kind offices of Tench, well, Hester had no illusions about how that was going to end. It was only a matter of when. She also knew she'd never enjoy the good fortune of actually marrying an officer or snagging a sea-borne trader bearing bags of gold. No, any husband for her was sure to be a soldier or sailor or perhaps in the more distant future an emigrant settler. That she would ever consent to marry a convict never crossed her mind. For the nonce, then, and until marriage became more of a necessity, the compressed strata of Sydney society offered a unique opportunity for Hester to live (and sleep) cheek-by-jowl with her wiggy better, and if her future was vague she wasn't particularly concerned. She'd enjoy this interlude as best she could and, as she said, "Let weightier matters wait."

And so they did. When at leisure under the same roof (albeit a tent roof), Tench and Hester dallied in gossip, shared an occasional meal, desserted on pleasant sex, and later drifted off in snuggled sleep. When Hester rose and hopped about in the chilly pre-dawn to mend the fire for morning tea, sometimes having to fish out an ice-pane from the water bucket, she followed a ritual of presenting the pane to the fading stars - an ephemeral offertory - before laying it carefully on a clump of grass to melt in the morning sun. Serving a breakfast of toasted bread and tea with sugar, she became aware of their relationship shifting as Tench dressed, for his clothes effected a visible metamorphosis. Tench the lover transformed into Tench the Marine officer. Even in breeches and freshly blackened boots he became more remote, and when she helped him into his coat and smoothed the shoulders and back, then patted her own hair in place and dropped her hands, the connection was broken and the separation complete. He stood before her as her better and master. Hester handed Tench his hanger and hat with unaffected deference, while a small part of her knew such deference was wrong.

Hester wasn't sure whether it was Tench's new hut or Dobber Thornapple's death that brought their pleasant times to an end. For certain she felt hurt when he finally moved into his new hut and chose not to invite her to share his space. Had he asked, she might have declined, or perhaps stayed only for a little while. But not to be asked at all was belittling. Was he tired of her?

To protect herself she began to withdraw in small ways. She began carrying out her duties perfunctorily. She spent more time away, found fewer laughs to share, and showed less enthusiasm in bed. Although he must have noticed the differences, Tench never mentioned the growing gap between them. Perhaps he wanted the affair to end anyway. In any case, after Dobber's death Tench began showing Malvina Thornapple what Hester called his "phony sympathy."

From her own experience, Hester realized Tench was an opportunist taking advantage of Malvina's vulnerability. She wondered if the attention Tench showed Malvina belied an insidious, deceptive game. Did this remote, charming Marine play at seeing how long it took to convince a fresh widow to spread her legs for him? Whatever the case, as soon as Dobber was dead Tench began paying special heed to his grieving widow.

But to be fair, perhaps Tench felt guilty about his own part in bundling Malvina off to this land of Dobber's sorry end. Or perhaps he felt sorry for the children. And who can say the captain's attentions were a dastardly manipulation? Perhaps they resulted from natural sympathy, or *noblesse oblige*. When Malvina finally did end up in Tench's bed, who can say it was a conquest? Or whose conquest? Perhaps they became lovers in a natural progression of their emotions, devoid of manipulation. True, Malvina wept and moaned and carried on in bed with what Tench felt was a rather odd, salty, clinging grief, which provided him the opportunity to be a comfort while slipping it to her, a juxtaposition of performance he found uniquely satisfying. But Tench didn't actually bed Malvina until several months after Dobber was hanged, until after Hester Thistlethwaite had more or less gone her own way. In fact, it wasn't until after Nell was raped, and Malvina's world literally collapsed around her.

After Dobber died of hemp fever (a gallows-humor term common in Sydney), Malvina must have wondered if life could get any worse. She was destitute. And she would get no widow's benefit, for why would the Service pay off a criminal's widow? The 18th Company chipped in, of course, and for a while Malvina bore up like a trooper. She girded her loins, put her

268

shoulder to the wheel, and worked her fingers to the bone to bring home the bacon. At times sounding semi-hysterical, she lectured Ned and Nell: "Heavens, we can do it! If we just put our minds to it we can do it! We can do it if we just try!" Bartering the last of Dobber's clothes, she hired convicts to clap up a little wattle-and-daub, and began soliciting day work from those with means. From first light to early night she and Nell washed and mended, cooked and baked, scrubbed and polished.

Tench was only one of several that Malvina and Nell labored for, but it was with averted eyes that Malvina arrived to serve the captain. Tench was special. Unlike those whispering wives who blamed her for Dobber's death ("'er bottomless gut's 'ut killed 'im, 'at's 'ut!"), Tench's solicitations were unreserved and gracious, and she came to believe that in Cape Town he'd only done what his duty demanded. After all, that was the way men thought. He had no choice but to kidnap her! And although he'd deceived her, she came to bear him no grudge. Quite the opposite. With the passage of time her memories of the elegant captain bowing as he sought admission to Madame Labonne's house, of his listening to her chatter in the kitchen with a hint of appraisal in his eyes, of promise, of his arms around her in a gesture of comfort - these memories became so selective, so gossamer and distorted, that in the bleakest hours of her life Malvina fashioned wonderful fantasies from Tench's manipulation and betrayal. And now that she actually served him she imagined Tench growing fond of her. She imagined him asking her on bended knee to marry him. She imagined his manly arms taking her and her children under his protection. It was only a matter of time.

Even Ned helped feed Malvina's hopes, for she found after Dobber's death that she couldn't keep the boy in line without Tench's help. Instead of helping with the work, the boy would sneak off to watch the soldiers drill, tag along to the brick kilns, ride down to Rushcutter's Bay in a boat going for thatch, or hang around the wharf or hospital or commissary. Ned worshipped Tench, however, and if Tench told him to get to work he got to work. But often as not when Ned faced odious chores Tench would wink at Malvina and take the boy with him on some captainly duty. And mother and daughter would be stuck with the Ned's chores too.

After Nell's rape it was a violent rainstorm one dark night that shook the last of the stuffing from Malvina, that saw her defeated form appear at Captain Tench's doorstep as a last resort, that saw her seek refuge from the disaster that had become her life. Of course Malvina felt she could have, should have, would have done anything to prevent that terrible rape! Even Nell felt guilty! To bring such shame on her mother! To cause poor Mr. Wright and his family to be packed off like convicts to Norfolk Island! But after Nell's rape, Malvina was shunned by women she once considered friends. She had nowhere else to turn.

Thus, on that dreary night when a violent rainstorm collapsed the roof of their slapdash hut on the sleeping Thornapples, Malvina felt she'd truly come to her end. Wet, muddy, frightened, utterly beaten down, she huddled with Nell in the dripping clutter, sobbing her capitulation. Adversity had won. Were her ruined hut a sinking ship she would gladly have slid with it to the bottom of the sea.

But Ned took charge. He led his sodden mother and sister to Captain Tench's hut, the only place he could think to go.

Could Tench turn the boy and his family away on such a night? Of course not! But Tench never imagined they'd end up staying and somehow become his wards. But as it was, the Thornapples began to share Tench's roof, and Malvina his bed.

And thus matters lay until John White one day teased Tench about Sydney's latest piece of gossip, something about "Tench's menagerie." For Tench that was the final straw!

White, of course, had his own menagerie - an assortment of humans, animals, and birds that lived in the hospital compound. At first the surgeon kept only a few domestic animals and young Broughton, his lithe servant, whose mincing ways caused endless smirks and sniggers in Sydney. The next addition was Nanbaree, the native orphan boy White nursed back to health from smallpox. Broughton displayed such affection for Nanbaree that White once said he kept Nanbaree for Broughton's sake. In truth, Broughton had the best ear in Sydney for Nanbaree's tongue, and by the time Hester Thistlethwaite drifted in to join White's household as his somewhat superfluous housekeeper *cum* mistress, the boys chattered constantly in a gibberish mixture of their two languages. Intent on compiling a dictionary of the so-called Port Jackson dialect, Judge Advocate Collins tried pumping Broughton for linguistic information. But except for the simplest translations the Judge Advocate's questions brought panic to Broughton's face. The lithe servant simply couldn't handle abstractions. To translate ideas like marriage and the hereafter was beyond him. While Collins waited with poised pen, Broughton would stumble, stammer, and pant in such agony that Collins would finally wave away his question and depart with a shaking head. But no sooner was he out of earshot than the two boys would be chattering away again.

For his part, after some rough early days in the new colony, John White had mellowed. The early stresses of rampant scurvy, diarrhea, and the unfamiliar maladies endemic to Sydney had taxed him severely, but those snappish, quarrelsome days were now but bad memories. Although he resented Sydney's continuing hardships, he viewed with some pride his hospital and hospital gardens and his dispensary under construction. Except for those 50 or so aged convicts who inexplicably were transported 14,000 miles to scrabble at menial tasks and fill Sydney's salubrious air with quavery complaints of old age, the colony's general health was good. Tending to victims of accidents, beatings, spear wounds, and assisting with an occasional birth comprised most of White's professional practice, and with considerable leisure at hand he indulged his natural curiosity.

He found the strange flora and fauna of great interest, and started a miscellany of curiosities; at first just a few birds captured with birdlime or wounded with shot and nursed back to health, but then he added a dingo puppy, a wallaby, and a possum-like creature. He loved hunting, and would often tag along with Marines sent on one errand or another to see what he might bag. He also accompanied Arthur Phillip on several explorations. With his offbeat servant and native boy, with Hester as his companionable mistress, with his comfortable quarters, easy work, and little menagerie, John White had a pretty good life. After the Governor's society, John White's was the most popular in Sydney.

When the lists were posted of convicts and Marines being sent to Norfolk Island, Sydney groaned with protest and complaint. Perhaps out of pique, those on the lists slaughtered their pigs and chickens for a feeding orgy. "Why not? If I can't take it, why should I leave it behind?" Community interest seemed an alien idea. Not enough to go around? Then look out for Number One! A panicky order from Arthur Phillip brought the butchering and feeding frenzy to a stop.

When Malvina appeared on the list she was devastated. Shocked, disbelieving, she hurried to Tench. She clung to his arm. "You can do something, can't you? Especially for her?" She indicated Nell, half-hidden around the corner of the hut, watching the grown-up drama with non-committal eyes. The young girl had reverted to thumb sucking. "The children like you so much, and we all get along so well! Heavens, you can fix it, can't you, Captain? You can fix it so we don't have to go, can't you! You won't let us be sent off, will you?" Poor Malvina never suspected that she herself was the issue.

And that was Tench's dilemma. In truth, he liked her plucky children. The more they were underfoot, the more he enjoyed them. But Malvina had become tiresome. For one thing, she was too skinny. He hated her grinding public bone. For another, she was whining again about food. He'd never paid much attention to Malvina's voice, but he was now acutely aware that as their rations shrank her whining grew, and her voice abraded his ears like a screechy violin. At times his fingers tightened with an urge to wring her scrawny neck.

For another thing, Tench thought it unseemly he should be perceived as living with a private soldier's widow, and certainly not with the appearance of a domestic relationship complete with children. Decidedly not. Even a convict woman would be better. No, John White's reference to "Tench's menagerie" struck home, and from then on Tench sought to be shuck of Malvina. For the children's sake he kept his peace, but it was Tench himself who suggested Malvina be sent to Norfolk Island. Thus, when she begged for his intercession, Tench commiserated and promised to speak with the Governor himself. But did nothing. That's the thing about a dilemma, he thought, no matter what you do, somebody gets hurt.

~~~~~~~~~~

A bottle of Tench's dwindling stock of slightly off-taste canary had appeared, been consumed, and then another. Then John White sent Ned to fetch a bottle of his own canary, along with Hester Thistlethwaite, whose arrival cheered Malvina enough so she too soon joined in the boisterous, rosy-cheeked laughter. Standing with one booted foot perched on the seat of his chair, his nose red, Tench finished his rendition of a rather long ballad with a flourish of his little stemmed glass:

> *He strack the tap-mast wi his hand,*
> *The fore-mast wi his knee,*
> *And he brake that gallant ship in twain,*
> *And sank her in the sea.*

Bowing to the applause, Tench sat and pulled Malvina onto his knee. He saluted her with his glass, "*Ma belle dame.*" Malvina raised her own glass. "Your health, too," she giggled.

Before long Malvina and Hester had exchanged places, and Malvina found herself on John White's knee with her own nose red and her speech slurred, oblivious to his exploring hand. "Why, Wat," the surgeon joked, "How can you send this beauty away, eh? How can you be so cruel? How'd you like to live with me, Malvina, eh? '*Come live with me and be my love*' eh? I'd never send you off! I'd beg the Governor on bended knee to keep my hands on you!" His hand slid under her dress and into her crotch.

"Oh, but he did, Excellency, he did!" Malvina said, "But, my Heavens, the Governor wouldn't hear of it. Said 'Can't be done.' Isn't that so, Captain? Said 'Can't be done.'"

When Tench didn't respond Hester turned on his knee the better to see his face. "Really, Wat?" Her voice had a subtle, mocking edge. Tench stirred and shot a look at White. The silence grew. Malvina pushed White's arm to remove his hand from her crotch. "Of course he did, Hester! Didn't you, Captain, and the Governor said, 'Can't be done!' That's what he said, Hester, you know he did!" Her voice broke and she covered her face. "That's what he said!" she insisted through her hands.

Hester came around the table to take Malvina by the elbow. "Come, dear, let's get some air."

After the women were gone, Tench exhaled a huge "Jesus!"

White was telling a story about a surgeon who sawed the wrong leg off a woman when Hester and Malvina returned. Without interrupting his story, White pulled Hester to his knee. Malvina stood looking down at Tench, her eyes teary. He glanced at her and made pretence of concentrating on White's story, but it was no good. White trailed off. In the heavy silence, her voice so low and serious she sounded like another person, Malvina said, "You didn't even talk to him, did you?"

Tench studied his glass, silent.

"And I trusted you!" she said in the same quiet voice. "I trusted you!" Tench sat mute, turning his glass by its stem. He would not look at her.

"Oh, come now, Malvina!" White said, "Of course he did! It won't be so bad! You'll all be back in a few weeks anyway, as soon as the relief ships get here. Sit down and have some fun!"

273

Malvina paid him no heed. She studied the averted face of Tench for another long moment, then put her hand on his shoulder and bent to kiss him softly on the brow. She moved around the table and kissed the surgeon with an identical soft kiss. Hester rose and the women embraced. "Do you want me to go with you?" Hester asked. Malvina shook her head. She caressed Hester's cheek with a sad, distracted smile, then leaned close to kiss her softly on the lips. Her hand lingered on Hester's shoulder for a moment. "I'll be all right. I just think I'll get some air. Thank you." At the doorway she turned. "I'll be all right," she said again, and disappeared into the dusk.

"You think she'll be all right?" White asked. Hester sat on his knee and said she thought so.

Tench remained silent, his eyes fixed on the glass of syrupy wine he still twirled absently. Suddenly he gulped it down and poured another. "Of course she will," he said. "John, tell us another story."

White launched immediately into a tale about scalawags who tried to gain entry to a country doctor's house by pretending to have a cadaver the old doctor ordered before he suddenly died. "But the maid wouldn't let them in because she was all alone." White sipped his wine. "So they left the body rolled up in a blanket on the doorstep and went away, figuring the girl would drag the cadaver inside. After all," White smiled, "who wants a dead body on their doorstep?"

Tench's laughter was too loud. White drained his glass and lifted it to Hester. She refilled all the glasses, her emboldened hand massaging the surgeon's neck, his own hand squeezing her waist. "Oops!" she laughed when his hand moved to her crotch.

"Well, now," White resumed with a wink, his left hand gesturing, his right hand exploring Hester's under-thigh, "Every few minutes she'd look out the peephole at that cadaver and wring her hands. She didn't want the body in the house, but she didn't want it at the door either. So after a while she screwed up her courage enough to rush out and drag the corpse away from the house a ways and rush back in and bolt the door almost before you could say your name." White drank, then bent his head to kiss Hester Thistlethwaite's bosom and wink at Tench. "Well, sir, try as she might she couldn't get that corpse out of her mind. She kept going to the peephole to look and kept wondering if she should drag the body beyond the gate. The moon was up then and she was about to run out again when...she saw it start to roll! Back and forth! Aye, back and forth!" White's hand slowly rolled back and forth on the table for Hester's benefit. "Lord! Her heart pounded so hard she thought she might burst! That corpse rolled back and forth, back and forth. Then she saw an arm." Oblivious to White's fondling, Hester was leaning back the better to see his face, her eyes rapt with fear and horror.

Having gulped down his sherry, Tench wore a foolish, drunken grin. He'd heard the story a dozen times. "Dear God in Heaven, the poor girl thought she'd die of fright!" White whispered, "but she couldn't take her eyes off that corpse as it slowly worked its way out of the blanket. Another arm, and then finally it threw off the last of the blanket and stood up! Aye, Hester, stood up! And it was a man! A stark-naked dead man!" Eyes wide, fingers over her mouth, Hester stared in horror. "Aye! A stark-naked dead man!"

"Well, now, the first thing he did was piss like a horse. And it was real piss all right - steamed in the moonlight! Then the dead man picked up the blanket, shook his fist at the house, and turned to go. But just as he passed through the gate he looked back once more. And she saw his face in the moonlight, clear as you see mine, Hester. Aye, and what do you think she saw?" White drank again, deadpan.

Still with her fingers to her mouth, Hester searched his face. She was about to speak when their attention was drawn to the frantic cries of "Cap'n! Cap'n!" and the appearance of a shaken Ned panting in the doorway.

"Cap'n," the boy gasped, suppressing a sob, trying to stand at attention, "Cap'n, can you come quick? Mama's hangin'!"

~~~~~~~~~~

Misery enough for all, all right! Hardship and disappointment leapfrogged in macabre play, threatening to crush the spirit of Sydney. With half the convicts destined for Norfolk Island with half the food, the uprooted bid doleful good-byes to huts and gardens, to longtime jail mates, messmates, and bedmates. Off they sailed to a new home a thousand miles from nowhere, a densely forested, surf-pounded dot in the Pacific, a dozen square miles of wilderness where a better life might be possible.

Of course the Marines sent along were angry. With their three-year enlistment expired, most wanted only to go home, to be done with soldiering. Patched, faded uniforms and bare feet were daily reminders of their forgotten status. They were constantly hungry. Morale bumped along the bottom. Barracks lawyers talked of involuntary servitude. "We're nothing but prisoners!" Porter cried. Half-true stories of "lost battalions" in remote postings were bilious sauce for their shrunken rations. For their involuntary service, for their dislocation and hardship, resentment festered.

But tiny Norfolk Island was like a ship at sea, vulnerable to mutiny. Already one convict insurrection had fizzled. A strong Marine force seemed prudent. Using his exile to Norfolk Island to bolster his self-interest, Major Ross approached his favorites to make the case for a better life, free of Arthur Phillip's secretive, capricious misrule.

But the resentment of the Marines was nothing compared the resentment of the native captive, Colbee, who'd volunteered for nothing. The older of the two natives kidnapped by Lieutenant Bradley, middle-aged Colbee intimated he was a chief. Imperious and disdaining everything his captors stood for, he made good his escape within a month by using a shard of oyster shell to saw through the rope spliced to his manacle, then leapt the Governor's palisade to freedom. Meanwhile, dozing around a corner, his convict keeper, as ordered, conscientiously kept the other end of the rope grasped securely in his hand.

But the younger captive seemed content. Bright, playful, talkative, in his early twenties, Bennelong was the perfect candidate for becoming Arthur Phillip's language teacher. Scrubbed head to foot to remove accumulated layers of the grease, clay, and dirt he wore as protection from insects and weather, Bennelong allowed his head to be shaved and his body to be clothed. After Colbee's escape, Bennelong stuck to Arthur Phillip like a burr, at first addressing him with the native word for "father," then calling him "father" in English. He learned to drink toasts to the King and to dine with manners at Arthur Phillip's own table. Within a few weeks he could caricature the gait, mannerisms, and voices of several visitors to the Governor's house, a talent that entertained Bennelong much more than his subjects. He was a bright light in a gloomy period, one of the colony's few successes, and after several months of his captivity Arthur Phillip ordered Bennelong's manacle removed.

Although scheduled for transfer to Norfolk Island, Malvina remained in Sydney. Intended to be her final act of rebellion (or defeat or despair, it's hard to say), her unsuccessful suicide left her an aphasiatic victim of a botched job, one more inadequacy in a long list. Cut down after Ned's cries of alarm, Malvina was blue and comatose. At Tench's hut they chaffed her hands and feet while trying to rouse her with ammonia spirits and rum. At last, after almost an hour, Malvina's eyes fluttered open in a bewildered stare. She began to breathe more naturally, and on the litter-ride to the hospital she slept soundly, even peacefully.

Late next day she awakened to a new world. If someone said her name, she looked. If someone helped her stand, she stood. If someone put a spoon to her lips and said "Open," she opened, and usually swallowed. Her larynx destroyed, her only speech was a soft grunt, mindful of the satisfied "ugh" of a pig. She could be led by the hand in a shuffling, flat-footed gait.

"Oh, God, I never thought she really meant it!" Hester cried. "Do we ever really mean it? I didn't think she really meant to kill herself. I thought it was just talk, making me promise to take care of Nell! All we ever do is talk! We never really do anything!"

It was John White who suggested that Hester tend to Malvina and Nell in a hut near the hospital vacated by convicts sent Norfolk Island. And Tench took Ned as his boy servant.

After a time Malvina had a regular place just outside the door of the hospital, which Hester taught her to open when someone approached. Led shamble-foot by Nell, Malvina sat on a three-legged stool beneath the overhang, expressionless, uttering her low, pig-sounds, rising to serve those coming and going on hospital business. If someone smiled, Malvina's eyes darted, as if looking for escape. Sometimes her mouth would twist into a grimace of smile as she lunged forward to be of service. In a few weeks her complexion acquired an invalid delicacy; her hands became blue-veined and soft. To hide the scar from the noose, Nell adorned her mother's neck with a broad red ribbon, which from a distance lent a touch of elegance to Malvina's ivory face and upswept hair, now tinged with gray. A closer look might discover a string of drool as she rocked to and fro to an internal rhythm. Some thought she looked saintly. Most thought her an idiot.

While spring days grew longer and life quickened in England, darkness fell sooner in desolate Sydney. But the gloomy dawn of March 28, 1790 was brightened with the birth of Mary B's second child, a healthy boy. The delivery was trouble-free. As Katie, Liz Bason, and Hannah tended Mary B, swaddled the child, and placed him at her breast for his first taste of mother's milk, Hannah suggested a name. "In olden times they had bad times like this, and Isaiah prophesied a newborn savior would bring an end to their troubles. He said the child would be called Emmanuel, which means, `God is with us.'"

Weak but happy, not sure of Hannah's meaning, Mary B smiled. "Lord, I hope this little one brings an end to our troubles. Is Emmanuel all right with you, Will, or do you want 'William'?"

Will had no objection to Emmanuel. With rare conviviality, he exclaimed, "A little man! That's what he is, isn't he? Emmanuel, a little man!" After service on Easter Sunday, Mary B's baby boy was baptized Emmanuel Trewardreth Bryant, his middle name a reference to Will's birthplace.

Misery enough for all, all right! Hardly 50 pounds of salt meat per person remained in the storerooms, and less than 200 pounds of flour and rice. Arthur Phillip cut rations again, to half. Convicts queued up each afternoon to receive their daily portion, a new system of distribution to safeguard the foolish from consuming all their food at once, like the gorbelly woman who died from stuffing herself on a huge pudding she made from an entire week's ration. When White's autopsical knife perforated her gaseous middle she literally exploded, spraying him with a putrefied paste of half-cooked flour. One dolt slowly starved to death because he paid out most of his food to have someone else cook his meals. The poor wretch knew nothing of cooking. Another convict died because he sold his food, the second who starved to death to get money for passage home.

Food became an obsession, and convict labor more and more inefficient. Arthur Phillip decreed that convicts need work only four hours a day. They were to use their extra time to fish or gather or work in their gardens. Some did. Others plundered the gardens of others.

~~~~~~~~~~

Emmanuel, God is with us! How did Hannah reconcile Isaiah's prophecy with the awful news that arrived the day after Emmanuel's baptism? At dawn the headland lookouts let loose a cry of giddy joy when they spied a long-awaited sail.

Salvation was at hand! The men jumped and danced and pounded each other though tears and laughter as the ship grew on the horizon. They sent up the signal flag with one eye on their deliverance and the other on the reaction from Sydney, the puff of a happy signal cannon and Arthur Phillip hurrying out for first news.

So slowly did the ship reveal itself that the first enthusiasm had already worn off when the young officer of the lookout realized with a sinking heart that the ship was the *Supply*. But why? With the realization growing that only bad news could bring the little ship back from Norfolk Island so soon, the lookouts now watched the approaching ship with a sense of dread. When they saw the decks crowded with men, certainty took hold that the *Supply* brought news of disaster.

Accompanying Arthur Phillip in his cutter, elated at the prospects, Tench was numb with disbelief when he realized the approaching ship's boat came from the *Supply*. Unable to disguise his disappointment, he croaked, "Sir, prepare yourself for bad news." They were still two hundred yards away when a voice carried the awful news across the choppy water: The *Sirius* is lost!

That afternoon Arthur Phillip called an emergency council to hear the report by Lieutenant Ball of the *Supply*. Driven by gale-force winds, the *Sirius* wrecked on the unfamiliar reefs of Norfolk Island. Luckily, not a single life was lost in shuttling passengers and crew to land on a hatch cover. While the ship appeared unsalvageable, they'd saved most of the cargo. Lieutenant Ball had watched them toss animals, boxes, trunks, and other paraphernalia overboard to be carried ashore by the surf. Salvaging the barreled provisions in the holds would be a different matter.

Incredulous, Arthur Phillip asked, "So we don't know what their food situation is?"

Lieutenant Ball answered, "Sir, I got my cargo ashore and the *Sirius* got some off before she struck, but I don't know how much."

The former commandant of the Norfolk colony, Lieutenant King, now replaced by Major Ross, shook his head. "If they don't get more they'll soon be hungry. Their reserves weren't much, and there's 500 people there now."

Arthur Phillip was silent. He stared at a sheet of blank paper, the eyes of a dozen silent men fixed on his fringed pate.

After the *Sirius* delivered the convicts and Marines to Norfolk Island, he was sending it to China for food, but now he had only the *Supply*, a ship of a scant 170 tons. If he sent the *Supply* for food, Norfolk Island would be cut off from Sydney, and Sydney be without an ocean-going ship.

A Navy man without a ship was like a surgeon without instruments, a carpenter without tools, a farmer without a plow. Responsible for the well-being of hundreds, Arthur Phillip felt naked, helpless, defeated. If he had even one answer to a storm of questions he could write it down, and that would be a beginning. But he had none. When would relief arrive? He didn't know. When would the Second Fleet arrive? He didn't know. How long could the people on Norfolk Island hold on? He didn't know. How long for the *Supply* to get to Djakarta and back with food? He didn't know; no one had ever sailed from Sydney to Djakarta. Jesus Christ! How could he make decisions if he didn't know anything?

In the long, awful silence Tench at last cleared his throat. After cleared it a second time Arthur Phillip slowly raised his eyes. "Yes, Captain?" he said quietly, relieved that someone had something to say.

Tench leaned forward, his demeanor earnest. "Governor, all we really know is that we have to hang as long as we can. Sooner or later, and we can only hope sooner, help has to come. But we have to hang on, and that's all I think we should be concerned with, hanging on. Everything we do should be aimed at hanging on." Heads nodded. Encouraged, Tench continued, "Now as I see it, the only thing we need to keep hanging on is food. If we have food, we can survive. I respectfully submit, Excellency, that we must bend all of our efforts towards food, and look to nature - nature's bounty - to see us through. Let the buildings go, let the roads go, let everything go except getting food. We can always catch up! But for now we must live like the natives and find our food in the forests and waters! We can do it, Excellency! I know we can do it, if we just try."

Heads nodded as faces turned to gauge Arthur Phillip's reaction. Nature's bounty! Of course!

Arthur Phillip lowered his eyes again to the blank paper. He sat silent in the gathering gloom, shoulders sagging, head wavering, tongue unconsciously massaging the gap of his missing buttertooth. Drifting in a breath of air, a few pathetic gray hairs on the fringed dome of his head caught a last gleam of light. A long minute of silence. Then another. The tick-tock of a mantle clock filled the room. The Governor seemed unable to rouse himself. Nervous glances began to flick around the table. When at last Arthur Phillip spoke, the collective sigh of relief was like the air let out of a dozen balloons. "I agree," he said softly, raising his eyes. "So say you all?" Heads that nodded unanimous agreement were shocked to see remnant tears in their leader's eyes. "Very well," Arthur Phillip said, "We shall consider Sydney under siege."

He spoke to the blank sheet of paper before him, as if he were reading his decisions there. He would send the *Supply* and Lieutenant King to Djakarta to charter a ship and buy enough food to hold out for another year. The *Supply* would return with as much food as it could carry, but Lieutenant King would find passage to England to give government a personal report on the state of the colony. Meanwhile, all private stocks of flour and grain were to be turned over to the commissary, his own included. No livestock was to be slaughtered except for purchase by the commissary. Hunting parties would go out every day. Every boat that would float would fish every day, with officers aboard to ensure the entire catch was turned in. But he wanted work projects continued. Convicts must always see they had to work to eat, even if they worked only two hours a day. And convicts would continue tending their gardens in their free time. Anyone malingering would lose a day's ration. Punishment for stealing food would get stiffer.

As Arthur Phillip talked, seeming to gather confidence, Judge Advocate Collins scribbled down the instructions. Others began to offer suggestions, modifications, cautions. A servant brought in candles, and for the next half-hour the men sitting around the table in "Phillip's folly" fleshed out a plan for survival. By the time they broke up, well past the supper hour, spirits had risen. As he stood to end the meeting Arthur Phillip handed his blank sheet of paper to Collins. "Here, Davey," he said, playing to the group, "so you want my notes?" To the anxious sentries outside, the burst of laughter was incongruous. As officers and officials went into the evening with smiles and good-natured chatter the soldiers looked at one other in bafflement. What on earth could be funny?

~~~~~~~~~~

But for the heroism of a strapping convict, Norfolk Island might have suffered mightily. Gale-force winds still blew when

Major Ross sent two convicts out on the hatch cover shuttle to throw off the last of the livestock. Better the creatures should take their chances swimming to shore than die of thirst or drown while trapped on the *Sirius*. The convicts pulled themselves to the wreck, threw the surprised pigs and fowl overboard, then looted the cabins for liquor. Time dragged as Major Ross waited for their return. He waited in vain. The two proceeded to get roaring drunk, and as evening fell those ashore saw lantern-light flicker here and there on the wreck and heard snatches of what sounded like drunken song. Suddenly the lights on the *Sirius* grew brighter. The fools had set the ship afire!

Efforts to get the attention of the drunken convicts failed. Volleys of musket fire loosed at the ship were lost in the crash of the gale-driven surf, and the balls had no effect. Even a three-pounder fired into the wreck produced no response. Perhaps the drunkards had fallen asleep. As the glow of flames brightened in the gathering darkness, men reeling with exhaustion regarded the 200 yards of roiling water and shrank from the obvious. With the hatch cover secured to the ship, there was only one way to reach the *Sirius*. Climb out along the six-inch hawser. Someone had to risk his life, or ship and cargo would burn to the waterline.

Why did John Arscott volunteer? He had less than six months to serve of his seven-year sentence. After losing more than six years of his life for stealing two silver watches and thirty pounds of tobacco, why risk his life with freedom so nigh?

Or was freedom nigh? Their terms expiring like his, how many convicts would seek berths as crew on homebound ships? Far more than needed. Perhaps John Arscott thought his chances would improve if he showed worthiness. Thus, after looping a line over the hawser and around his chest, after Major Ross shook his hand and wished him success, after the muscular convict knelt to cross himself, he struggled a few steps into the surf and launched himself into a head-high wave.

Several times people ashore feared John Arscott lost. Surf and undertow played him like a cork, sucking him under, popping him up, once tossing him completely clear, like a fish in the water fleeing a predator. Ashore, their lips moving in prayer, shoulders rolling in unconscious motions to assist the struggling swimmer, Marines and convicts alike urged him onward.

John Arscott disappeared for what seemed an eternity. Hearts sank. Then the dim blotch of his body dragged itself onto the hatch cover and he turned to wave that he was all right. Enthusiastic cheers broke from a hundred mouths.

Now working in darkness, Arscott extinguished the fires and drubbed the delinquent convicts before sending them back on the hatch cover to take their punishment. He remained aboard the wreck all night.

Following Arscott's heroic feat, the gale abated, and several days of calm weather permitted almost all the cargo on the *Sirius* to be salvaged. Once the food was safe ashore, the ship's captain, a seasoned Navy officer named John Hunter, kept the crew busy bringing ashore whatever they could pry or cut loose - iron, rigging, fixtures, wood, even some of the ship's guns. At last the vessel rested as a skeleton on the reef. While the men picked away at the carcass of his ship, Hunter awakened each morning secretly hoping to see it gone. The hulk was a taunting symbol of a blemish on his record, a reminder of the inquiry he'd face in England. Hardly a triumphant homecoming for good service!

After almost three weeks, the *Sirius* thankfully disappeared in the dead of night.

~~~~~~~~~~

Who was England to believe? Letters the *Supply* carried to Djakarta were as variegated as Sydney's sunsets. Arthur Phillip wrote: "*Dismal accounts will, I make no doubt, be sent to England, but we shall not starve, though seven-eighths of the colony deserves nothing better....*" From his post at the head of the harbor, young Lieutenant Daniel Southwell wrote: "*We are now at less than half allowance, and...on the brink of going three on one man's dividend; and a few weeks must, if nothing arrives, put us on a quarter allowance.*" Even John White, usually sanguine, complained of "*a country and place so forbidden and so hateful as only to merit execration and curses.*"

The salt beef was gone. Adults now received one-half pound of flour or rice a day, and four ounces of salt pork. And a few dried peas. Ned and Nell got half that. After the rum ration was also halved, Marines found even less solace in their grog.

But Arthur Phillip promised "*we shall not starve,*" and so the fishers fished and the hunters hunted and the laborers labored. And the thieves thieved. William Lane, convicted of stealing 13 pounds of bread from the storehouse, was sentenced to 2,000 lashes. Thomas Halford robbed a garden of three pounds of potatoes and was sentenced to 2,000 lashes. Sentenced at Exeter with Mary B, William Chaaf was hanged in late April for robbing a convict's hut. Chaaf was the sixth of Mary B's shipmates on *Charlotte* to be hanged in Sydney.

Joseph Elliot stole 1¼ pounds of potatoes from the Reverend Johnson's garden and received 300 lashes, had his flour ration stopped for six months, and was fettered to John Coffin, who was also being punished for theft. A sailor named Tom Paul stole six cabbages from Arthur Phillip's garden and was sentenced to 500 lashes. When John White's goat was caught eating a cabbage in John Fuller's garden, the convict wanted the goat flogged. Although White made restitution to the convict, Fuller crabbed that the goat, *sans* flogging, had learned nothing from its brush with the law.

Despite Arthur Phillip's plan to defeat hunger's siege, disappointment and discouragement soon prevailed. Nature's bounty proved illusory. The kangaroos and emus, the possums and wallabies, the ducks and swans had all vanished. Day after day the hunters tramped themselves to exhaustion, bringing back little more for their efforts than big appetites. After a time the hunters were called in from their fruitless task.

The boats did better, but not much. In May the biggest day's catch was only 200 pounds. Most days the catch was 40 or 50 pounds, the caloric equivalent of a dozen pounds of meat. With the accumulated failures of his plan to live on Nature's bounty, Tench's spirits plunged. He took to his bed with an owlish complaint and refused to rise. If Ned looked in, Tench waved him away. Free to forage, Ned augmented his own diet with grubs, mice, ants, and other tidbits that Nanbaree showed him could be tasty. And good-natured Bennelong, deciding he'd had enough of the starving colony, leapt the Governor's palisade and escaped in the darkness.

As May drew to its hungry end, convict efforts became desultory. Brick-makers and road-builders went through the motions. Trying to grow grain was considered a waste of time. The entire grain harvest the previous season was but 300 bushels, a modicum stored in a room just ten feet square. And they had to reserve every bushel for seed. Those who toiled at turning the soil with their broad-bladed hoes cursed the hard ground, muttered at their purgatorial labors, and tried to ignore their growling stomachs.

Some convicts were more fortunate. As sawpit work was too strenuous for men on such skimpy rations, sawyers were freed of their labors. With the slowdown in his masonry labors, James was sometimes assigned to a fishing boat. In his off time he tended the garden and livestock he and Tartop shared.

The unlikely friendship of James and Tartop had proved good for both. Inexplicably, Tartop had a green thumb. Most convicts grew bored and discouraged with garden work - weeding, watering, picking off potato bugs, mulching, thinning, squashing fat grub worms, propping, staking, chasing off chickens - that was the way to garden. Tartop was happy with these mundane tasks.

His customary work was as a beast of burden, harnessed to a sledge that pulled bricks from the kilns to the settlement, a labor reserved for the most miscreant. But Tartop accepted this daily drudgery without complaint. While straining at the harness he thought of his onions and cucumbers, worried about his cabbages, made plans for potatoes and peas, and mentally ticked off his pumpkin blossoms. From the Reverend Johnson he got seeds to try, a cutting to root, and advice on staking his beans. From others he got less useful advice. One night the quarter guard found him turning in a circle and reciting a charm as he buried each section of a seed potato by the light of a full moon.

Tartop also liked animals, and enjoyed helping James with their growing inventory. From a single hen and rooster, James had increased his flock to almost two dozen. From the garden Tartop collected beetles and bugs and weeds to feed the chickens, which in turn made manure he fed to the garden. In exchange for sawing lumber on his time off, James had acquired a pair of piglets from Lieutenant Creswell, which later produced a litter of their own. Before times turned really bad, James had traded vegetables and eggs for tobacco and spirits from Marines, and once even managed to acquire two gallons of rum in exchange for a piglet. But he learned Tartop was not to be trusted with liquor. James returned from work one day to find a rum-soaked Tartop snoring in the shade with their sleeping pigs. Tartop also wanted sheep, and made James promise that if the Second Fleet brought sheep James would try to get a ewe.

As the bad times settled in, James regularly dispatched Tartop to Mary B with eggs and vegetables and a little flour.

Mary B accepted the gifts with a smile and a kind word and would ask after Tartop's garden and offer sweet-tea, which Tartop always declined with a mumble and yank at his forelock. On rare occasions he might play awkwardly with Charlotte for a minute or two. If Will was around he teased Tartop about coming to court Mary B, which Tartop denied with red-faced embarrassment. Sometimes Will asked in all seriousness whether Tartop had seen any Chinamen in the woods or how many bricks he'd hauled that week. Tartop had no idea about the bricks, but Will would keep after him, "Well, do you think it was a hundred? A thousand? Five score? Four thousand? Six dozen? A dozen dozen? More than ten?"

"Oh, yes, Will, more than ten!"

Later Will began teasing Tartop about Tahiti, which Will described as an earthly paradise of easy living and naked, golden women. "Wouldn't you like to live on Tahiti, Tartop? Beautiful naked women feeding you grapes and loving you? Wouldn't you like that, Tartop? Wouldn't that be better than fucking your hens?"

~~~~~~~~~

Under surveillance for several months after his reinstatement as the colony's "official" fisherman, Will became even more a cynosure as Sydney's desperation for food grew. In the early days of short rations he'd brought in several bounteous catches, and strutted with a superior air that irritated Judge Advocate Collins. But who could argue with success? With such catches Sydney could stretch its salt meat reserves indefinitely. But now Will's luck had turned bad, and Arthur Phillip himself sometimes met the returning boats to measure their success. Will was failing, his failure tormented him, and he was unable meet the Governor's eye.

Determined to prove his mettle, he drove himself harder. He suggested keeping boats out around the clock with rotating crews. "I'll keep 'em going," he promised. And he did, but his bad luck persisted. He grew haggard. When his crews begged him for rest, he exhorted them to greater efforts.

Several times Mary B joined him in his boat, which was usually the Governor's cutter. A large rush basket in the bow served as a bed for Emmanuel and playpen for Charlotte, and a canvas lashed to the stem-post and forward tholes protected them from raw winds and foul weather. When he drew Will's boat as his assignment to watch over the catch, Tench learned

286

from Mary B that she rode along so she might eat the raw hearts and livers to better nurse Emmanuel.

Mary B thought the healing process between Will and James began the day James showed up and Tench motioned him into Will's boat with Mary B and the children. His voice hoarse with weariness, Will said, "All right, Jimbo, you're a charmer! See if you can charm the damn fish!" So a chary James spent the day under Will's command, pulling an oar with three others, while Tench browsed an oft-read book in the bow near Mary B and the children. Tench never noticed how Mary B avoided even looking at James, so fearful was she that Will might explode. The day passed uneventfully, however, and at the end of their labors Will marveled at how easily James had handled his oar and the nets. "Jesus, Jimbo, your arms are something! Sawing and bricks do that?"

~~~~~~~~~~

On Norfolk Island, meanwhile, life was strict and punishments quick as Major Ross imposed martial law. The margins of survival required no nonsense, but martial law suited Ross. Fortunately, the boobies appeared for their annual nesting. Convicts and Marines marched nightly across the tree-tangled island to ascend Mount Pitt's barren reaches. There they clubbed bag after bag of the stupid, nesting birds. For weeks the island feasted on fresh, fishy birdmeat. Lieutenant Creswell later figured the colony killed and ate some 200 birds for every man, woman, and child. Unlike Sydney, Norfolk Island found itself able to live on nature's bounty. Indeed, the abundance of fresh fish and boobies allowed Major Ross to stop the salt meat ration completely. And with fresh meat in her diet, even Seedy regained a measure of enthusiasm for living.

~~~~~~~~~~

"*We shall not starve,*" Arthur Phillip promised, but few in Sydney believed they were not starving. True, they had something to eat every day, and no matter their station they received equal shares (except for children), but the rice was now so vermin-infested it literally crawled, and they lived largely on flour and greens. Meat-eaters especially suffered. Those who'd grown up gnawing juicy joints of beef and succulent rib racks and smoked hams and roasted fowl and who gorged themselves on savory mutton stews and spicy sausages - those unfortunates felt they had nothing at all to eat. Marines whose regulations assured them seven pounds of meat a week groused constantly. Convicts found their skimpy portions of shriveled salt pork simply a return to a familiar past of daily bread and rare meat.

As May drew to an end, convicts and Marines alike were walking scarecrows...thin, gaunt-faced, in rags. Like everyone, Mary B found her joints painfully swollen, and she had to rise carefully to ward off dizziness. But her milk continued to come, and Emmanuel, God's promise of deliverance, fed with gusto.

After long days of peering seaward from his lookout on the headland for a relief ship, young Lieutenant Southwell often perceived in a distant cloud a hull-down ship in the glow of late afternoon sun. How convincingly the shifting vapors reported an invisible crew at work. See how they alter sail to bring their ship to Sydney! How often the young officer prepared himself to espy through his telescope the distant flutter of a top-gallant pennant that announced, not another cloud, but a ship from home! Safe harbor! And salvation for Sydney!

But the sight of sun-washed sails on the horizon would not announce relief. The fateful evening in early June was foul-weathered when Lieutenant Southwell was sure his imagination did not deceive him. Through heavy mist and low, scudding clouds he saw a distant ship. Then the light of day failed. Rising before dawn, already rehearsing his first words of announcement, he scrambled up the path to the top of South Head, convinced daybreak would reveal a ship awaiting light enough to work into harbor. His disbelieving eyes were greeted with the familiar expanse of an empty sea. Once again his spirits plunged.

Were the gods playing with Sydney? Southwell would later learn he truly had seen a ship, the *Justinian*, fatted with provisions to release the colony from its bondage of hunger. But within sight of the harbor entrance a stiff land wind and contrary currents carried the ship away again, carried it north like helpless debris. For days the *Justinian* fought unfamiliar currents and southerly winds that drove her farther and farther north. Finally, almost 200 miles north of the entrance to the Sydney estuary, imprisoned between on-shore winds and threatening reefs, the captain let go his anchors to ride out several days of terror. The *Justinian* was unable to work back to Sydney for more than two weeks.

So it was the *Lady Juliana*, freighted with its well-used women, that finally made her leisurely way into Sydney Cove to break the famine. This time when Southwell ran up the flag to announce a ship, Tench had no doubt relief had arrived, for through his telescope he spied the tiny figures of the lookout dancing around the flagpole. Again every boat in Sydney made for the arriving ship, thoughts of food jostling eager anticipation of news from home, the first news from the outside world since the *Sirius* had returned from Cape Town more than a year earlier.

"The King is recovered!"

"Hurrah! and God save the King!"

"Revolution in France!"

"No!"

"Aye, mobs and fighting in the streets! Fat Louie is toppled!"

"No!"

"Aye, God save England and good King George!"

"Aye!"

"The *Guardian*, your relief ship, was holed by an iceberg!"

"No!"

"Aye, she limped back to Cape Town. We've got a bit of her cargo, but she looks beyond saving."

"Good God, no wonder! She would've been here months ago. We knew England wouldn't forget us!"

"Aye, not England!"

"Aye!"

"And I suppose you'd be interested in some letters from home?"

"Aye, and any newspapers.... By the way, where's the rest of the Second Fleet?"

"The rest? Why, I expect they'll be here any day!"

Oh, the rejoicing! No matter that the lubricious *Lady Juliana*'s leisurely voyage from England had taken nearly a year! As soon as John White inspected the health of the convicts and crew and cleared her to anchor, a full week's food was served out, Marines and convicts alike drank the King's recovery. Smiles and handshakes prevailed! No more hunger! Sydney is saved! The King is recovered! God save Sydney and good King George!

~~~~~~~~~~

Misery enough for all, all right! But at least there was respite. The spectral *Justinian* anchored at last to disgorge food by the ton for Sydney's storehouses. The *Lady Juliana*'s women settled into Sydney's empty huts, convict gangs resumed their labors, and Arthur Phillip read and reread letters from England. All official. Not a word of his wife. He'd expected bad news. But no news? Was she still hanging on? My God!

Sydney's respite was brief, for the deathships of the Second Fleet arrived soon after. First the *Surprise*, followed two days later by the *Neptune* and the *Scarborough*, back again with another cargo of convicts. A report the *Surprise* was dumping corpses in the harbor was the first sign of something amiss. John White boarded the ship to find ranks of sick, emaciated men needing immediate treatment for scurvy, diarrhea, and even dreaded jail fever. Some died as White and his assistants examined them. The convicts had been kept in irons and confined below-decks for most of the half-year voyage.

Emergency tents raised near the hospital isolated the sick, who were shuttled ashore as quickly as possible. More died in the boats bringing them to land. With tears in his eyes, the Reverend Johnson surveyed the misery of the hundreds with an overwhelming sadness. Oh, the cruelty! Oh, the heartless greed! His ministrations kept him too busy to record the deaths of those first terrible days, but his register in July picked up the doleful tale.

"July 2, 1790 Buried this day Thomas Delling, William Budge, a man named Owlit, and another who is nameless before God, convicts.

"July 3, 1790 Buried this day Thomas Carter, William Glover, William Dubbell, John Shellick, Daniel Jones, William Richards, Alexander Aspernal, convicts.

"July 4, 1790 Buried this day James Robbenot, John Gray, convicts.

"July 5, 1790 Buried this day William Bead, William Pointon, John Bland, Frances McGurk, John Cumson, convicts.

"July 6, 1790 Buried this day Benjamin Creamer, John Lufbridge, Edward Glin, Edward Bonnock, Cornelius Broad, Ann Hardyman, William Tilbrook, convicts.

"July 7, 1790 Buried this day Richard Johnson, John Williams, convicts, William Elivin, sailor.

"July 8, 1790 Buried this day Joseph Chant, James Jones, George Admes, convicts."

Day after day, at least two or three died.

Arthur Phillip wrote to voice his displeasure to William Wyndham Grenville, the new Home Secretary. Despite the colony's long months of hunger, "*when the ships arrived we had not fifty people sick....*" Now 488 convicts were under medical treatment. Out of more than 1,000 convicts sent out, only 60 had not died or been put ashore too sick to work. He summarized the sad history of the three ships:

	Males		*Females*	
	Embarked	*Died on Ship*	*Embarked*	*Died on Ship*
Neptune	424	147	78	11
Surprise	256	36		
Scarborough	259	73		
Died in hospital		122		2
Totals	939	378	78	13

The horrors of the Second Fleet cast a gloomy pall over Sydney, and gave Will pause in his plan to seek passage home.

~~~~~~~~~~

Some thought the trunk of the yellow gum tree resembled the English walnut for the first dozen feet, but then it sprouted a mass of long spiral leaves that hung like hair. Each year from the middle of its hair the yellow gum sent up a single, straight reed that natives harvested to make their spears. Softening the gum that oozed from the lacerated trunk with fire, the natives affixed hardwood tips to the shaft and then adhered small chips of oyster shell or shark teeth or sharp chips of rock with the same yellow gum. When the gum cooled it effectively welded the barbs into the tip. Fitted into a throwing stick and cast with wondrous skill, the native spear was a lethal weapon that exceeded the range and accuracy of most firearms. Quint was lucky when he hit the native's spear at sixty yards, for even the stump was at the limits of his musket's accuracy. Time and again, however, Sydney saw the natives cast their slender spears 80 and 90 yards with impressive skill. They had nothing like birdshot, however, reason enough to fear the white man's weapon.

In the earlier, less suspicious days of the colony, the English noticed that when natives parleyed they laid their spears near their feet as a sign of non-aggression, but kept them close, shifting them with their feet and toes as conversational groups eddied and flowed. They used their feet and toes with marvelous dexterity. When approached with bread or hatchets or beads, the natives would subtly alter their positions, always on guard and always keeping their weapons near. After Arthur Phillip captured Arabanoo, his first kidnap victim, natives didn't suffer one of their own to go near a colonist without several protectors with spears at hand. With the kidnapping of Colbee and Bennelong, they fled outright at the sight of whites.

But Arthur Phillip had instructions to treat the savages with kindness and respect, which he was able to do with Abaroo and Nanbaree, who after all lived in Sydney, were learning English, and could be reasoned with. But the others in their hundreds, divided into at least a half-dozen family groups, each with its own domain in the Botany Bay-Port Jackson area, were inaccessible. Thus, in early September, when John White sent word to the Governor that his hunting party had encountered a large body of natives feasting on a beached whale and that Colbee and Bennelong were among them, Arthur Phillip immediately set out to try once again to establish meaningful contact. Upon reaching the now-deserted whale carcass, Arthur Phillip entered the nearby woods alone. "Colbee," "Bennelong," he called, his voice growing fainter and fainter. Tension on the beach grew. At last Arthur Phillip returned to the boats to say he'd encountered both Colbee and Bennelong and wanted to take them presents. His subsequent trips into the woods seemed to go well. Soon Arthur Phillip called for others to join him; the convivial Bennelong wished to see some of his English friends.

When Judge Advocate Collins and others arrived at a small clearing with wine and bread, they found the Governor with Colbee and Bennelong, surrounded on three sides by a score of armed natives, clearly on guard. Three or four others ranged between the Governor and the safety of the boat. Bennelong shook hands with the new arrivals, drank to the King, and promised to visit Sydney soon because he'd received the promise of his white "father" that he'd be free to leave when he wished. Highly pleased with the success of his mission, Arthur Phillip turned to go. The English party had retreated only a few steps when Bennelong began to introduce some of the other natives. Arthur Phillip turned back and, hands extended, approached a native.

Perhaps the native remembered how Colbee and Bennelong were seized by their wrists and dragged into a boat. Perhaps he thought of Colbee and Bennelong's ordeal as captives and of how they'd had to hammer at Colbee's iron manacle for two days before it finally gave way. Perhaps he thought Arthur Phillip intended to seize him. Suddenly the native's foot flipped a spear into his hand and he assumed a threatening posture. Arthur Phillip stopped to drop his hanger and short sword on the ground. "Warra, warra," Arthur Phillip said, wanting to indicate that he didn't like being threatened. He advanced with empty, outstretched hands. In an eye-blink the native launched his spear. Ducking, Arthur Phillip felt the blow as the spear entered just above his right collarbone. Those behind saw the barbed tip laden with bloody shreds of flesh emerge below his shoulder blade. "Ooof!" Arthur Phillip cried.

Both sides panicked. As the English turned with cries of alarm to run for their boats, Parthian spears hissed from natives yelling in their own flight. Not yet feeling the pain, Arthur Phillip ran too, but stopped with another "Ooof" when the tail of the spear caught a tree. "Help me break it off!" he cried. He drew his pistol from behind the shelter of a tree, where one of his men tried unsuccessfully to break the shaft. The man saw the flash of a spear that grazed his hand and a spout of his blood. With a rush of adrenaline he snapped Arthur Phillip's spear. Arthur Phillip discharged his pistol to let the natives know he was still armed, and then broke for the boat.

It took the English party two hours to row back to Sydney with their pale, panting leader. He complained of chills and was alternately groggy and hyperactive. Given that the spear protruded both front and back, they propped him up in the bow as best they could, his legs awkwardly elevated on the Judge Advocate's knees. Collins tried to keep him covered, gave him wine, and tried to reassure him. With the spear serving as a plug, the wound bled little, but that was the only hopeful sign. The anxious faces in the boat showed their fear that Arthur Phillip would die before they reached Sydney.

But he was still alive when the cutter ground onto the beach just below the hospital, and as the boat crew leapt out to carry him up the steep slope, Collins ran ahead to alert the surgeons, his shocked, pale face a signal of disastrous news. By the time the rest of the party reached the hospital with their semi-conscious patient a crowd of buzzing on-lookers trailed behind, a disconcerting hubbub that sent Malvina Thornapple lurching to her feet with a drooling grimace to swing wide the door. While White's assistant examined the wound, Arthur Phillip came to enough to instruct Collins with matter-of-fact calm, "Get the word out. This was an accident, and no pretext for revenge."

Perhaps the native who panicked at Arthur Phillip's outstretched hands played a larger role than he knew. Perhaps his yellow gum spear, inflicting a wound that even the natives thought would prove fatal, was a chosen instrument. Perhaps Arthur Phillip's suffering was at least partial expiation for the sins of the colony. In any event, Arthur Phillip's wound seemed to open the door to better relations with the natives. Several days later, when John White spotted smoke in a familiar stopping place for natives. He rowed across the harbor to investigate, and found a small group camped near the water. Approaching cautiously, he tried to make them understand; the Governor was not dead; the Governor wanted peace. The natives agreed to remain until afternoon, when White promised to return with a sign of goodwill.

Arthur Phillip dispatched Tench and a small party as emissaries bearing gifts, with Abaroo as translator, the native girl who lived with the Reverend Johnson. Bennelong stood among the waiting natives. Struggling for English words, Bennelong asked whether "father" was dead. He seemed much relieved to hear Arthur Phillip was alive. Bennelong then asked by a combination of words and gestures if "father" was angry. "No," Tench assured, "Father not angry. Father send presents. Father want see Bennelong."

Bennelong then indicated he would like to see "father" too. Meanwhile, Bennelong's wife implored Abaroo to quit the white man's world. In the end, the natives agreed to meet again to exchange more presents. Forsaking her people, Abaroo returned to the white man's world with Tench.

Arthur Phillip made a remarkable recovery. He soon walked with little pain, and rendezvoused with Bennelong to give his personal promise that the native had nothing to fear by visiting Sydney. Before the end of the year Bennelong and his wife and dozens of others were regular visitors. They slept in the Englishmen's houses, ate their bread, drank their wine, and exchanged weapons and crude utensils for bits of clothing and glittery trifles. Fully reconciled with "father," Bennelong asked Arthur Phillip to build him a white man's hut, which the Governor erected on the east side of Sydney Cove, past Will and Mary B's hut. Soon the natives were beaching their canoes at the hut to stay for a day or two in a constantly shifting assortment of husbands, wives, sons, daughters, uncles, aunts, nieces, nephews, grandmothers, grandfathers, betrothed, agemates, and tribal and intertribal links the English never really understood. Soon some of them learned to say "hungry" and "bread" and to suck in their stomachs as a sign they wanted something to eat. Soon some were filching from gardens and making off with chickens and pigs because they didn't seem to grasp the idea that something growing from the earth or foraging around in the undergrowth could actually be the property of someone they couldn't see. Soon some sought the authority of white men to intercede in their family disputes and intertribal quarrels. Soon the young women were granting sexual favors for small presents.

~~~~~~~~~

Will was steamed when Judge Advocate Collins threatened to have him flogged if he persisted in the issue of his freedom. But what was Will to do? The Second Fleet arrived with the mind-boggling news that the court records of the First Fleet convicts had been left behind again! By then dozens had approached the Judge Advocate to report they'd completed their sentences; what was the process for getting back to England? Most wanted to return at first opportunity, which meant going back on the ships of the Second Fleet.

Collins chided Will, a man he disliked for his ego and bluster. "We can't let every lag in Sydney tell us when his sentence is up, Bryant, or you'd all take years off your sentences. Do you think we're such fools? Besides, you're married! You can't go off and leave your wife."

"My wife's not the point! The point is, my sentence is soon up, and you don't have a right to keep me here!"

"Watch your step, Bryant! You don't have any say about your rights. The point is, you're a convict! The point is, you're here! The point is, you're married. The point is, you're responsible for your wife! The point is, you're going to be here till the Governor says you can go, and that won't be till we get your records from England! And then you can't leave before your wife's sentence is up. You can't leave without her! So get back to work! I'm sure you have fish to catch!"

A few weeks later, getting headache powders at the dispensary, Will cocked an ear. The boom of the signal cannon meant a ship arriving! He joined a hugger-mugger throng running towards Observatory Point to get a better view down the harbor. Hurrah! It was the *Supply*, returning from Djakarta like a long-absent friend! Within hours its news of Bligh's remarkable open-boat voyage from Tofoa to Timor would set Will to thinking.

~~~~~~~~~

"Hi, Jim, look at me!" James leaned on his hoe and wiped his brow to look at Tartop. Sure of his friend's attention, Tartop stood ramrod straight, peered at the sun, and turned a little to the left. He looked along his left shoulder and raised his left arm to point straight out from his side. "North!" he announced. He turned his head to the right and raised his right arm, "South!" He wheeled a quarter turn to the right. "East, West!" He threw a glance at James. "I'm a compass, Jim! I can tell directions! Will showed me how!" He dropped his arms, turned to get his bearings again, then wheeled through his litany of directions, raising and slapping his arms to his sides enthusiastically. "North, South, East, West! North, South, East, West! Just like a compass, Jim! Will says I can be the navigator!"

James regarded Tartop with an amused, somewhat puzzled expression. "Navigator? Where you going?"

A little dizzy from his wheeling, Tartop staggered and looked surprised. "Why, Tahiti, Jim, Tahiti! Will says he's going to Tahiti, and I can be the navigator! Lots of naked women on Tahiti, Jim! Lots, and they feed you bread that grows on trees!" Tartop sprawled at the edge of the garden, wiped sweat from his brow, then raised himself to his elbows. "Just think, Jim, cool shade, naked women feeding you grapes and bread and rubbing you all over." He closed his eyes, opened his mouth, and closed it around an imaginary delicacy. He chewed the morsel with obvious pleasure as he rubbed his genitals. "Ummm! Just think, Jim, naked women and bread on trees!"

James wore a little smile. "Will says he's going to Tahiti, eh?"

Tartop scrambled up. "Right, Jim, quick as his time's up! That's what he said. And I'm going to be the navigator!" Arms flapping, Tartop wheeled through his directions again. James watched the gullible youth. What did Will have up his sleeve? Weeks earlier, convicts from the Second Fleet stole a rowboat at Parramatta, sneaked past Sydney in the dark, then stole a larger boat from the lookout post and sailed silently away, bound for Tahiti, according to friends.

Upon hearing their destination, Sydney shook its head. For anyone to attempt Bligh's remarkable voyage in reverse was certain death, especially for a bunch of convicts. Despite Sydney's envious speculations about the idyllic lives of the *Bounty* mutineers, an impossible 3,000 miles of ocean lay between their penal life and the Eden of Tahiti. No, even for a voyage much shorter, the odds against success were staggering. James chopped around their potato plants while he mulled over Tartop's information.

~~~~~~~~~

When Arthur Phillip came back from a tour of Parramatta, he learned his favorite convict huntsman had been speared by a native. He boiled over. Enough was enough! Struggling to control his temper, he grilled Sergeant Motherwell, "Are you sure McIntyre did nothing to provoke this incident?" Motherwell was certain. He explained that McIntyre's hunting party was dozing the afternoon away in preparation for stalking kangaroos by night when they became aware of natives creeping closer. McIntyre recognized one of them and tried to engage him in conversation, but the native kept edging farther and farther into the woods. Perhaps it was foolish for McIntyre to follow, unarmed as he was. Finally the native assumed a defensive stance on a windfall, and when McIntyre advanced the man sent his spear through the convict's ribs.

"I'll never forget his words when I came up on him, Excellency. 'I'm a dead man!' he said, 'I'm a dead man!' He was on his knees, surprised-like, holding the shaft like this..." Motherwell dropped to his knees to demonstrate. He shuddered as he stood again. "We thought he'd be dead before we got back, but he's tough."

Arthur Phillip fought off a wave of nausea at the memory of his own wounding. He rocked to and fro on the balls of his feet, deep in thought, then smacked a fist into his hand. "Goddamn it!" he cried. Enough is enough! His patience had run out. It was time to teach the savages a lesson. He summoned Tench.

Tench had never seen Arthur Phillip so angry. "Look!" the Governor shouted, shaking a sheet of paper, "Seventeen! Seventeen murdered or seriously wounded since we arrived! Seventeen! And we have never, ever countenanced a single act of violence against them! Not one! Ever" He ranted about his efforts to live in harmony, to engender good will, to bring the benefits of civilization to the savage band. But apparently all to no avail. His voice dropped. "Well, maybe we have to show them for once and for all that we can destroy them, utterly, any time we want. Maybe then they'll see what's in their best interests!" He ordered Tench to bring back the heads of ten men, plus two live prisoners to be publicly executed, hopefully in front of tribal brethren.

By the time Tench emerged from the Governor's house he'd managed to ameliorate his superior's gruesome objective. He was now instructed to bring back six prisoners, if not six heads. He was not to harm women, children, or destroy possessions. And his force would not to try to gain an advantage by pretending amicability, but would act in a determined and warlike manner.

At last, a military operation! This is what the Marines had sailed halfway around the world for! After almost three years they'd prove their mettle! At 4 a.m. next morning a force of fifty set out for Botany Bay under the command of Captain Tench.

Three days later, having pushed through thickets, woods, and endless marshes on some of the hottest days in the colony's brief history, having suffered through sucking bogs and swarms of mosquitoes, having engaged in one brief, fruitless foot-chase of a few, shadowy enemy, the Marines dragged back into Sydney, their food and water gone and the men exhausted.

They returned in time to help welcome the *Waaksamheyd*, a Dutch vessel Lieutenant King chartered in Djakarta to deliver the rest of his purchase of rice and salt meat to the little colony. Eight months earlier, when the *Supply* on its desperate mission for food, Sydney was on the verge of starvation. Although the colony was no longer starving, the *Waaksamheyd* was welcome anyway, for Arthur Phillip had once again cut food rations to two-thirds, convinced that future supply ships would fail to arrive any more dependably than past ships. Prudence dictated that Sydney endure a little suffering. On its voyage to Sydney the Dutch ship had also suffered. Sixteen crewmen died on the three-month voyage.

The master of the *Waaksamheyd* ('wakefulness') possessed an unflattering character; he provoked in the English both dislike and scorn. For one thing, except for a dickering pidgin, he couldn't speak English. Indeed, some thought he wasn't Dutch at all but Malay, or at best a half-breed. For another thing, he frustrated Arthur Phillip with the two letters he brought from Djakarta, one written in English, the other in Dutch. The letter in English, well composed, set forth in meticulous detail the terms of the charter of the *Waaksamheyd*. The letter in Dutch (that not a single Englishman could decipher and that the cloddish Dutch captain could only roughly translate) appeared to tell of a state of war between England and Spain. Just like the Dutch - take pains with the guilders and never mind the politics!

Too, the Dutchman appeared to be a cheat. When the English calculated the weight of his rice he was several thousand pounds short. Shown the evidence, Captain Detmer Smit agreed to make up the difference with some of his other cargo, butter so long absent from the colony. He was also willing to consider the further charter of his vessel to return the shipless men of the *Sirius* to England. To open the negotiations for this service he asked a price ten times what the English thought reasonable.

Negotiations with Captain Smit continued as Tench regrouped his Marines for another expedition against the wily native. In fact, Tench rested for only a couple days before assembling a smaller, more mobile force. He let word slip out. This second foray would be north towards Broken Bay to apprehend the villain who wounded Arthur Phillip. The story was a ruse. Tench set out by moonlight, then doubled back to Botany Bay to catch the natives sleeping.

Three days later, Tench again staggered back into Sydney with no more to show for his efforts than empty food packs and filthy uniforms. In oven-like heat, his Marines had again scoured Botany Bay for the phantom natives, executing feints and forced marches to catch their quarry unawares. Once they'd bumbled into quicksand. Sunk to his waist, immobilized, Tench stupidly watched his brass buttons disappear one by one, his mind suddenly blank. When his men called for instructions he looked at them with such bewilderment they wondered if he'd lost his senses. Luckily, Pottle remembered hearing how British soldiers in America escaped quicksand, and cut branches to fish Tench and the others to safety. Last to be rescued was Sergeant Motherwell, nearly sunk out of sight. It took a rope to pull out the squat sergeant. If the Marines hadn't brought rope to bind their elusive enemy, Motherwell would probably not have lived to recount the thrilling story of his Botany Bay campaign under the command of Captain Watkin Tench.

As 1790 drew to an end the unnatural heat continued unabated. Once again gardens shriveled and died. Livestock panted for air. The stream that quenched Sydney's thirst shrank. At Parramatta, thousands of giant bats blown seaward from a baking interior fluttered down dead from trees.

Would life ever be bearable in this end-of-the-world purgatory of violent rainstorms, drought, bone-chilling cold, and unremitting heat? Memories of hunger and the terrible death toll of the Second Fleet haunted soldiers and convicts alike. To be going home to England aboard the *Waaksamheyd*, quit of this hellhole called Sydney, the 130 sailors of the lost *Sirius* were deemed luckiest of men.

Kupang

1791

*H*e saw himself near Bennelong's hut. He was alone, but felt someone might be watching. Tench? Then Gooreedeeana emerged from the hut and strolled toward the beach. He followed her silently, seeming to glide, not sure why he followed. She was naked except for a short skirt of strap-like leaves that swayed as she picked her way through the undergrowth to the water. She waded to her knees, paused, waded deeper, and bent to splash water on her face and arms and front. Not like other natives, she was taller, more slender, and her breasts were firm and high and gleaming with wet. He could feel her firm breasts in his hands, her protruding nipples nested between thumbs and forefingers, cool to his lips.

Her face was regular, almost European, her nose narrow, her lips well formed. When she turned and saw him watching, she smiled. Even, white teeth. He glided closer, captivated by the most beautiful eyes he'd ever seen. They were brown and lustrous and looked out from beneath long, dark lashes with a hint of teasing, of coyness. His heart ached at the loveliness of her eyes.

She backed into deeper water, laughing softly, inviting him to join her. Her skirt floated petal-like around her hips. She was a floating flower. He followed, not thinking of his clothes until he looked down and saw his uniform. He followed her deeper. When the water reached his groin he felt a delicious coolness. She kept backing away, smiling, speaking words of invitation he couldn't understand. He wanted her with the most powerful ardor he'd ever known.

Then she was on the beach, her wet hair shining and sun-spangled, her body shimmering in water-mirrored sunlight. She began a slow dance, both arms swaying above her head. She reminded him of a graceful palm tree swaying from wind. She crooned a repetitious little melody of nonsense. Through her hair he felt bumps and scars and welts. She turned to him and indicated he should raise his arms. His arms felt heavy and awkward, and when he raised them he saw they were bare and that he too was naked. He tried to follow her graceful dance. At first he was clumsy, but then he began to feel the languorous rhythm and his movements became smoother. He could make out the soft words she sang - her name, "Goo-ree-dee-ah-ah-nah, Goo-ree-dee-ah-ah-nah." She indicated that he should sing too, and when he began to sing her name she shook her head. "Pee-der-pah-ah-ah-dul," she sang for him. "Pee-der-pah-ah-ah-dul." She raised her fingertips to his mouth as he tried the sounds, and he saw the end of her little finger was missing, the three remaining pads the softest things his lips had ever touched.

At first his voice was rough and cracked, but as he sang through her fingertips the sounds and his dance came together and he sang with pleasure, his heart swelling. He swayed with the rhythm, following her light, joyful steps, wanting the music to go on forever as they sang their names in the dance of the palm trees.

Then he was by the roots of the big tree that had lost its footing at the very end of Bennelong's point. The roots still nourished foliage, so a boat always had to get beyond to see down harbor. Gooreedeeana sat and scraped away sand, revealing a treasure of clamshells, oyster shells, and whitened fish skeletons. She selected a shard of oyster shell and indicated he should sit between her legs. As he scuttled back against her he wanted to be behind her screen of leaves, to smell the sweet pungency of her sex. He knew she would smell like flowers. He sat back, her smooth inner thighs cool against his sides. He felt the softness of her cheek on his, and she kissed the lobe of his ear, crooning his name. He felt her pinch his shoulder but didn't feel the shard slice his skin.

He watched her slice and lick until she'd cut parallel tracks down to his biceps. Then she began tattooing the other shoulder. He licked his salty blood and traced the small ridges of the lacerations with the tip of his tongue. Then he found himself sucking her knee, her skin smooth and cool, salty. He knew Quint was watching.

He looked back and saw Quint sitting against the roots of the tree. He was crying. "I can't find her," he said. "I've looked all over, but I don't know where they put her. Help me."

"They took her to Norfolk," Pottle said.

Quint put his tongue in the gap of his missing tooth. He looked perplexed and unhappy. His eyes were red from crying. "I can't find her," he said again, "Help me."

302

"They took her to Norfolk. You have to go to Norfolk."

"I can't find her." His head sank and his shoulders shook. "I don't know what she looks like. Help me."

Pottle had seen Quint cry only once, the night before he was hanged. Now he put a hand on Quint's shoulder and noticed that Quint's uniform looked as new as when they left Plymouth. He saw he was in his own uniform again, but it was faded to an ugly pink. Quint's shoulder felt warm and solid.

"You mean your baby?" Pottle asked.

Fisting his eyes like a little boy, not looking up, Quint nodded.

Now he was across the harbor where they buried the dead from the Second Fleet. He watched the convicts finish a grave in the gray mist, a big hole. When the boat came they stripped the wasted, stiffened corpses and tossed them into the hole. The grave-workers wore dirty cloths over their noses and mouths and handled the white, naked bodies roughly. He knew now he was dreaming, because the dream kept coming back of masked convicts in a mist, tossing stiff, naked bodies into a hole. After the convicts filled the grave he'd dream of their snarling squabbles over the salvaged clothes.

Every day Pottle had stood guard on a rock above the gravediggers, watching them dig, watching them squabble, keeping them from harm by the natives. He began to cry at the endless, stiffened bodies, at the waste, at the stupid machinery of punishment, the indifferent cruelty. Pottle wanted to go home.

~~~~~~~~~~

James took a seat near Will on a slab of rock and wiped his face and neck. "Listen, Will, what're you up to?"

Will wiped his own brow and resumed knotting. "What does it look like? I'm making a seine."

James watched Will knot the new seine, admiring his deliberate, dexterous skill. "The talk is you're slipping out."

Will worked for a time before he looked up. "My time's almost up, Jimbo. They got no right to keep me here."

James nodded. More and more convicts insisted they deserved freedom, but the unspoken system was clear. If you said your time was up, they asked if you planned to settle. If you said "Yes," they were all sweetness and smiles. "As soon as we hear from England you'll get land and tools and livestock." But if you said "No" - end of discussion. So far only one convict had been emancipated, James Ruse, and he was an experiment. Along with food for a year, Ruse got an acre of cleared land, plus the tools and livestock. Arthur Phillip wanted to know how soon Ruse could become self-sufficient. Freedom for James Martin was not yet an issue. He had two years more to serve.

James wiped his face and neck again. The heat was infernal. Unless he and Tartop watered their garden every day it shriveled overnight. The Parramatta grain crop looked like a total failure.

"What about Mary B and...and the little ones?" James asked, stumbling over how to handle paternity.

Except to wipe his brow and mouth with a shirtsleeve, Will gave no indication he heard. He stretched out his work and eyed it, his project a commission from one of Tench's fellow captains. "I'm not a farmer," he said.

James chewed his lower lip. Suddenly he put a hand over Will's spindle. "Goddamn it, Will, talk to me! This is my business too!" Will looked at the intrusive hand. James lifted his hand. "Come on, Will, I don't hold you a grudge. We ain't enemies."

Will studied James, then took a huge breath. His words sounded rehearsed, as if he spoke a practiced a speech. "My time's up. There's nothing for me here. I'm not a farmer, I'm a fisherman. I don't know farming. And even if I stay they won't let me build a boat big enough to get outside the headlands. They think someone will steal it and sail away. Well, that's not my lookout. I've got to have a good-size boat. I'll be poor forever if I don't have boat enough to get to open water."

Then he said his marriage wasn't even legal. The banns weren't published three times according to church law. They weren't published even once, so none of the early marriages were legal. And they still used Mary B's name at the commissary. According to the commissary agent, she wasn't Mary Bryant, she was Mary Broad. So how could she be his wife if she still had her own name? And they never asked if he had a wife in England. This last he said with a tone of disgust.

"Listen," he said, "You know why they don't want to know? Because if I had a wife back home I couldn't marry here, could I? Sure, they'd believe me if I told 'em I had a wife in England; then they wouldn't let me marry here. But they won't believe me when I say my time's up, will they? They won't let us go home, will they?" He shifted his work and began knotting again with abrupt, angry movements. "They only believe what they want. They only do what they want. And we're the ones to pay." He paused to hawk and spit. "Well, I'm not paying any more! They took my boat back home! They took seven years! But now my time's up, and they ain't getting more. The law's the law!"

James was silent as Will poked and yanked the spindle. Finally James put his hands to his knees and pushed himself up. He took a couple of steps before turning back. "Don't you love her? Don't you care what happens to her, to the babies?"

Will did a couple more knots. His head bent to his work, he said, "My time's up and the law's the law."

~~~~~~~~~~

When he was still some distance from the Bryant's hut, Stork heard voices raised in argument. Their little girl played in a dirt patch under a tree. He felt a twinge of regret when he saw her stiffen at the sight of his uniform, and although he smiled, she remained impassive. He heard something about "no right to go off" before he coughed to make his presence known. There was silence in the hut, then Will appeared in the doorway. His eyes grew hooded when he recognized Stork. The soldier flushed.

He was surprised at Bryant's haggard appearance. Even long seasoning by sun and wind didn't compensate for his invalid skinniness. His naked, white torso was a register of great hunger and the return to two-thirds rations. Stork was momentarily shaken by a revisitation of pale, emaciated, grave-bound convict bodies.

Stork raised a hand to his mouth and coughed again when Mary B appeared behind Will. "Afternoon, mistress," he nodded. She was holding a baby. A brief shadow crossed her face at the sight of the uniform, but she brightened when she saw it was Stork. Conscious of his ragged coat and ill-fitting, convict-issue shoes, Stork leaned on his staff, his eyes searching the ground right and left, his Adam's apple working violently. Finally he said, "I wonder if I...look, I brought...." He fumbled in his side-pocket as his eyes roamed the ground. He looked up. "It's about Quint, Quint's baby, I mean."

Their confused expressions flustered him further. He was going to speak again when he was distracted by the little girl sidling past him, then darting between Will and Mary B to cling to her mother behind her dress. "I don't quite know...."

"Won't you come in out of the heat, Mr. Pottle," Mary B offered, pushing past Will. "Perhaps you'd like a drink after your hot walk. Will, fetch Mr. Pottle a chair." Stork deflated with relief.

After a cup of water and small talk of the heat, Stork fumbled out his bottle of rum and presented it to Will, and after they drank several toasts to health he was finally able to explain his mission. He chose not to tell of his dream, or that Quint now appeared regularly, sometimes in dreams, but also during the day, when Stork was sure he watched from somewhere near, a presence so real Stork constantly caught sight of Quint's grin just as it ducked behind a tree or a barracks corner. Unable to tell them of Quint's presence, he introduced a story about Quint wanting something on his marker, something about "father of...."

"But she didn't have a name," Mary B explained. "She was stillborn, you see, so she got no name."

"Well, yes," Stork agreed, of course that was true, but Quint hadn't known she'd be stillborn, so he would've thought she had a name he could put on his marker. In fact, Stork thought Quint would want to know her little grave was properly marked too, now that she was dead, and when you consider....

Mary B sipped her grog. "Well," she said, "I'm not sure what the name was going to be. We didn't know whether she'd be a boy or a girl, so Seedy was thinking about several, and then when the poor thing came dead, well, it didn't seem to matter."

"Well, that's all right," Stork said. He could put all the names on Quint's marker. He thought that would be fine with Quint too, because then Quint could choose which name he liked best. In fact, he thought that's what Quint would want when you consider....

The Bryants nodded, and Will refreshed grog all around. Then Will asked if Stork was going to put all the names on the little one's marker too. "You'll need a pretty big marker for such a little grave, Mr. Pottle."

Stork looked non-plused. He thought for a while, drank, thought some more, then drank some more, and finally said, well, Quint actually might not mind if he put 'father of Seedy's daughter' on Quint's marker and 'daughter of Quint and Seedy' on the little one's marker. The important thing was to keep it all straight. That might even be better, Stork said. In fact, that's probably what Quint would want when you consider....

"Well, yes," Will agreed, "'Father of Seedy's daughter' sounds good to me." They clunked their pewter mugs at Pottle's solution to this knotty problem.

After a minute, Mary B said, "Cedelia! Seedy's real name is Cedelia, not Seedy, so it should say 'father of Cedelia's daughter.'"

Will reared up a little. "Well, if everybody calls her Seedy, why put Cedelia on the marker?" He said that seemed like a dumb idea to him.

"Well, it's her name," Mary B said. "Seedy isn't her proper name, and if it's going to be on a proper marker she ought to have a proper name."

Will snorted, "Well, nobody will know who Cedelia is. It could be anybody, even you!" He laughed and winked at Stork as he stood to refresh their mugs again.

Mary B rose to place her now-sleeping baby in his crib. The light in the hut was dimming with the sinking sun. She stood reflective at the crib before addressing Stork. "Well, I don't know how you'll do it, but I can show you where we buried the little one so you can put up a proper marker. But she isn't on the hill with the other ones, being born dead and never baptized and all."

Stork nodded, his bleary eye distracted by the foreshortened little finger of Mary B's left hand. Memories of Gooreedeeana flooded his mind. The "Botany Bay Beauty" they called her, discovered on the second futile mission formed by Captain Tench to punish the natives. They'd come upon her probing for shellfish, immersed to her thighs, a sunlit goddess rising from the sea. Although her companions fled, she remained to enthrall Tench with a bold self-assurance that set the captain back on his heels, a clearly smitten Marine. Tench later speculated that someone from Cook's voyage of exploration might have been fathered her. Now Stork remembered the dream-touch of Gooreedeeana's fingertips and the missing tip of her little finger. In a ritual at birth the natives removed the last two joints of the little finger on the left hand of infant females by tying them off with thread braided from the mother's hair. After a time the end of the finger shriveled and dropped off like a second umbilical cord. Now he remembered how Mrs. Bryant's finger was bitten nearly through and how Sergeant Motherwell snipped off the mangled end.

He looked up after a long introspection to see Will and Mary B staring at him expectantly. Embarrassed by his reverie he regarded Mary B's expression and suddenly realized she'd asked him a question. In the dim light he saw the lustrous eyes of Gooreedeeana. "Whew," he said, swirling the grog in his cup, "Strong stuff!".

The hot walk must have dried him out, Mary B said. She asked again if he'd heard anything from Norfolk about Seedy Haydon. No, he said, he'd heard nothing of Seedy, but now that the rest of the men from the *Sirius* were coming back there might be news.

"I hear they'll all go to England on the Dutch ship," Will said.

"Aye," Stork answered. A pause. "And I'd go too."

"Aye, and many like you, Mr. Pottle."

"Aye." Silence. The men drank, Stork draining his mug. "Might as well finish it," he said, reaching for the bottle to share out. He shook the last drops into the fisherman's mug. Will said, "Thank you, Mr. Pottle, you're a good man. Here's to what's right." They clunked mugs and drank.

"So you won't be staying to settle, Mr. Pottle?" Will said.

"No, I'm going home, Mr. Bryant, going as soon as I can." A pause. "This is no place for me."

"Aye, it's a hard life." A long pause. "Too bad about Mr. McIntyre." Will referred to the death of Governor's convict huntsman, whose spear wound had festered and caused a lingering death

"Aye, too bad for sure." Silence.

"And too bad for a woman with children to be left alone," Mary B said with surprising archness. She smoothed Charlotte's hair as the little girl leaned against her knee while sucking her thumb, the child's eyes unwavering from the military man. Stork looked at Mary B, at Will, then studied his cup.

Will rushed in. "Misery all the time," he said, "One good month and then two bad. Seems to get worse all the time."

"Aye. Misery, all right. Misery enough for all. And at the end of the world to boot."

"Aye, you're right there, Mr. Pottle, well said. Misery enough for all, and at the end of the world to boot."

They heard a distant roll of thunder. The three exchanged looks. "Maybe we'll get some relief from the heat," Mary B said. The men agreed.

"Will you be staying a military man back in England, Mr. Pottle?" Will asked.

Stork swirled his cup. He suddenly found the play of light in the sharp, aromatic turmoil of his liquor quite profound. "No, Mrs. Bryant, I've had enough of Army life."

Will nodded. "Well, it's not an easy life."

Stork agreed.

More silence. Then Mary B spoke. "Will the military men leave their wives behind, Mr. Pottle?" Stork looked blank. Will made a sound in his throat. Mary B paid no heed. "Well, don't soldiers have wives among the convicts...and children to boot?"

Stork's shoes scraped the rough planks. "Well, yes..."

"But that's mostly officers," Will interrupted, "Not private soldiers like Mr. Pottle! Besides, what's a man to do, eh, Mr. Pottle? When a man's served his time he's served his time, ain't that right? After all, they ain't real wives!"

Again a roll of thunder, closer. They glanced toward the door as if the coming storm might announce itself there. Stork gathered his legs to rise. Will beat him to his feet. "Well, awfully nice of you to stop by, Mr. Pottle," he said, extending his hand. Stork was able to half-rise, then fell back, and took Will's offered hand to pull himself up. He grinned foolishly at the fisherman.

Mary B asked, "What about the grave? Shall I show you the little one's grave?"

"Grave?" Will said, "Oh, yes, the grave! Well, maybe we all better go. I haven't seen it yet myself. How would that be, Mr. Pottle, we'll all go!" A momentary gleam of lighting in the hut, then silence, then another long rumble of thunder. Indecision hung in the air. Befuddled, Stork looked from one to the other. "Well, maybe when the weather's better!" Will said, advancing to the door. "A man might get soaked the way it's looking."

Mary B extended her hand in farewell. "Quint was a good man," she said. Stork nodded. He turned to leave, but Charlotte detached herself from her mother to grab his finger. She pulled on him, indicating she wanted to tell him something. "What is it, child?" Mary B asked. "Do you want to tell Mr. Pottle something?" Smiling, Mary B explained, "She's so shy around strangers, she must like you."

Stork smiled back and took a long time to bend his lanky frame until his face was near hers. He swayed. "What is it?" he asked. Charlotte put a hand between her mouth and his ear and whispered semi-intelligible words. Stork straightened a little with mock surprise. "A secret? You want to tell me a secret?" He winked at Mary B and bent again, a hand on Charlotte's head to steady himself. "All right," he stage-whispered, "Tell me your secret!" Charlotte whispered in his ear. "No!" Stork said, "No!" Charlotte nodded up at him solemnly. "A coney hole!" he said as he straightened. "Well, isn't that something!"

Will strode over to pick up Charlotte and kiss her roughly on the cheek. "Well, aren't you the little tattletale!" he said, giving her a spank on her butt. He looked at Stork with a fierce grin. "She's going to give away all my secrets. Pretty soon a man won't be able to keep anything safe."

Stork smiled back, at a loss for what to say. Finally he blurted, "Well, best have a hiding place."

"Aye," Will said, "A man can't be too careful with all these thieves around!" He laughed loudly at his own joke, which Stork joined as Will steered him to the door.

Outside they regarded the storm cloud, a blue-black mountain threatening to tumble down on their heads. "Jesus, we're going to get a hell of blow!" Will said. "I better check the boats." Distant treetops whipped in the vanguard wind. Mary B carried out the baby, who fussed at his nose with a fist. She joined the men to look across the cove at the approaching storm. A blast of wind, blessedly cool, pressed her dress against her body, streamed her hair behind. Stork couldn't help looking at her with admiration. They all enjoyed the relief of the wind until it strengthened and began to swirl dust and debris, then Stork briefly shook hands with Will and headed back to barracks, his coattails flapping, his long, rapid strides eating up the path.

It was a year later, on his way back to England aboard the *Gorgon*, when time hanging heavy nurtured memory and rumination, that Pottle thought again of that strange, rum-befuddled afternoon with the Bryants and of their later escape from the prison colony. It was then he realized he'd been surrounded by hints of Will's plan.

"Why was I always so sure of things back then," he wondered. "I didn't even see the obvious!" He wondered also how Quint could know that Seedy's stillborn baby was a girl.

~~~~~~~~~~

Reseating the broad-brimmed straw hat he'd plaited himself, the Reverend Johnson pocketed his handkerchief and leaned on his hoe. His face was florid. His eyelids fluttered in confusion. "Why, of course," he said, sputtering a little, "I married you myself! Why would you ask such a thing?"

Mary B shifted Emmanuel to her other hip and looked at Charlotte staring at Abaroo. In the almost daily traffic of natives going to and from Bennelong's hut along the path in front of Will and Mary B's place, the women were nearly naked, scarified, sometimes painted, and wore pieces of shell and pretty stones or bits of ribbon stuck to their hair with yellow gum. But Abaroo, costumed in a dress and bonnet, clean-faced and unadorned, was the first dark-skinned woman Charlotte had ever seen who dressed like Mary B.

Mary B said, "Well, I just want to make sure."

"Well, do you doubt you were married?" the preacher asked.

Eyes to the ground, Mary B didn't answer. She watched the progress of a beetle and raised a bare foot. The beetle froze. Mary B let the preacher's beetle live. When she put her foot down again she felt the heat of the earth creep through the thickened sole of her bare foot. The beetle scuttled beneath the foliage of a turnip.

"Well, if a husband's time is up is he free to go?"

The preacher wiped his brow again and watched Abaroo patiently pick potato bugs and drop the squashed insects into a bucket for the chickens. "Well, he's a free man, of course, like any man, but he can't just go off and leave his wife. That wouldn't be right! At the very least he'd have to come back."

"But what if he says he won't come back? Back home men go off all the time and don't have to promise they'll come back, so what would the law say here?"

The preacher hesitated, "Well, I'm not sure. You come as individuals, of course. Just because you get married doesn't change that, I suppose." He studied the ground. "Actually, I never thought of this before. Interesting problem." He pondered, then looked up. "But of course none of that changes the fact that you're married. In the matter of marriage you're bound forever, even if your sentences don't run concurrently, ah, you know, together."

"So he has to stay till his wife's sentence is up?"

"Well, he's the husband! Of course he should stay! He can't just go off and leave a wife, and especially with little ones! That wouldn't be right!"

"But what if he says the government has to feed his family anyway? That his wife gets her own ration and doesn't need him to provide for her?"

"Really? Is that what Mr. Bryant is saying?"

Mary B shuffled. She hadn't wanted to get so deep into the matter. After all, what if Will was all talk? What if he backed out of his plan? Just like him to brag and make big plans and do nothing! She forced a smile and shifted Emmanuel to let her side cool a little. "Oh, no. Some of us women were just wondering. You know, there's talk all the time now with so many men getting ready to leave. But like I said to Will, when we go back to England we'll all go back together, as a family, a First Fleet family."

Once more the preacher removed his hat to wipe his brow. "Why, yes, yes, of course," he smiled, "That's the only way. A family has to stick together. Like the Bible says, seven years will seem like seven days, if he loves you."

Mary B nodded. She realized later that she'd forgotten to ask about the banns.

~~~~~~~~~~

Will's manner was urgent. "Listen, Jim, I want you to come."

James Martin looked at the fisherman with surprise and glanced at the receding figure of Tartop, off to fetch water to keep their garden hopes alive with two wood buckets dangling from a shoulder pole. Water-dark circles of soil surrounded their wilted potato plants. Storms promising rain brought only wind and a few spattering drops. Even Sydney's water supply was drying up. The once-vigorous stream that quenched Sydney's thirst, boiled its beef, washed its clothes, and hid its alligator was shrunk to a polluted trickle. Stonecutters chipped reservoirs in its sandstone bed to store the frail trail of water that trickled in from nearby marshes. As daily temperatures reached the high 90's, sometimes breaking 100, drought destroyed all but the most carefully nurtured gardens.

James said nothing, but led the way to the overhang of his hut. He leaned his hoe and the two hunkered in the shade. "Why?" James asked, his voice unnaturally loud.

Will looked around nervously, picked up a piece of bark, and began shredding its edges. His answer was a half-whisper. "Well...I need you."

James took up his own bark to shred. His voice again too loud, "Why?" he asked.

"Jesus! keep it down, will you?" Having shredded his piece of bark, Will got another and set to work again. "I need your arms," he said in his half-whisper, "We need good arms if we get chased."

James tossed aside the remnant of his bark. "We? That the only reason?" His voice was a little softer.

Will shredded the last of his bark and whisked his hands clean of the debris. He scratched in front of his ear, picked his nose with a thumb, and snorted twice. Finally he said, "Well, she wants you to come too."

"Mary B?"

Will nodded. He leaned forward, hawked, and let his spit fall between his feet.

James felt his heart swell a little. "She wants me to go with you and her?"

Studying his phlegmy blob in the dirt, Will nodded again.

James shifted as he thought for several moments. "What if I don't?"

Will spit again between his legs. "She won't go."

"Well, what's wrong with...ahhh, I get it...if she don't go then you don't go, is that it? She won't let you go alone."

"She says we're married."

"Well, you are!"

Will said nothing. After a long pause James asked, "What's in it for me? We get caught and it's a flogging, maybe worse. I've never been flogged, and I'd like to live my life without the honor. Besides, you haven't got a chance in hell of making Tahiti. Christ, Will, that's open water! And you're taking Tartop?"

Will snorted and regarded James with mock disdain, a gleam in his eye.

"All right then, damn you!" James said, rising to his feet, his voice a harsh whisper, "Tell me! Tell me what you've got planned! Tell me why I wouldn't be a fool to climb in a rowboat with you and Mary B and two babies and Tartop! Come on! Are you going out past the heads to catch a ship to England? Is that it? You going to find a whaler? What, you think there's ships out there just waiting to welcome you aboard? Listen, this is a penal colony, Will, and they'll know you're a convict! Besides, if they ever catch you back in England it's life, Will, life, and probably right back here, too. If you live to see the day!"

"You through?" Will asked, rising to his feet with lackadaisical ease, "You got your spleen out now?" He faced James with arms akimbo. "Listen, you said she wasn't yours, that you had a wife! She asked you to marry her and you said you couldn't! It was you who decided, Jimbo, not me! So, don't blame me!" He dropped his voice. "You could've, Jim, you could've married her and no one would've been the wiser!"

"Like you?"

"Aye, like me! I did what I did, but that's over now, and now I'm asking you for help. She wants you to come and I need you! So now are you Goddamn satisfied?"

"So it's her!"

"It's you, Goddamn it! You! I need your arms!"

"Will, you'd say anything! This is just more of your bullshit, trying to sucker me into your hare-brained...."

"It's not hare-brained!"

"Well, tell me then, Goddamn it!"

"Not here."

"Why not he...?" James interrupted himself at Will's gesture. He turned to see Tartop whistling up the path under his pole. They joshed with the boy a few minutes before starting up the hill behind the huts, Will talking earnestly.

When James was alone again with a chance to think Will's plan through he marveled at its thoroughness, its timing, the logic. He now saw he was the last piece of the puzzle, the piece that completed the picture and locked it all together. With him joining the escape, Will secured the last pair of strong arms he needed, plus Mary B's silence and cooperation.

~~~~~~~~~~

To William Wyndham Grenville, Secretary of State for the Home Department:

"*Sydney, New South Wales, March 5, 1791*

Sir,

In my former letters I have requested instructions relative to those convicts who say that the terms for which they were sentenced have expired, and who, refusing to become settlers, desire to return to England. To compel these people to remain may be attended with unpleasant consequences....

"The language they hold is, that the sentence of the law has been carried into execution, that they are free men, and wish to return. I have no means of knowing when the sentences of any of the convicts expire who came out in the first ships....

"...though there has been no very great impropriety in the conduct of any of those who say the time is expired for which they were sentenced, it is more than probable that they will become troublesome as their numbers increase...."

Arthur Phillip laid aside his quill, removed his spectacles, wiped his face, and massaged the sore bridge of his nose. He felt headachy, and the throbbing in his side had returned with a vengeance. His left shoulder burned where the spear skewered him. For the hundredth time he silently cursed the fools half a world away who caused him such daily aggravations in this hellhole. Every week another convict or two declared his sentence completed, and most wanted out by the first ship. What was he supposed to do?

He turned his little hourglass to watch the trickle of fine, white sand dance on the base and begin forming a mound. *Omnia tempus revelat.* Someday perhaps the fools will wonder why they ever thought they could people this distant desert with jailhouse jetsam. No convict in his right mind would choose to live here. Any convict in his right mind would rage to return to England rather than live here one day longer than he had to. But most will remain, he thought. Lacking intelligence or skill or strength or means or youth or determination, they'll accept this half-life as the best that life can be. Failing at their one chance as farmers, most will live as beasts of burden, as inept servants and wage slaves. And get in trouble again and again, and some will be hanged, and most will go to their graves with sighs of relief.

Arthur Phillip roused himself from his grim brooding and reread a letter from the Reverend Johnson. The preacher thanked Arthur Phillip for his grant of four hundred acres for church property and asked again for convict labor to help build a church. The tone of the letter made clear the Reverend thought it unseemly in the eyes of God that he should be holding holy services under trees and in half-finished storehouses and in the fishy boat-house for three years. How many pestering letters had this tiresome preacher written about his Goddamn church? With everything Arthur Phillip had to deal with, Johnson bedeviled him like a Goddamn whining mosquito about his Goddamned church!

The Governor dipped his quill and scrawled across the bottom of the letter, "*As soon as conditions permit. A.P.*"

~~~~~~~~~~

Katie handed Mary B a little packet, "I hope this helps, Mary B."

Shifting Emmanuel and slipping the packet into her apron pocket, Mary B said, "Me too. You know how Will is about letting on he's sick, but I can't stand seeing him suffer those awful headaches. If this doesn't help I hope he'll see the surgeon."

"Well, maybe these powders will do the trick," Katie said, and caught Hester Thistlethwaite's eye behind the dispensary counter. She indicated she was going outside with Mary B. Hester smiled and nodded, then smiled at Mary B. "I'll watch her if you like," she said, pointing with her chin to Charlotte, who was trying to rouse a sleeping cat by yanking its twitchy tail. She scolded the cat, "Get up, you razybones, get up, you razybones!" Mary B smiled her thanks to Hester.

Outside, Katie chuckled as she tossed her head back to the door. "I wonder where Charlotte learned that?" Mary B smiled sheepishly.

The heat was already building as the women strolled toward the new battery on Observatory Point, a makeshift little fort rushed into construction with the *Waaksamheyd*'s news of a possible state of war with Spain. No obstacle to a ship of war, the fort would serve little better purpose than to poop out a shot or two of symbolic protest before Sydney surrendered. But appearances were important. If England must bow to Spain, let her bow with dignity! Hear, hear!

Katie fished in her apron to give Mary B a small vial. Mary B slipped the vial into her pocket with the powders. "Thanks, Katie. It's just what we need."

Katie nodded. "Just a few drops now. They're little kids. Do you need anything else, anything at all?"

"I don't think so. Have you heard anything, any talk?"

"No, just the usual."

They strolled in silence, their progress so slow that Katie's limp was scarcely noticeable. Mary B stopped. "Katie, do you know what day this is?" Katie thought as her eyes searched the skyline, then shook her head. Mary B said, "Four years ago today we were put aboard the *Charlotte*. Do you remember? Do you remember that day?"

"Hardly anything. So much seems like a dream now, like I was somebody else."

"I remember parts. Do you remember Captain Tench chasing Blackpool? So angry about how dirty we were! He wanted to smack him with his sword, remember?"

Katie laughed with Mary B, then added, "And do you remember how scared Seedy was we'd all drown? We thought she'd piss herself she was so scared!" They continued laughing as they recalled scenes from their early days aboard the *Charlotte*, the hulk, the Exeter jail, and finally the fateful day that saw them arrested for Katie's fight with Agnes Lakeman. Their laughter died. They grew silent, pensive. Katie began to cry.

Mary B set Emmanuel down and took Katie's hands. "Listen, Katie, I've wanted to tell you something for a long time, but I couldn't, and now is my last chance."

Her head bowed, Katie shook with silent sobs. After all this time, she thought, I'm finally going to hear it.

Mary B lifted Katie's chin so she could smile directly into her eyes. She spoke with a catch, "I love you, Katie. I want you to know that I love you. I really do. And for a long time I hated you because of this." She indicated Sydney. Mary B paused. "But I've learned a lot about myself, Katie, by what we've gone through.... Part of it was thinking I deserved this and had to pay. I don't know why. Sometimes I thought it was something I never got caught for, something bad, and what happened in Plymouth was just my past catching up." She paused. "But I don't feel that way anymore. Now I know I don't deserve to be here, not anymore. Maybe some do, some bad ones, but I don't, and you don't, and that's why I'm ready to go. This isn't where I belong. If I thought I deserved this I'd stay, even alone."

Katie nodded. She looked as if she might speak, but remained silent.

"But I'm ready to go now. Really, I'm ready." Mary B paused for a long moment. "Look I want you to have this." She'd reached into her apron and now extended her *Book of Common Prayer.* "It was my Grandma's, in Little Bill's box. Remember? It's the first book I ever owned, and I want you to have it. Maybe it can give you something, some comfort, some words."

Katie fondled the little book, gently riffled the pages. "Thank you," she said, her voice small. She rubbed the cover as she struggled. "Thank you," she said again, "I...I'll keep it forever." Mary B nodded.

On their way back to the dispensary Katie became expansive. She chattered about her work and how she and Hester Thistlethwaite were planning to open a shop. "Nobody's got money now, but when people from England start settling here things will be different. Even if we start with just a little grog shop, that'll be something, people with money coming to pay." As Sydney revealed itself through the trees, Katie exclaimed, "Isn't that something, Mary B! All these buildings! Do you remember how it was nothing but trees and darkness?"

~~~~~~~~~~

"Goddamn!" Tench spoke with quiet amazement. A soldier had just uncovered the hiding place beneath the plank floor of Will and Mary B's hut. The cavity was almost two feet deep, the size of a coffin, lined with dried grass still bearing the impressions of several objects.

Judge Advocate Collins peered into the emptiness. "That son of a bitch! I knew he was up to something! He got too quiet. That son of a bitch!" He paced the cabin, then kicked the bed. "I knew it! I told Henry a couple weeks ago we ought to tear this place apart, didn't I, Henry, a couple weeks ago! Too much back and forth out here I said, didn't I, Henry? I knew it! Who was supposed to watch him anyway?"

Weary and bilious, with a massive hangover, Provost Marshall Henry Brewer shifted his feet, but said nothing. Several weeks earlier, overhearing raised voices, the quarter-guard sneaked up on Will's hut after curfew, sure of a violation. But instead of breaking up a card game or drinking party, the constables had found several convicts gathered around Will's table drinking sweet tea. The men were from the Second Fleet, come down from Parramatta to fetch supplies. They intended to return to Parramatta at first light, they said. Despite a search the quarter-guard found nothing incriminating. Summoning Will to his hut *cum* office next day, Henry Brewer warned the fisherman about the curfews, sentence up or not. But if the Provost-Marshall shared Collins' suspicions, he'd done nothing. Badly hung over then too, desperate to lie down again, he'd dismissed Will summarily.

Now Brewer raised a sheepish eye to Tench. Bryant the bully, the braggart, the boaster, as puffed up as one of those ridiculous hot air machines - gone!

Red-faced from the heat and exertion, Sergeant Motherwell appeared at the door. "I think we picked up their trail, Captain." Tench nodded and turned to follow the others, then paused. Something nagged, something unusual about the hut, a feeling that came to him as soon as he began looking around, some incongruity. The plank floor was an anomaly of course. Nearly all huts had earthen floors, but Arthur Phillip had treated Bryant well. No, it was something else. The hut was stripped clean except for the crude furniture - a rope bed with the ropes gone, a table, two chairs, a stool, a crude bench, a crib. Two kids in the crib? He put his hand on the rough bed-rail. This was where she slept, made love, gave birth.

He failed to find whatever nagged him, but as soon as he stepped into the early morning heat and waited for his eyes to adjust he found himself looking at it. The geranium! In morning shade by the waterbarrel, a still-dark circle of soil. He imagined her juggling the children, pausing in the darkness to dip a cup of water from the barrel to put on the flower, Bryant hissing, "Hurry up!" Why did she bother? Was the little flower a token for posterity?

Soldiers speckled the area around the hut. Ragtag, they peered and poked, commented and speculated, scratched their heads and stubbled jaws. What were they supposed to find, an escape plan? They poked bayonets in the ash-dump - decaying fish-bones, oyster shells, lobster shells, shards of clayware. Tench passed the remnants of a withered garden, and close by, a sagging, rush-seat chair near the child's shady play place. Tench imagined her sitting on the chair nursing one while watching the other at play, a scene of domestic tranquility.

He followed Motherwell along the path that led past Bennelong's hut. A month earlier Bennelong's hut had swarmed with people, and Bennelong's sister and her children had joined Bryant in the Governor's cutter for a few days of fishing, a new era of joint economic cooperation. When a freak storm foundered the cutter, everybody swam for their lives, and natives ashore helped to pull the half-drowned occupants to safety, then rescued the oars and fishing gear. Pleased with the helpful spirit of the natives, the Governor announced to his table guests that at last Sydney was becoming a community.

Bennelong's hut was empty. Perhaps today, perhaps tomorrow, perhaps next week or next month, crude boats would unexpectedly appear bearing twenty or thirty or more familiar and unfamiliar faces. They'd crowd into the hut jabbering in their rapid, slightly guttural language, a spectrum of generations, a tangle of relationships, a mystery of affiliations and alliances. Tench realized he'd never understand the complexities of the native culture.

Past Bennelong's hut, Motherwell pointed to a sprinkle of rice. Tench imagined hoarse curses as the bag slipped and a pale gleam of rice streaked the path. Frantic fingers scraped up the precious grain. Two seconds! Three! No more! Then hurrying on again. More streaks of rice. A hole in the sack? As Tench followed Motherwell along the escape path he began to perceive the larger trail, the trail of Will Bryant's grievances and resentment.

Bryant had never hidden his feelings. From the first he'd carped about government unfairly seizing his boat. He probably thought the Governor's cutter was just recompense! Now, his sentence up, his seven years paid, he simply sailed away. He didn't steal the cutter, he reclaimed his boat! He didn't escape, his sentence was up!

Behind the screen of the big fallen tree at the end of the point Tench saw where they'd waded out to the cutter. The cutter, for Christ's Sake! Bryant stole the Governor's cutter!

After the cutter was staved in getting it ashore in the freak storm, Bryant worked patiently at the repairs. Like so much in Sydney the boat was a patchwork of previous fixes. Worse, it was punky. "Tell the Governor this boat's about done for, hardly even safe! Look!" He demonstrated his conclusion to Judge Advocate Collins by driving his knife an inch into a mushy wale. Deceptive bastard! Tench thought. Bryant already knew he was going to steal the cutter. He'd repaired it for himself!

Motherwell showed Tench the cutter's seine, one of several items abandoned behind the tree. "We figure he took the new seine he was making for Captain Hill," Motherwell said. Tench nodded. They showed him a broken saw and a handmade balance scale, the kind used to portion out food. Whose carpenter's saw? He imagined the scene. "Jesus Christ! A broken saw? What good is that? Throw it out!" A moment for its owner to feel the familiarity of the tool, to remember the pleasure of its good performance and the frustrations of its bad, then tossing it with a too-loud twang. "Christ! Get in before you wake the whole town!" Pushing off. But who? Who besides the fisherman and his family and the man with the broken saw?

They must have gotten away about midnight. A perfect night. No moon, cloudy, just a little wind to ruffle the water and muffle the oar sounds. They would've rowed across the bay and hugged the far shore. After a couple hours they would've come up on the lookout huts at Camp Cove. Easy now, keep low. Easy. Easy. Quiet.

Safe past! Then it was through the headlands, and out! Out! OUT! For the first time in three years they saw the limitless expanse of the sea!

But out to where? Goddamn! The man was smart! He'd waited until the harbor was empty of ships - the *Supply* gone to Norfolk, the Dutch ship just sailed, the Lump useless for pursuit. Not another boat left was big enough to catch the cutter. The Governor's cutter, for Christ's Sake! Bryant stole the Governor's cutter!

Corporal Baker and Pottle approached prodding a snuffling Tartop. The boy's eyes were red-rimmed. Hangdog, he shuffled his dirty bare feet in the hot sand. "Muster for Sydney shows Martin is gone, Captain, Martin and Cox and someone named Bird, the younger one. A man's gone horseback to get a muster at Parramatta." Tench nodded and tried to remember the convicts. Martin he remembered: good-humored, a fellow who stuck to business. He'd never been trouble before. Why now? He remembered Cox's weasel features, his clear tenor voice. One of the *Mercury* mutineers! So this was his second escape. Cheeky bastard! The younger Bird drew a blank.

There must be some from Parramatta. Figure the six-oared cutter with Bryant at the rudder, then Martin, Cox, and Bird, all younger men, all strong. Probably three more at least. Bryant would want six men on the oars to make a dash. What would she do? How would she keep the children quiet? Tench suddenly realized Baker had spoken. "What?" Tench said.

"Tartop here says they're going to Tahiti, Captain. Says he was supposed to go along as the navigator. And we found this up by the trees where Cox worked." Baker handed Tench a folded paper. Judge Advocate Collins stood behind to see. Tench considered the disheveled figure before him. "You were supposed to go too, as the navigator?" His head down, Tartop snuffled and nodded. "To Tahiti?" Tartop bobbed his head, then burst into tears.

The folded sheet was a letter addressed to a Sarah Young. The sun glared so bright on the paper that Tench couldn't make out the writing. He handed the note over his shoulder to Collins. His eyes could get no relief from sparkling blue water and white sand that seemed ablaze. A steady wind from the southeast blew too warm on his face.

"Listen to this," Collins said. He'd turned his body to read in its shadow. "It's from Cox." He looked up. "I know this woman, this Sarah Young. She came on the *Lady Juliana*. Just a girl. Came with a bastard child." Across the harbor the dark forest invited Tench into a cool, quiet solitude. Collins read: "*My dear Sarah, I regret that I must leave you without a better farewell, but when you learn what we have done you will understand. God willing, when you hear these words I will be a free man. I must try my freedom while I can. A life sentence means that I can never hope for freedom.*" Collins raised his head. "Well, what else would a life sentence mean? Silly ass!" He continued reading to a gathering knot of men. "*It grieves me to leave you, for you are kind and sweet.*" Collins raised his head again, "And many a man will vouch for that!" Listeners chuckled. "*As poor amends for my sudden departure, I give you all I leave behind, but wish I could also leave you strength to resist the evil of this sinful place, and the will to lead a life of virtue.*"

"Jesus!" Collins said, "We ought to give this to Johnson for Sunday service." The gathering chuckled again.

Suddenly Tench felt depressed. He knew enough of Cox to know that his prattle about God and virtue were hypocritical. Why did men always want to reform whores only after they used them?

Tench placed an avuncular arm around Tartop's shoulders. "Let's find some shade, son. What's your real name?"

~~~~~~~~~

322

Day one. The rising sun had barely tipped the blue-black mainland when Will Bryant, muffled against a fresh breeze from the southeast, peered past the faint foamy history of their progress through a small telescope. Mute, he lowered the instrument, rubbed his eyes, then raised the tube to again scan the southern horizon and the line along the mainland. When at last he dropped his hands he continued looking south for long moments before turning to the anxious faces in the cutter. But he was unable keep his pretence of solemnity. He suddenly threw wide his arms, his face split with joy, and he tilted back his head to release an exuberant cry of triumph. Men rose to pound backs, to hug, to shake hands. "Nothing!" Will cried, "Nothing! We did it! We're free!"

"Free! We're free!" the boat echoed.

The celebration woke Mary B's children from their opiate sleep, their cries of distress a piercing counterpoint to the noisy men. And the young ones prevailed. Their jubilation supplanted by tolerant smiles, the men chuckled their concern about the "little nippers." Mary B quieted Emmanuel with a breast and comforted Charlotte with a drink and a piece of sugar. She smiled at the curious faces of the rough men.

They all wanted to talk at once, eager to recount the night's adventure, but the jumble of food and equipage tossed aboard in haste had to be sorted, inventoried and stowed, the space in the cutter had to be more clearly defined, personal belongings had to be secured, duties assigned, routines established. "Then we'll eat," Will said, which set them all to work with alacrity. Although Will was "Captain," cooperation and goodwill were essential for his punky command, an undecked, lug-sailed twenty-three feet cutter, six feet at its broadest beam. The craft was crammed with nine adults, two children, six oars, two masts, two seven-gallon water casks, an antique quadrant and precious pocket-compass, two muskets, sails and gear, axes, a sand-filled firebox, kindling wood, a few scraps of lumber for repairing the boat, carpenter's tools, a couple hundred pounds of rice and flour, fourteen pounds of salt pork, cooking pots, Little Bill's battered box, personal bundles and boxes, and Captain Hill's seine. By wit and luck and determination they'd escaped their sentences, Sydney's oppressive discipline and hard life, their dependency and subservience, and their hopeless futures. Aye, freedom was their hard-won object, and now they were free!

But were they? Bearing north in bright sunshine, rigged wing-on-wing, their little cutter was an inspiring sight. The bustle of activity in the cramped boat was good-natured and playful. But let the skies darken, the winds moan and the swells build; let the boat plunge in dangerous troughs and a cold rain lash these ill-clad bodies; let night fall and the glare of lightning reveal mountainous seas and frightened faces bailing frantically; let the sail shred in a fierce, howling gale and this aging boat buck as if intent on tossing them all to watery death; let the boom of wild water crashing on unseen rocks turn innards liquid with fear - then the issue was not freedom, but survival. Aye, to be truly free they must first survive.

After a breakfast of bread and dried fish, already several miles north of Broken Bay, their prospects for success seemed strong. They were bound for Timor, some 3,000 miles distant, a large island in the Indonesian archipelago, Bligh's objective after he was deposed from the *Bounty*. On the south coast of Timor lay Kupang, the closest settlement to Sydney with a permanent European population. Of course the captain of the *Waaksamheyd* didn't tell Will that Europeans at Kupang numbered scarcely fifty and that Kupang was a sere, desolate little failure the Dutch were ready to abandon. No, Detmer Smit happily sold Will a compass and quadrant that had belonged to a deceased ship's mate, a cloudy little telescope, two of the *Waaksamheyd*'s muskets, some ball and powder, and had copied a chart and carved the air with his hands as he explained, "Kupang big, *hein*? You make...big *bezeigheid*, Kupang, *hein*?" to which Will nodded his understanding, not understanding at all. So Kupang was their objective, where, by way of Djakarta, they'd seek passage to Europe or America by working for passage on ships engaged in the China trade.

Will's plan was simple. Go north and double the northernmost reach of New South Wales, skirt the shoreline of the vast Gulf of Carpenteria, then sail west along Arnhem Land until they must strike across the Timor Sea. Except for the last dash across the Timor Sea they could hug the coast, get fresh water, catch fish, repair their boat, and rest. If discovered by a whaler or trade ship, they'd say they'd been bound for Sydney when their ship foundered. God help them if their rescuers were bound for Sydney too, but in the seas north of Sydney such prospects were unlikely.

Will had picked his crew carefully - some for their knowledge, some for their bodies, some for their gold (the only exchange the Dutchman would accept). Each man had to pledge absolute secrecy, everything he owned, and be willing to risk life itself. Each collaborator and each detail compounded chances for betrayal. And even if the conspirators were trustworthy, what of their women, their hut-mates, and enemies? Jealousy, spite, revenge, advantage - a swarm of motives threatened Will's enterprise. Jangled nerves gave him raging headaches. Sleepless, even skinnier, those who saw black bags grow under his eyes must have wondered if an inner worm fed, if Will's wasted frame would rest in a Sydney grave.

But all that was past now. If success had yet to work its effect on Will's body, his spirit knew he'd triumphed against authority, had risked all and won. He was giddy with pride. He smiled an irrepressible smile of satisfaction as he surveyed the crew making his cutter shipshape. How strange! he suddenly realized. After weeks of planning, surreptitious meetings, and whispered confidences, this was the first time they were all actually together.

Bill Morton was essential. In his late-twenties, from a solid, seafaring Exmouth family, Morton had learned compass and sextant as a "young gentleman" in the West Indies trade. But rum proved his downfall. From tippling to binges to blackouts, he pursued his dipsomaniac career to its inevitable end, and by age twenty-five was harbor flotsam, his body a museum of drunken injuries. For a while he labored on the southcoast docks - Torbay, Plymouth, Falmouth - but a back injury did him in, and his slide from a life of petty crime to a life of punishment was swift and unobstructed, greased as it was by drunken incompetence. He earned seven years for sneaking off with a traveler's trunk.

Sobered by imprisonment, by the time Morton was put in chains aboard the *Neptune* he'd weaned himself from the bottle, but the deathship was a nightmare undeserved. Six more years of abuse seemed intolerable. America beckoned. America meant a fresh start.

When word got out in Sydney that family had entrusted the *Neptune*'s captain with money for Morton's return passage to England, Will looked him up. Did he know anything of Bligh's voyage to Timor? Aye, Morton knew about it. All England knew about it. Did Morton think Bligh had worked a miracle? No, Morton said, Bligh was lucky with the weather and kept his men in line, which was what it took.

Before long Will and Morton were plotting in a caricature of courtesy, all "Mr. Bryant" and "Mr. Morton." Morton promised, "You get the boat and I'll get you to Timor, Mr. Bryant."

"And stay off the rum, Mr. Morton?"

"And stay off the rum, Mr. Bryant, don't you worry!"

His eight guineas bought the Dutchman's instruments, the muskets, and the crude charts, with money to spare. Of average size, with mousy brown hair and the blotchy complexion of a boozer, Morton mused with an out-thrust underlip, blinked habitually, and seldom laughed.

Another essential man was James Cox, a *Charlotte* shipmate, now 35, briefly a ship's carpenter on a West Indies trader, whose obsession was escape and whose fate was failure. He'd failed in the mutiny on the *Mercury*, failed to get away on the French explorer ships that put in at Botany Bay, and was unable to get aboard returning transports of either the First and Second fleets. To his credit he didn't join Tartop's plan to seek China just across the Blue Mountains. But Cox had run out of ideas for escape, and he worked at Will to make something happen. "What're we going to do, Will? Just tell me and I'll do it. I know a pigtail from a pig's ass, Will, you know that. I've got my tools! If Bligh can do it, we can too, Will!"

While the authorities kept a careless eye on Will, Cox quietly maneuvered men and materials to Bennelong's point. Now, reclined midships with a self-absorbed air, Cox sang a repetitious refrain to himself:

Around-o, around-o, around-o,

And the world it goes around!

He paused, lost in wistful memories of Sarah Young's moist allure. He caught Charlotte staring at him, winked, pulled a face, and went back to his self-absorbed singing.

The oldest and most experienced seaman was another prisoner from the Second Fleet, William Allen, a man they called "Allie." Allie had given Will details on the escape of the five convicts from Parramatta. "I could've gone with them, Will, but they didn't have a plan. Just get up the coast is all. Hail a whaler. Hell! What kind of plan is that! By now they're all prob'ly ate by savages."

Allie was age 54, lanky, weathered, heavily tattooed, toothless, a sailor whose repeated brushes with the law finally meted him a life sentence for stealing a dozen handkerchiefs from a Norwich shop. If he got back to England he planned to seek the south coast to go see his eternal, bent-backed mother in Norwich. It was years since he'd seen her to show off his latest tattoos and entertain her with tall tales of his life at sea

John Butcher (AKA William Butcher, AKA Samuel Broom, AKA John Brown), a shambling six-footer, age 49, was a man for an oar. He'd spent his life moving bits of Kidderminster from one place to another, whether muck from a ditch or stones from a field or wood from a woodlot, but always a shovelful, an armful, or a backload at a time. Had he not intervened in porcine nature's way, Butcher would still be moving bits of Kidderminster from one place to another, but he rescued three newborn piglets to which he assumed title, reasoning thus: as their mother had already rolled over, crushed, and eaten the rest, she'd eat the last three too, and then there'd be none. If Butcher ate them it came to the same thing. But the little pigs squealed, and for getting caught he earned a seven-year sentence at the Worcester assizes. A Second Fleeter, he had almost three years more to serve and no particular skill, but Allie recommended him as a good, dependable man. Butcher had sandy hair streaked with gray, and wore a large copper earring in his left ear. He kept a sunny, youthful disposition, and handled a heavy oar like child's play.

The last from the Second Fleet was Nat Lilley, age 38, stocky and dark, another oarsman. Lilley hated Sydney's sere wilderness. His birthright was the cool, green hills of County Cork, and he longed for its gray, still, misty mornings and dark, lush forests and the peaceful, dewy damp of the gloaming. Fleeing the poverty and repression that were also his birthright, he ran afoul of the law in Bury St. Edmunds in a dispute with an employer. "Arrh, and why should I be backing down?" Nat still asked, the injustice a metallic, bad taste in his mouth. To make up for day-labor pay he thought due, he pilfered a couple of spoons, a watch, a small seine, and some other items just to make sure he got fair share, and was serving a seven-year sentence as a damn Irish troublemaker. He'd ridden with Butcher and Allen in the prison wagon to Woolwich, there to be loaded on the *Surprise*.

Sam Bird (real name John Simms) was a youthful 29. With only a year left to serve, an anxious dread had overtaken him that he'd never see his home in Surrey again. Quiet, preoccupied, Bird was a slight but willing helper Will got to know during the great hunger, when he was taken off his work at the brick-kilns and assigned to Will's boat. As they made ready one day, Bird asked in an uncharacteristic rush of speech, "Do you ever think of just sailing out, Will? I mean to hell with it and just sail out, no matter what, even if they come after you and shoot? I mean, what the hell! they're going to kill us anyway! If they don't work us to death they'll starve us to death!"

Will was amused by Sam's intense, anxious face. "No," he answered finally, "I think about it smarter." Only Bird had neither great strength nor skill nor experience to recommend him, but the haunted, desperate look in his eye led Will to include him. Perhaps such desperation would be useful.

And then there was James, age 31, two years remaining on his sentence, and with the most entangled motives of all. Former lover of Mary B, father of Charlotte, enduring a bruised friendship with Will, James wasn't sure why he was escaping: was he is escaping with Mary B and his daughter, or returning to his wife and little boy in Exeter? For certain his escape was costly. For only the second time in his life (the first being when he was but a youth) he found it difficult to gather up his bundle, hitch up his breeches, and set off down the road. In Sydney James left much behind that he valued - his convict hut and little trove of pigs and chickens, his garden and faithful, doltish companion, the stuff of a kind of contentment James hadn't known since boyhood. But at least he'd secured a ewe from the Second Fleet for Tartop, a gift that brought tears to Tartop's disbelieving eyes. Now James sat on the seine and reclined against a thwart, his outstretched feet nearly touching Mary B's, who sat facing him, her own back near Will's knees. He watched her amuse Charlotte with the finger game of church, steeple, people, and smiled at Charlotte's wonderment. When he happened to catch Will's appraising look and his smile and wink, he smiled back and shifted to look forward, his expression thoughtful.

Mid-afternoon the wind veered to the southwest and the skies grew overcast. A chilly drizzle settled in that sent the escapees seeking shelter under a makeshift awning. Slicker-clad, Will and Allie heard the laughter of those snug under shelter, entertained by a hiding game with Charlotte. Allie commented on the cutter, "She's leaky as hell, Cap'n."

"She's old, Allie, old and tired, but she'll get us there. We just got to keep her tight." He ran a hand along a beaded seam and noted water sloshing beneath the floorboards. A burst of laughter from under the awning darkened his face. "All right! I want the bottom dry! Sop it up!" he ordered. "Mr. Morton, do you have the log-line rigged?"

Morton emerged with a stick of firewood weighted with a stone. He silently presented the device for Will's inspection. "How did you set your markers, Mr. Morton?" Will referred to the light line attached to the stick with small scraps of cloth tied at intervals.

"Three double arms, Mr. Bryant."

Morton tossed the piece over the side and began counting aloud as he paid out line to the receding marker: one halibut, two halibut, three halibut.... At a count of thirty he wound in the line, counting the cloth flags as he retrieved them. "Just under six flags...a bit less than three knots." He stuck out his underlip and frowned and blinked. Will nodded slowly. "You might be a pinch slow on your count, Mr. Morton. And then there's the current."

"Aye, Mr. Bryant...so maybe two knots." He blinked at the water sliding by for perhaps a half-minute before he said, "Fact is, I learned my count from a slow talker."

A bead of rainwater depending from the tip of Allie's nose fell as he shook with a spasm of shivering. "Cap'n, will we be putting in to dry off?"

Will smiled tolerantly. "No, we'll stick to plan, Allie. We won't put in till tomorrow night. Go below and warm up. Jimbo! get a fire going and tell Mary B to make a pudding. Two pounds of pork. And make some tea. Nat, you and Sam come out! I want you to learn something about the tiller and compass."

Light rain continued into late afternoon next day, when the air cleared enough for Will to satisfy himself again that no sail pursued them. He steered toward a gap in the headlands, and by evening they'd crossed a shoal to beach the cutter several hundred yards up a tree-fringed river at the juncture of a plashy brook. As everyone tumbled out, stretching cramped muscles, unlimbering creaky joints, groaning, smiling, laughing, the realization began to grow that this serene, hidden glade, softened by the tree-filtered light of evening, offered their first real freedom. "Hurrah!" James suddenly cried, and picked up Charlotte to toss her in the air, "Hurrah! hurrah!" Others joined his celebration, and once more there was hugging, jumping, and joyful dancing. Even Charlotte danced in the luxuriant grass until she tangled her feet, fell flat on her face, and began squalling.

Soon they were all breathless, and after a bustle of coming and going to take care of bodily functions, they arranged themselves around a smoky fire and a cauldron of heating water. Drawn with weariness, their flame-lit faces were nonetheless animated as they took turns recounting once again their parts in the escape and their plans for the future. But as darkness gathered there was a brief, tense exchange between Will, Morton, and Mary B about Quint's book. As a result the mood around the fire changed. The convicts stared quietly into the flames and waited for the sweet-tea to finish brewing. After Mary B served out the soothing liquid they sat around the snapping fire in silence, too exhausted to feel hungry. As night drew on they began yawning.

With Emmanuel in her lap, Charlotte asleep at her side, Mary B inhaled the warm aroma from her cup of sweet-tea and absently stroked the head of her daughter as she turned over Will's latest surprise. Over the fire he'd announced that he'd use Quint's book to begin a log. "But that's for his mother!" Mary B protested, "I promised Seedy!"

Will shrugged. "We've got to keep a proper log, isn't that right, Mr. Morton?"

"A log is good, Mrs. Bryant," the navigator answered. Blinking rapidly, Morton said, "Odd things happen at sea. A man loses track." He sat silent with out-thrust underlip, but when Mary B's expression told him she wanted a better explanation, he continued with a sigh. "Too much time and too much all the same, Mrs. Bryant. After a while a man can't remember what his heading was the day before, or the day before that, or the day before that. He can't remember when the wind veered or picked up, or when he noticed a different current. A week goes by and everything still looks the same. You start to wonder if you've gone anywhere, if you're moving at all. I've seen it, Mrs. Bryant. A black doubt starts to grow." Morton quaffed tea. "To know where he is, a man's got to know where he's been. Seems odd, but that's the way it is. To know where you are you have to know where you've been." He drank again. "Got to have a log, Mrs. Bryant."

Mary B bit her lip. "But won't we follow the coast? Won't we know where we are by the land?"

With a glance at Will, Morton smiled. "That's our plan, but you never know. Things happen."

"Look!" Will said, determined to put the matter to rest. He'd fished out Quint's book from Little Bill's box and was holding it open to the picture. "Here's Quint." He flipped the book upside down. When he opened it again he revealed a blank page at the back of the book, "Here's me."

Mary B gave up. At least Will should have told her before!

For a long time the dying fire would brighten with a breath of moving air, then continue expiring to a bed of hissing coals. The convicts were stretched out two by two and, except for Will, would rotate a guard through the night. Although several had never fired a musket, they were assured that raising an alarm would be sufficient protection. Musket training would take place next day.

Mary B lay between her sleeping children. She was put out at Will, who lay an arm's length away. Then the silence was almost imperceptibly penetrated by the gentle sweetness of Cox's disembodied tenor soothing the cool, night air. His song brought a shiver of pleasure to Mary B.

Drink to me only with thine eyes
And I will pledge with mine
Or leave a kiss within the cup
And I'll not ask for wine....

When the song was finished, Mary B became aware of the peaceful breathing of her children and the sigh of the fire, the gurgle of the brook, the sough of trees, the whispering grass. Directly overhead, a hole in the clouds was spangled with brilliant stars. Her thoughts mellowed. Will had got them here! Aye, it was his planning and attention to detail, his keeping them going that did it! She had to give him that. She was here...they were all here and enjoying their freedom because of Will. After all this was over, after they reached Kupang, maybe everything would work out between them after all.

She reached out and gently squeezed Will's arm. He grunted, changed position, and gave her his back.

~~~~~~~~~~

Three days and two nights at sea, two nights and one day ashore, that was Will's plan. Three days at sea gave them each almost two quarts of water a day, with a small reserve. Two nights ashore let them refill their water butts, tend to their leaky boat, dry their clothes, gather wood, forage for edibles, and draw the seine for food. The sea had to provide their main sustenance, for after their first celebratory pork Will reserved the meat and flour for harder times. They'd live on fish and rice, sometimes shellfish.

Ashore, Morton would take careful readings with the Dutchman's quadrant, a clumsy instrument made obsolete by the sextant, but a device still used by old-timers. When pitched about in their little boat the quadrant was impossible to use, but ashore Morton could plot their progress in general terms. From Sydney near the 33rd parallel they'd follow the continent's bellied coast to the 11th parallel, then turn west for Timor. Their crude, copied chart (still based on Captain Cook's work of a quarter century earlier) was on a rough scale of 100 miles per inch, with only the most prominent landmarks indicated, and often placed carelessly.

Three days out, one day in. The schedule worked for eight days as fortune favored the convicts. At that first riverain rest they feasted on broiled fish and hearts of palm until they could eat no more. They cooked over coal discovered in the riverbank, and hacked out several bushels more to load in the cutter. To a handful of shy but curious natives who approached their glade they presented scraps of clothing as if manor-born.

On their second cycle at sea they made steady progress for three days, but the cutter leaked badly. Putting into a commodious harbor for repairs, they were driven off by hostile natives and had to sleep in the boat. Next day natives followed their progress up and down the harbor as the escapees looked for a safe landing place. Whenever they touched ashore the natives advanced with shouts and brandished spears. Instead of fleeing at Allie's warning musket shot, they charged! Amid flying spears, the convicts scrambled back to the boat to row for their lives. They spent the day on a barren island in the broad harbor, *sans* fresh fish, *sans* hearts of palm, *sans* shy natives. Next day they got safely ashore to find greens and fresh water, then it was back to sea. With a brisk wind from the southwest propelling them at a good clip, they felt confident on day nine.

Then Fortune's wheel turned. That night a gale from the west baffled their attempts to stay within sight of land. Dawn found them riding huge swells in driving rain, the wind still westerly, land nowhere in sight. All day Will tacked to keep from being blown farther from land. Progress north was poor. "We just have to ride it out!" he encouraged, "Keep bailing and we'll just ride it out!" By late afternoon the wind shifted and he was able to steer for the mainland, which they sighted in failing light with cries of relief.

False security! Wild surf prevented a landing next day, or the day after, or the day after, or for two weeks after that. Each approach revealed impossible reefs or thunderous surf. Several times they barely escaped the dangerous clutches of currents racing through hidden reefs. They used the last of their coal, the last of their wood, the last of their flour and salt pork. Rain refilled the water casks, but without fuel they were reduced to eating raw rice. They pounded it into more edible form, which Mary B chewed into a warmed grainy mush for Charlotte. But the child would not eat, and cried for her mother's milk. Catching Allie's look, whose toothless gums caused much suffering with the raw rice, Mary B passed him what she had in her mouth. His eyes were the most grateful she'd ever seen.

After seventeen days at sea, rains failing, water exhausted, the cutter so leaky someone had to bail constantly, they decided to risk landing. Spotting a wide-mouthed bay where the treacherous surf seemed quieter, they secured what they could, lashed Charlotte and Emmanuel to the water casks, unshipped the oars, and headed for land.

James said later he was sure they were all going to die. The cutter began to rise and plunge like an insignificant chip in the building swells. Up and down, up and down they rode the increasingly wild water as they struggled to stay in the boat and clawed at the water with their oars, often clawing at air in the bucking craft. "Pull!" Will cried, "Pull, damn you, pull!"

"Rocks!" Morton screamed through the roar, pointing from his bow lookout, "Starboard!"

Will swung the tiller. "Pull, damn you all, pull!"

The cutter reared nearly vertical and seemed as if it would tip backwards. It teetered. It plunged forward to smack the trough with a sickening shudder. The following wave sent a cascade into the boat. Will heard screams of terror as he struggled with the tiller. Bird's oar was torn from his grasp. Mary B lunged half over the side to grab the oar and restore it to Bird's hands. She snatched a hat and knelt to fling desperate jets of water over the side, one eye on her screaming children. Shaken, Bird rested on his oar. Mary B yelled, "Row! Damn you! row! Don't give up!" When Bird raised despairing eyes, she grabbed his oar, thrust him from his seat, and picked up the rhythm. "Bail, damn you! Bail!" she shouted at Bird. Bird began bailing with the hat.

Mary B's strength gave a boost to Butcher, Nat, and James, the only oarsmen still effective. Arms swollen with the blood of their exertions, breaths rasping, they strained against the violent sea. Utterly exhausted, Cox and Allie could only wave their oars feebly. The boat struck with a sickening crunch, then was lifted free. Oars pulled, the cutter crunched again, and again was lifted free. Oars pulled. "Back! Back!" Will yelled. They backstroked desperately to free themselves from entrapment.

At last, the safety of the bay! Riding gentler swells, they hung over their oars, utterly drained. A few hundred yards distant they saw a sandy beach fringed with palms and the inviting green of a forest paradise. They skirted the south side as they looked for a quiet landing place. They found none. Finally, reckless with thirst, Nat and Butcher volunteered to swim ashore with the waterbutts. Stripped to the waist, they hugged the empty casks and thrust themselves to shore with leg-power. No sooner had they staggered onto the beach than Nat turned to wave excitedly. Over the top of a dune a half hundred natives fanned out to fence in the swimmers with spears poised. Their voices were a threatening uproar. Allie uncovered a musket. "Belay that!" Will commanded. He stood at the tiller, yodeled to Nat and Butcher, and pointed at the nearby whitened skeleton of a storm-tossed tree. The two picked up their casks and worked toward the tree, indicating to the natives that they wanted only wood by stooping to toss occasional pieces of driftwood into the surf. The fence of natives moved with them, their hubbub undiminished. At the tree Butcher and Nat hurriedly broke off branches and tossed them into the surf, one eye on the steadily encroaching natives. The cutter was within fifty yards. Will called "Musket!" Allie dropped his oar and checked the priming. Judging the boat within reach, Nat and Butcher bolted for the cutter, awkwardly burdened with the casks. With renewed war cries the natives advanced a few steps, feinted with their spears, taunted, advanced a few steps, feinted, taunted. Nat and Butcher plunged into the surf.

"Jesus!" is all Nat could gasp as they were hauled to safety, "Jesus!" Will maneuvered to retrieve a few symbolic pieces of wood while the natives milled on the beach, some dashing forward to fling words, some wagging their buttocks indecently. They laughed and taunted.

Thirsty and exhausted, the convicts sat sag-shouldered, watching the antics in silence, their hopes dashed of slaking their fierce thirst and spending a comfortable night ashore. When the cutter had retreated a hundred yards, the natives broke into triumphant celebration. Allie fired his musket, a pop and a puff. The natives were quiet for a moment, then one who'd been particularly demonstrative ran to water's edge and launched his spear with a hoarse cry. The dumbfounded convicts watched the spear arc unerringly towards them, a deadly needle piercing the gray sky, straight for their hearts, watched it swish over their heads and slup into the water twenty yards farther out.

"Cocksucker!" Will cried, "Let's get out of here!"

It took an hour to cross the wide bay, and another to explore the fringe of a vast treeless swamp before they discovered the dubious outflow of a small, weed-choked river. The badly leaking cutter now rode so low they strained to move against the meandering flow. The water kept its briny taste. At last they escaped the sea and put ashore on a hummock of more-or-less solid ground, convinced they'd find nothing better. Mosquitoes swarmed, but flint and steel got a fire of marsh grass going, and its smoke provided relief from the pesky insects. Dazed by their long ordeal on the unrelenting sea, the endless afternoon, they crowded the smoldering fire and savored the warmth and smell.

They found little to eat. No fish, few shellfish. But they enjoyed the freedom to stretch and move and the luxury of all the water they could drink. They ate steaming rice and boiled waterplant roots with relish. Because they had nothing better, they payed the cutter's seams with soap. Then they set off again, because they had to. They explored the vast bay to forage for food and fuel, and when they faced the dangerous surf to put back to sea, they'd experienced their freedom for twenty-seven days. And were still 2,500 miles from their goal.

~~~~~~~~~~

The next two weeks were cause for saying "No more!" In working out of Danger Bay (as Will named their swampy refuge), seas broke repeatedly over the bow, drenching their belongings and making the craft dangerously logy. They'd already jettisoned everything aboard but essentials. Will ordered the last sacrifice they could offer - their extra clothes. Dolly's heavy wool cloak waved a brief farewell in the raging waters. Now they had only necessities.

As they slept offshore in the cutter a few days later, unwilling to risk a dangerous surf, the anchor-line parted and they waked to find themselves being swept in darkness towards a thunderous peril. They scrambled to unship the oars as they were flung sideways and backwards, the boat tossing like a woodchip in the roiling waters. The cutter began to fill. It banged into unseen rocks in its wild gyrations. Cox's oar snapped. Only extraordinary efforts enabled Mary B and the others to pull to safety. Once ashore, James hoisted aboard a heavy rock to use as their new anchor, but natives soon appeared to harass them, a regular occurrence now. Most times a musket shot would send them away, but whenever the convicts put ashore two had to stand guard night and day.

Nighttime of day thirty-four. A howling wind blew them into black, mountainous seas that rolled in from the south. No one slept. Will struggled to keep the bow to the wind as teams of two traded off rowing and bailing. Two rowed, two bailed, two rested. Prayers for relief went unanswered. With the wind undiminished next day, 30-foot seas heaved and plunged the little boat mercilessly. They rowed, bailed, napped, rowed, bailed, napped. They ate the last of the cooked rice. They made do with a pint of water morning and evening. Soaked, shivering, bone-weary, they dreaded the approach of darkness. Overcast, spumy daylight at least offered hope for sighting land.

Pitted against this wind-lashed maelstrom, Will worked without relief, unwilling to entrust the tiller to another. Single-minded, he wiped stinging sea-spray from his eyes, clamped his jaws, challenged yet another wave, rode it higher and higher until it seemed they would touch the swift-scudding clouds, then fell in a sickening plunge. A wrong move, an error in judgment, and he'd steer them all to watery death. His eye constantly swept the sea and the sky. He wiped his face and steered into another relentless wave.

As night fell the crew began to break down. Unable to manage the heavy oars any longer, the weaker ones traded off on bailing. Their lips blue, shivery, feverish, thirsty, Charlotte and Emmanuel cried themselves to exhaustion, their sailcloth swaddling useless for warmth in the constant dousing. They slept fitfully, and waked to cry themselves again to exhausted sleep.

Day thirty-eight. Ninety straight hours of raging sea. Mary B huddled under a scrap of sail. Her ankles and feet were painfully swollen, her hands numb, her fingers stiff and clumsy. She offered suck to Emmanuel's clammy face. Her own face was wind-raw. She tried not to let the thought creep closer that lurked at the edge of her weariness, that it would be better if life for the two little suffering innocents ended, if they drifted off in a peaceful final sleep. Aye, if they all drifted off and let the unforgiving sea win. What had these babies ever done to deserve such punishment? What had she ever done? She stirred her legs. Pain. Everyone was breaking out in open sores from the constant wet of salt water. She imagined herself standing up to shake a fist at the heavens. Damn you, God! Why are you doing this? She saw herself hold her babies up to a bearded tyrant in the clouds. You want them? Take them! Take us all, damn you!

Struggling simply to survive, Mary B grew angry. Was that why she stood firm when others retreated? Was that why she fought when others surrendered? She drew Charlotte closer and tried by will alone to summon a little more heat from her anger to warm her children. No use. Everything in the boat was wet and cold and clammy and sucked the warmth away.

After a time she stirred from her stupor, willing herself to do battle with despair. She tucked her children under the thwart and chafed her ankles, feet, and legs. Salt spray stung her raw hands and open sores. She worked her way midships. "I'll row!" she shouted in James's ear.

Late that night the wind finally abated. The swells slowly subsided. Low in the water, half-reefed lugsail filled with a steady gale from the southeast, the logy cutter plowed slowly through gentle glassy swells. The exhausted convicts sprawled in sleep, even Morton, who had the watch. Already they'd sorted themselves out in their dreams - survivor, victim, survivor, victim. Perhaps some already sensed an infection of doubt sapping their will to survive. Was a vision of freedom worth this much misery?

Mary B hitched up her dress and eased herself back in the predawn half-light to relieve herself for the hundredth time before this intimate assembly of sleeping strangers. Scant modesty lived aboard the cutter. Most was gone long before with other excess baggage. She smiled with weary, tender regard as she contemplated flush-faced Charlotte asleep in the crook of James's arm. She hoped Charlotte would keep sleeping, for soon Emmanuel would stir from feverish sleep to demand her attention. Only Will was awake this shrouded, gray-shine morning. His stare was glassy. He was like a statue, the tiller bar and his hand and arm and body carved from gray stone. Did he see Mary B? Perhaps after four days at the tiller he slept with open eyes.

This was not the same Mary B who clambered aboard that propitious night of escape, a special day of hope she later told Tench, for it was Emmanuel's first birthday. God will deliver us! Since then she'd grown stronger and more confident. In the unspoken pecking order of the boat she'd risen above Bird and Cox and Allie, an unspoken transition that occurred when she snatched Will's hat to bail the cutter, when she saved Bird's oar and helped pull them to safety. By the time they escaped from Danger Bay, the entire boat wordlessly acknowledged her rightful place at the oars.

Her place in the councils of decision-making remained more tentative. She still tended the children and gave them her milk, but others tended them too. If Allie's cold had him wheezing and sweating in his sleep, it wasn't Mary B who nursed him, but his snoring shipmates from the *Surprise*, Butcher and Lilley, now sprawled in their bow territory.

Because he commanded the stern, his quarterdeck, Will still commanded Mary B, but less so as his wife and the cutter's cook. She now taught others the kettle and skillet.

Command sat well on Will, a revelation to Mary B. He conferred with Morton, issued orders, checked measurements and soundings, and made notes in Quint's book. But if Will in a Great Cabin bid Mary B to his bed, she would not have obeyed. That change too had been wordless.

Perched on a wale for her morning pee, Mary B raised her head and cocked her better ear. Tentatively she half-raised a hand for silence, but the snorers snored and Will sat oblivious. She listened, peering through gray mist. Nothing. Listened hard. There! She waited to be sure before she pulled herself to her feet and began waking the crew. With an excited cry she informed them, "Land! I hear land!"

~~~~~~~~~~

The first prospects looked bleak. They braved a dangerous surf to land, and after the morning fog cleared they saw no sign of the mainland. An exploration of their refuge, scarcely a mile around, revealed sparse vegetation, no ponds or marshes, no fresh water at all. The waterbutts held less than a gallon. And Will looked awful. Helped ashore from the boat, he stood wooden and vacuous before he collapsed face down in the sand. They stretched him out beneath a rocky outcrop. He lay stiff, his eyes open, his lips working, and after awhile sank into a sleep of the dead.

But the barren isle Mary B discovered that misty morning was the turning point of their long journey. They were safe on *terra firma*, and clouds promised rain. They found driftwood aplenty for fire. Prospects for food much improved when the receding tide revealed an extensive reef. Searching for shellfish along the reef, James and Nat spied a ponderous green turtle emerging from the sea to lay her eggs. They staggered back bearing their huge treasure between them with exciting news. They'd put four more of the big turtles on their backs! Spirits soared. Fresh meat! Flesh!

Before long they'd butchered the turtle, built a fire in the shelter of the cutter, and boiled sea-water and rice to make a savory turtle-meat stew. Roused with difficulty, Will ate a few spoonsful, then fell asleep again with a half-finished bowl in his hands. Allie laid him back and sat in faithful watch.

As the assembly reclined in sated pleasure that evening, wishing perhaps only for a mug of real tea and a pipe of tobacco, the skies opened to release a torrent of warm rain. Perhaps Heaven sought to make amends! Hastily rigging water-catchers from the sails, the convicts slaked their thirst with the downpour. They caught enough to fill the water casks. They filled empty cups and bowls. They filled the skillet and kettle. They filled every vessel that could hold water. Then the men doffed their salt-encrusted clothes and rinsed them in the gush of water from the rain-catcher. As the rain let up they began to cavort around the hissing fire in near-naked leapfrog. Laughing at each other's silly antics, they circled the flames in follow-the-leader. James began an atavistic, guttural punctuation to his steps, "Ho! Ha! Ho! Ha!" Others joined. "Ho! Ha! Ho! Ha!" They sped the tempo of chant and feet, more and more abandoned, thrusting this way and that, thumping their chests, slapping their stomachs and thighs, leaping, yipping, yowling, wilder and wilder, "Ho! Ha! Ho! Ha!" At last, exhausted, one by one they dropped out to retrieve their clothes and retire to the shelter of the canvas. There they quaffed cup after cup of steaming sweet-tea. His heaving torso glistening, James caught Mary B's look over his mug. His fire-lit eyes crinkled in a warm smile, which she returned.

With Charlotte and Emmanuel warmed, fed, and fast asleep, Mary B slipped away with a dab of soap. The landward sea was gentle. As she shed her salt-encrusted clothes the rain picked up again, and she gently cleansed herself, wincing at her sores. She lathered her hair, inhaling the simple pleasure of soap smell. Freshened by her bath, she walked by the water and saw another green turtle emerge to lay her eggs. She stopped to watched the slow, deliberate digging, feeling an affinity with this female of the sea. Mary B raised her arms in a languorous reach and swayed to the gentle rhythm of a secret memory. Eyes closed, lulled by the rain-softened rhythm of the waves - shush, shush, shush - she luxuriated in the soft, warm fingers of rain on her upturned face. Alive! Alive! For the first time in years she felt hope, and the promise of life. Her senses heightened by rivulets of baptismal rain streaming down her body's pathways, she suddenly yearned to share this moment, to enjoin her rediscovered promise of life. Perhaps she willed the presence of James and the dream-like tentative touch of his hands. She shuddered at a gentle kiss on the back of her neck and felt the warmth and reassurance of being enfolded in powerful arms. She swayed to the rhythm of the waves, caressing herself, until she found the center of her pleasure, and in the gentle rain achieved a wondrous and complete sense of well-being.

~~~~~~~~~

They recuperated for almost a week on their bountiful little island refuge. Sun-warm, dry, well-fed, Charlotte and Emmanuel bounced back from their fevers and were soon digging in the endless sand and venturing farther and farther to explore a child's-eye world of sandy mountains and grassy forests. Allie began to recover from his cold, and Will, after thirty hours, awakened from his sleep of the dead, and after a groggy hour or two was soon issuing orders again.

They lived well. Propitious rains kept the water casks full. They discovered a pepper-like plant to spice their turtle stews. They clubbed nesting boobies, harvested crabs and oysters, caught a few fish, and turned seven more turtles. They feasted on turtle until they could eat no more, and smoked some two hundred pounds for future food. They payed the cutter with turtle fat, and Mary B rendered some into soap from a lixivium scented with sweet-tea liquor.

Morton took careful readings. After conferring over the Dutchman's charts, Will announced their island lay just south of the 26th parallel. Faces fell. Five weeks, and they were not yet halfway up the coast! At this rate they'd be six months reaching Timor! "All right! All right! I know this isn't good news, but listen!" Will said they were through the worst part. Past Sandy Cape a hundred miles north, they could work their way up the coast protected by the great reefs, just as Captain Cook had done. "We've been lying around all day with a good south wind blowing. We could cover eighty or a hundred miles a day with wind like this. We could make Cape York in three weeks, and then it's fair way to Timor. Allie, you're just getting fat here anyway! I say, let's get stowed away and tomorrow head for the mainland, then it's go like hell for Cape York!"

The response was enthusiastic agreement.

At first light next morning, the wind holding, they worked around Mary B's little island and headed west. They made land by mid-afternoon. Turning north, with clear skies and a steady south wind, they gobbled up the miles.

The next several days saw progress beyond their highest hopes. The freshly payed cutter seemed reborn. They doubled Sandy Cape and slipped in behind the reefs. Here the southerly current they'd battled since Sydney was scarcely noticeable. Hour after hour they skimmed past myriad islets and reefs. Will and Morton checked their speed with the log-line. Will was delighted. Four and a half knots. Four and a half knots. Five! A bowman took soundings. Twelve and no bottom, twelve and no bottom.

Day fifty. As Bird sat with Will at the tiller, Morton asked with a serious mien if Bird could see a goat grazing on an island off the port bow. Bird studied the island. Cox looked up from shaping an oar to replace the one they'd lost. "A little goat sure would taste good," Cox said. Soon some of the others saw goats. "There's one, I see it!"

"No, that ain't no goat!"

"That's a goat!"

"No, that's a rock!"

"Well, then, that's a goat!"

"That's not a goat either!"

"Is so!"

"Is not!"

"Is so!"

"Is that the goat you mean, Mort?"

"That isn't a goat, damn it!"

Before matters got out of hand, Morton announced he was only joking; he thought the island straddled the Tropic of Capricorn. Then he had to explain that Capricorn was Latin for goat horn. The men in the boat stared at him. Bird put fingers to his head like horns and playfully butted Will. Morton looked at the blank looks of the convicts, then at Will. "Never mind," he finally shrugged, blinking rapidly.

To avoid further confrontations with natives, they stopped only at islands that looked uninhabited, and then only long enough to replenish water and wood, scour the shoreline for shellfish, and draw the seine. They got little for their efforts, and lived most days on their reserve of turtle meat. Except for a few pounds Will held back, they consumed the last of their rice.

Just a week from Mary B's island, Morton announced they were at the 20th parallel and just five days later they were crossing the 15th! At this news they cheered and whistled and clapped. They were doing it! They could feel it! They were going to make it! As they chewed turtle that night bathed in the light of a brilliant moon, James moved back to sit at the tiller by Will. "Goddamn, Will, we're going to make it, aren't we? We're going to make it!"

Will's grin was white in the moonlight. "Goddamn right, Jimbo. I'm a free man now. You couldn't hold me back if you triple chained me! Not me! Not me, you couldn't!"

James nodded. He looked at Mary B in the bow, where Allie played cat's cradle with Charlotte. After sitting silent for several minutes, James drew a big breath. "Will?" he said. Will said nothing, He raised his eyes to the pigtail to check the wind and reached out to assure himself a stay was secure. "Will?"

"You still want her, Jimbo? You have all along, haven't you!" He busied himself with a minor adjustment of the tiller rope. "Well, you're the Catholic, Jimbo, not me." He sat wordless for so long James thought he was finished. Then Will said, "When the time comes, Jimbo, when the time comes, you'll know."

Day fifty-seven. Two weeks from Mary B's island the cutter threaded past the bleak prospect of Cape York, its occupants dazed by the glare of a nearly vertical sun. A hot, gusty wind blasted them from the northeast. After eight weeks of struggling up the coast, the forbidding landscape and a snappish Will dampened the excitement of turning west. He cursed the endless shoals and sandbanks that impeded passage through the island-entangled strait. In his haste to double Cape York he'd let their water shrink to a couple gallons, and in the parching wind and heat all eyes scanned the barren, forbidding hills of the peninsular tip for signs of water.

Mary B held a corner of the makeshift awning she'd rigged for the children. The two little ones were endlessly fussy in the heat. Will had given up trying to keep a larger shelter secured in the blasts of the oven-like wind. When the wind died between blasts the sun lay like a heavy blanket on Mary B's head and shoulders. She felt it burning her forearms. She looked at the tense, sunburned face of James studying the coast, saw him lick his lips and swallow. Perhaps feeling her eyes, James turned and smiled a crooked, unhappy smile. Mary B smiled back and held his look for a long moment before she shifted her gaze to the sere hills of the cape.

"Mr. Bryant, if that big island west nor'west is Prince of Wales, then that's Booby Island just past that. Past Booby Island we're through the straits!"

"Very good, Mr. Morton. We'll follow the coast till we find water."

"Cap'n, over there! Natives!"

Heads swiveled to follow Allie's pointing arm. They made out tiny figures on the lee shore of a small island. Will swung the tiller. "Are we going after water, Cap'n?" Allie asked. Will nodded.

When James later told Boswell of this incident in Newgate Prison, he said Will's recklessness had almost cost them dearly. But given their relentless push for the Cape and their befuddling thirst, Will's presumption that a few men with muskets were more than a match for savages was understandable, if incautious.

"Jesus! Black bastards, aren't they!" Butcher remarked. Mary B shrank lower. Something about the situation made her nervous. These natives were so different! Sleek-skinned, black as coal! One who was larger than the others wore some kind of fancy chest ornament.

Will ordered the sails struck, oars unshipped, and muskets readied. James, Nat, Butcher, and Cox pulled for shore. They were within a hundred yards when Morton hissed, "Bows and arrows!"

"What?"

"They got bows and arrows!" Heads swiveled, astonished. Natives around Sydney didn't know what bows and arrows were. They'd advanced the development of their weapons no further than clubs and spears.

"Well, I'll be Goddamned!" Will said, "Lay on your oars." He stood on the thwart, braced against the tiller bar. As the cutter drifted closer he doffed his hat, bowed, made a scooping motion, and pretended to drink from his hat. The natives remained silent and impassive, arrows notched in their short bows. Will repeated his motions and called out, "Water!"

His voice evoked immediate agitation. The throng began to jabber and threaten and the man with the chest ornament strode back and forth as he orated with defiant gestures. "Water! Water!" Will repeated more insistently, "We want water!" But his words only caused the natives to become even more vociferous and more belligerent. "All right, Allie," Will said, "Give 'em a shot over their heads. See how they like that."

The belch of smoke and noise from Allie's musket froze the natives for a moment, then arrows flew. Flit! Flit! Flit! Tiny streaks of light and shadow headed towards the boat straight for their eyes! Will dove for cover. "Back! Back!" he cried unnecessarily as arrows showered the retreating boat. Allie fired off the second musket in a panic and ducked behind the protection of the stem-post. A sound like "tuck!" and Mary B stared in disbelief at an arrow quivering in the wale, inches above her head. "Fool!" she screamed at Will with sudden fury, "Fool!"

~~~~~~~~~~

Lieutenant Bligh took nine days to make the 1,100-mile voyage from Booby Island to Timor in a longboat crammed with starving men, but he didn't enter the vast Gulf of Carpenteria.

He knew that in early summer the current of the shallow Arafura Sea flows swiftly from east to west and was augmented by monsoon winds and that progress across the Gulf would be swift. Neither Will nor Morton possessed that knowledge. They planned to follow the long scoop of the Gulf of Carpenteria, then hug the coast of Arnhem Land until it was time to cross the Timor Sea. But their plans changed.

After the natives drove them away from the island Will named Danger Island (actually Possession Island on charts of the day), the convicts sailed several miles down the coast before they spied the swampy outflow of a river. Desperate with thirst, they discovered several recently deserted huts only after landing. Nervously filling their bellies and waterbutts, they retreated to the cutter and spent the night drogue-anchored by their stone anchor a couple miles from shore. By morning they'd drifted considerably south. They began rowing back to replace water consumed during the night.

"God Almighty!" Morton gasped. From behind a small island perhaps a mile distant emerged a magnificent fifty-foot war canoe. Single-masted and square-rigged, the vessel had the wind and was making straight for them. A few moments later a second giant canoe appeared. Will came about, ordered the oars shipped and the sails raised. When perhaps a half-mile distant, the first canoe dropped its sail. Will tacked.

"Jesus! What're you doing, Will?" Cox cried, "Let's get out of here before we all get ate!"

Peopled with black bodies wearing painted faces, the war-canoe carried several figures atop a raised fighting platform on the bow.

"I don't know if we can outrun the bastards! We've got to see what they're going to do. Maybe they just want to scare us off."

When the canoes had come nearly abreast each other, the first hoisted its sail again and the imposing vessels bore down on the cutter together. "Goddamn!" Will howled, swinging the tiller, "Make all sail, Allie! Break out the oars!"

Compared to the progress of the huge canoes the cutter moved like a log. The gap between the convicts and the canoes quickly narrowed to a quarter mile. The enemy attacked in tandem, keeping a couple hundred yards between them. Each held thirty or more warriors, with another half-dozen positioned on the platforms, their bows and arrows and spears clearly visible. The warriors pranced and threatened with their weapons.

"We can't outrun 'em!" Bird cried, "What're we going to do?" Will remained wordless. He seemed frozen by fear. Suddenly he turned. "They'll go past and cut us off. We'll wait, then come about and row into the wind. We might be able to out-row them into the wind. Listen, pretend we're trying to pull away! Allie, when I say 'Drop the sails' you and Bird drop the sails as fast as you can. Then the rest of you start rowing like hell!"

Will's analysis was correct. When the rowers appeared to increase their efforts, the powerful canoes pinched in, but when the cutter failed to gain any advantage the canoes maintained a cautious distance, perhaps because Allie and Bird kept their muskets pointed. As the canoes came abreast they pinched in more, and the gibberish war cries and sight of their grotesque war paint loosened convict bowels. After what seemed an endless time, Will suddenly swung the tiller. "Drop the sails! Now row, my beauties, row for your lives!"

The oar Cox fashioned to replace the one snapped in the surf may have proved the difference. Will caught the war canoes by surprise, and they nearly collided in coming about to pursue the cutter in the opposite direction. By the time the enemy regrouped, struck their sails, and picked up the chase under the power of their paddles, the convicts had gained a half-mile and Allie had the crew settled in a steady stroke as Will concentrated on gauging any change in their situation.

Soon Will was looking for some advantage. Although not losing headway, they seemed unable to gain. After a half-hour he saw no perceptible change. Allie and Bird heaved the stone anchor overboard to lighten their weight. Will ordered the masts unstepped and altered their course to head northwest. He now had the wind almost broadside.

Being a hundred pounds lighter and altering their course helped, for after an hour, when everyone had changed off on the oars several times to keep their foaming progress strong, they thought they might have added a couple hundred yards margin. But they were wearing down. They cursed the cannibal savages aloud and sobbed with exhaustion, thirst, and rage. After two hours they were gaining distance slowly but steadily, but still the savages pursued. In her anger Mary B began growling to the rhythm of her stroke. "Give UP you bastards give up! Give UP you bastards give up! Give UP you bastards give up!" James picked up her chant, then Nat, then every voice in the cutter was defying the savages, those not rowing shouting at the tops of their lungs, "Give UP you bastards give up! Give UP you bastards give up!" The cutter took on new life. Each powerful stroke of the six determined oars sent the craft surging.

Did the savages sense defeat in the cutter's defiance? Did they see the difference between oars driven by fear and oars driven by angry determination? Who knows? At last, after almost three hours of pursuit, the enemy gave up the chase and turned away.

Too exhausted to speak, the rowers collapsed. Later they'd talk of their victory, their pain, and how they'd survived by working together. More than one would credit Mary B's show of defiance as the final edge that saw them through. But now, stuporous with fatigue and fearful what a return to land might mean for their safety, they hoisted sail beneath a tropic sun on a westward course across almost four hundred miles of the Gulf of Carpenteria. With eight gallons of water.

~~~~~~~~~~

Day sixty-two. They made the Wessel Islands in the morning. The islands meant water at last, as much as they wanted to drink and pour over sun-blistered faces and necks, to cool bodies. For four days they'd prayed for rain, their thirst fierce. Despite the dehydration caused by their exhausting escape from the war canoes, Will nonetheless rationed them to a pint each morning and evening. He allowed Mary B an extra pint for nursing milk. But no rain fell.

They found no water in the Wessel Islands. Hour after hour they snaked through a disheartening confusion of sun-baked islands and salt-water swamps, their thirst growing more and more desperate. Their water casks contained less than a half-pint for each. They'd eaten nothing but cold food for days. But food seemed not to matter. They'd sucked the juices of raw fish and drunk the blood of a hapless booby, but the salty fare only increased their thirst. As the day waned they retreated from their fourth false hope, another bay rich with waterfowl and mysterious slitherings and sloshings among the watergrowth, with clouds of insects but devoid of discernible fresh water. They glared at the hostile land. They muttered curses and pressed on. Only a cup or two of water remained.

The sun was low when they emerged from an island channel and Allie dipped his finger. His face brightened. "God be praised!" he cried, tasting the influence of fresh water.

"Are you sure, Allie? Are you sure?" Fingers dipped, cracked lips sucked. God be praised! Fresh water near by! Spirits rose. They worked along the swampy verge with keen anticipation. They peered ahead for a hint, an opening in the vegetation. Five minutes. Ten. Their parched throats could barely swallow. The sun dropped behind distant trees and shadows thickened in the tangled vegetation. Then, at last, there it was! There! Around a bend the unmistakable sign of a rippling outflow through the tidewater. In the cheering no one heard Will croak "All right!" As if they'd just surmounted their last obstacle to success. "All right!" he said again as he regarded the grinning, raw, emaciated faces of his charges. He stood on the thwart and shook his fist at the rose-gray western sky. "There, Goddamn You, there!"

Djakarta
1791

*A*t the southern edge of a vast archipelago once called the Moluccas lay an almost-forgotten island in the Far East trade. Although of dubious value to either Portugal or Holland, one nation ruled the eastern half and the other the western half. This divided island was Timor.

On its southwest corner, suffering a long, rainy winter and a baking, dry summer, the Dutch East India Company carried on a shrinking trade in beeswax and sandalwood from an aging fortress at the edge of a shallow, mangrove-fringed bay in a place called Kupang, a native name meaning "village."

Kupang too was divided. Wandering through an incongruous tropical savanna, a river separated the Europeans from the natives. Non-whites lived on the west side, and Europeans and a few Chinese merchants inhabited the fortress side.

The river was the center of life for Kupang. Although a fresh-water well served the fortress, servants and slaves from the European side joined much of Kupang morning and evening to bathe and do laundry.

Near the river's mouth stood a small, red-columned Chinese temple with a shady veranda from which a dozen loungers one day watched a small craft creep across the broad glare of Kupang Bay. They observed its progress for some time, speculating on the odd lugged sail. Although not a familiar craft, it steered directly for the wharf, ignoring the formality of asking for permission to land. But as the battered craft scraped to a halt against the brief stonework, the loungers saw a story. Its rotting sail was patched and badly frayed, its faded paint a mere remembrance of red and green, its bearded European occupants barely covered by remnant clothes. The interest of a Company corporal quickened when he recognized among the sunburned figures a woman and two small children. Disaster must have befallen the strangers! He hurried forward. The dazed, emaciated company looked like a revisitation of Captain

Bligh's sorry seamen who landed at this very wharf just two years earlier. Were these unfortunates also victims of mutiny?

As the strangers stumbled ashore, their appearance foreshadowed their tale of how a dry wind that drove their craft across the swells of the Timor Sea had dehydrated bodies already punished by too much sun, too little food, and never enough water. Of how, even after landfall, water casks nearly empty, they skirted Timor for a day and a half to locate Kupang Bay. But sixty-nine days after escaping from Sydney, Will Bryant landed his charges safely at this dusty, somnambulant, disappointing little town. Free at last, they staggered and weaved under the tropical sun of June 5, 1791.

By the time a sleep-groggy, red-faced officer arrived from the fortress, a melange of Chinese, Malays, Papuans, and mixed-bloods crowded in a semi-circle of curiosity. They wondered in several languages and dialects at the gaunt strangers, at their open, oozing saltwater sores, their feverish eyes, their skin burned to blisters. Knowing little English, the red-faced officer managed to deduce that the occupants were English and were seeking mercy. A boy fetched water. The English greedily sucked an entire bucketful dry, then a second. They were giddy, and kept drinking water.

Soon the strangers were escorted to the sanctuary of the fortress, from whose rubblework walls of age-old construction a half-dozen small, brass cannon peered through deteriorating embrasures. The Governor would grant them an audience soon, but meanwhile His Excellency ordered the refugees to be refreshed. Surrounded by the amazed examination of most of Kupang's white population and a rag-tag fringe of servants and slaves, the survivors gorged themselves in the shade of the church portico on bread, cheese, bananas, and oranges.

Although they were nervous, the convicts relished this first taste of real freedom. Winks and nods confirmed the fulfillment of their oft-told fantasies. They basked in the clucking sympathy of the solicitous Dutch and raised grateful toasts of sweet Kupang water to their success and salvation.

The convicts couldn't have wished for more sympathetic benefactors. Although His Excellency's understanding of English was poor, Timotheus Wanjon offered the hirsute, emaciated Will Bryant a chair and listened attentively. He grasped the essentials of Bryant's long-winded story of how the *Goodfellow* (Captain Robyns), an English trader working north from Port Jackson and bound for the coast of Coromandel, came to grief on a treacherous reef off New South Wales. Describing himself as the supercargo, Bryant explained that he'd planned to settle in Port Jackson with his wife and children, but when he saw the destitute, discouraging colony of convicts he decided instead to look for passage to America. He told how Captain Robyns, fearing he couldn't save his wounded ship, entreated Bryant and Morton (another disappointed settler) to lower the cutter and try to reach the mainland.

The Governor traded glances. How selfless!

Will then told how they'd nearly given themselves up for dead in a tempest that soon followed, a storm that tossed and drenched them mercilessly for almost two weeks; they lost nearly everything trying to keep afloat. Thirsty, starving, and disheartened, through God's deliverance they were finally set on a small island where they were able recover their strength. There they decided chances were better of reaching Kupang than of returning to Port Jackson against the currents, so they worked their way up the coast of New South Wales and followed the track of the heroic Lieutenant Bligh across the Timor Sea to Kupang.

Such a voyage! And with a woman and two babes! Despite his sun-ravaged body, his hollow eyes and shrunken frame, Will couldn't help preening at the lavish admiration of the Dutch. And why doubt him? Although Bligh left a complete description of the *Bounty* mutineers with Kupang authorities (which the Governor prudently reviewed while the English refreshed themselves on the church portico) the presence of the woman and children made it highly unlikely these sun-baked survivors were associated with the scoundrels from the *Bounty*. And surely they couldn't be escapees from Port Jackson, not in a cutter! Not 3,000 miles! Impossible! And with no other English outposts....

So Governor Wanjon accepted the sad history of the unfortunate strangers and extended them the traditional courtesies of the sea. After they were examined for contagious diseases, the Governor issued a notice of credit for clothing and such incidentals as the English might require, a debt to be honored according to custom by sending the bill to London on an English ship. Then the convicts were escorted to the very house in which Bligh and his men recovered, a small affair of three rooms with a little courtyard and a detached cooking hut, a house once a storehouse, the only vacant building within the fortress. The Bryants took the front room, the rest shared the long room and a small room at the rear, and the escapees settled in. Soon members of the Governor's staff brought the first of many meals from His Excellency's own kitchen.

What luxury! In three little shops that comprised Kupang's European commercial district the escapees discovered real tea, sugar, tobacco, soap, preserves, colorful fabrics, shoes, ready-made clothes, even china from England! Bathed and shaved and trimmed, chattering like children, the convicts outfitted themselves with happy abandon. Transformed from wretched escaped convicts to well-clad, well-fed guests, could they have any doubts about their gamble for freedom? Only once were they required to maneuver through treacherous shoals, when His Excellency invited Will and Mary B to his table. And Mynheer Morton, the other passenger, *ja*?

Will and Morton thought they carried the affair off in good form, but the evening was tense, and interminable. After Will plodded through yet another telling of their voyage up the coast, the Governor queried them about the health of the King, for news of Bligh's reception in England, of affairs with Spain, the turmoil in France. And what are the prospects for Port Jackson? But pretending difficulty in understanding the Governor's English, Will and Morton were vague. After many evasions, expectant smiles, and awkward silences, the Dutch retreated to conversation among themselves, and only occasionally acknowledged their guests with nods and smiles.

Later His Excellency's wife protested, "Why must we share our table with whatever riff-raff blows this way? Mynheer, the woman is low-born! A lump! Ignorant of fashion and news! You'd think she was three years at sea, not three months! The pitiful creature! And so frightened, the poor thing, too terrified even to eat! And did you see how the men gulped their wine? Did you see? God grant them pity, but spare me their company! Ach, so tedious! My head is pounding!"

So Vrouw Wanjon suffered no more dinners with Mary B, and the convicts were left to pursue their daily lives in peace. Nonetheless, like strayed sheep they left wispy clues as they wandered on the fringes of Kupang society. Perhaps not all was what they said.

What was it about them that after several weeks found the Governor increasingly uneasy? Not their language, for who in

Kupang could pick up hints of their criminal cant? And what conversation might have chanced to reveal inconsistencies or improbabilities in their stories? None, for they had few conversations, and Morton had rehearsed them endlessly about the imaginary voyage of the *Goodfellow* around South America's Cape Horn, knowing it was a route the Dutch were unlikely to know firsthand.

But Surgeon Zimmers, who examined the unfortunates on their arrival, noted scars on Morton's ankles he thought might have come from fetters. And what of the scars on Will's back, scars he laughed off with a joke about younger days? Those were flogging scars! And what of their body language? What history with authority evokes that anticipatory cringe, that evasive cast in the eyes, that nervous licking of the lips? What secrets bound them together like peas in a pod and made them seem ever on guard and drop their voices when a soldier or official happened by? *Ja*! and what of the quarrel that followed Morton's fall from grace.

Cox was in fine form that night, entertaining his fellows with sea songs and ballads, lubricating his voice with a steady consumption of the Dutchmen's fiery arrack. Well into the evening he began to badger Morton, "Come on, Mort! are you going to drink water for the rest of your life? Have a go!" Cox struck a pose, winked, and directed himself with a waving cup,

In Plymouth town there lived a maid,
Bless you, young women
In Plymouth town there lived a maid,
Oh mind what I do say....

Poor Morton! He'd struggled for days with his resolve. For more than a year he'd touched no hard liquor, but at Governor Wanjon's table had drunk wine with no ill effects. So perhaps.... He scanned the shadows for signs of Will, feeling hurt and ill-used, ready to break out and assert himself. Really, Will had become such a recluse. It was one thing for Will to withdraw into his shell like a tortoise, to brood in his room and drive the others away with his growling distemper, even his wife. But to shut himself off from Morton too...well, damn it! that wasn't right. Will had needed him. He'd implored and promised him. But after Morton helped get him to Kupang, Will acted like their partnership was dissolved, emerging from his ill-humored shell only to order them all about or complain. Ahhh, go to hell, Will!

At last, after repeated urging and encouragement, feeling unmannerly in continuing to resist the convivial fellowship, convinced he could imbibe a drop or two without harm, to the cheers of his friends Morton cleared his throat with a modest draught of the burning liquor and joined in song,

In Plymouth town there lived a maid
And she was mistress of her trade....

He fell off the wagon with a thud. The liquor rushed to his head, filled his voice with gusto, and his chest with glee. He drank and clowned through a first cup, then a second, then a third. He drank with relief he was drinking again, for their freedom, for the blessings of fellowship and the good god Bacchus. He drank into the night until, gesturing with his cup to make some drunken observation, he keeled backward off his seat, passed out cold.

Just after dawn in a purl of moist smells - vomit, spilled liquor, urine, rotting fruit - Morton opened his eyes to discover a barely distinguishable trail of ants marching across Butcher's outstretched hand. In the crepuscular light he watched the ants carry rice away from Butcher's bowl, two busy, bumbling columns, like insect convicts clearing rocks. He raised his head to follow the wriggling dark line over the straw-strewn floor - under the table, around the legs of an upended stool, then disappearing in a slit of gray light under the double-door. His head pounding, Morton lay back. He became aware of Butcher's horrible breath, and turned away to survey the others sprawled amidst the debris of the night before. Allie let go a massive fart, rolled onto his back, and began a violent, open-mouthed, toothless snore. Trembling, so weak he could scarcely rise, Morton got himself upright and weaved outside to piss against the wall. Oh, what punishment! His throbbing head felt split in two! Fingers guarding his limp penis, he rested his head against a forearm on the wall and closed his eyes, weary beyond belief, his fluttery stomach vaguely comforted by the faint warmth of the rising sun on his back.

He lurched back to consciousness when Mary B emerged with a chamber pot. Morton bobbed his ass to send his penis into trousered retreat, forced a sickly smile, and weaved back into the house, heavy with remorse. Goddamn Cox! Morton lay down on the straw and fell asleep.

That afternoon Morton wouldn't suffer Will's carping. He told Will to go to hell and said he wasn't their captain any more. The quarrel escalated, and after Will stalked off the continued argument provoked Cox and the rest to a noisy brouhaha that saw knives drawn. At the sounds of raised voices and curses and a shattering bowl, two soldiers materialized. They glimpsed the disappearing knives and the threat of English snick and snee, but in the presence of authority the convicts closed ranks. They assured the soldiers all was well.

The Governor, however, was not amused. Struggling in English, he reminded Will that they lived in Kupang by his leave. He could send them on their way at any time, and he would not abide brawling. Another breach of the peace and he would personally see that the English suffered the consequences of the law. Will bowed, apologized, promised the good behavior of his party, and thanked the Governor for his many kindnesses.

The chastisement made him boil. Innocent of wrongdoing himself, the disturbance fed his growing frustration. Now safe in Kupang, he regarded the other escapees as nuisances. He resented his entanglement with them. He'd needed their help, yes, but once he brought them to freedom he no longer needed them. He owed them nothing! Now they threatened his freedom! After all, he'd completed his sentence and had a right to be free. But they were escaped felons. And here he was stuck, linked by the isolation of Kupang to the bibulous instruments of his possible ruin, and their stupid brawl only made his precarious situation more obvious. Bitter, helpless, he retreated again into seclusion. Morton also secluded himself, and stayed drunk.

Ja! and what of Mr. and Mrs. Bryant?

Several Dutch wives had noted the feigned domestic harmony of the English couple. More than one had caught Mary B's smoldering look, Will's resentful, dismissive growl, his half-raised hand. *Schandaal*! Mynheer Fruey, the lieutenant-governor, speculated that the woman's intimate ordeal in the cutter might have turned her head. She seemed attentive to the toothless sailor and the one called Martin. Eyes atwinkle, he said to his wife, "Perhaps there was a little," and here he waffled his hand, "After all, those English...."

"Poof!" Vrouw Fruey rejoined, "What foolishness! With two small children in a boat of farting, belching animals! You men!"

Then, fearful she may have overstepped her bounds, Vrouw Fruey hurried on, "But consider, dearest husband, if he raises his hand in public, what does she bear when they're alone? Poor woman!"

Mynheer Fruey agreed. *Ja*, a sad thing. The man might be a remarkable mariner, but he was strange. Perhaps too much sun. Perhaps *es wrak*.

But what most provoked the Governor's growing unease was the inertia of his guests. He was well aware of the European disease, an endemic lethargy in the topics born of high sun and crushing heat and boredom, of leaden limb and strangled ambition. But Lieutenant Bligh had been dogged in his inquiries about ships that might call or vessels he might hire to carry him back to his responsibilities. After a few early questions, however, Bryant had secreted himself while the rest sank into sloth. Except for James Martin, who was soon employed in overseeing the lieutenant governor's sawpit, showing the Dutchman refreshing qualities of English industry. And Cox would carpenter odd jobs if he felt like working. But with Morton and Allie, with Butcher, Bird, and Lilley, it was always a bad back, a sore finger, a touch of fever. To this little community of Dutch, the English unfortunates were revealing themselves as indolent oafs, and most agreed Kupang would be better off with them gone.

~~~~~~~~~~

Captain Edward Edwards of *HMS Pandora* wasn't happy. For Lieutenant Oliver to miss their rendezvous was just the latest frustration to plague his search for the *Bounty* mutineers. All this bother for a pack of mutinous dogs!

Edwards regarded the forlorn figures in the round prison on the foredeck. "Pandora's box" his men called it. "Pandora's box" held fourteen of the *Bounty* mutineers, handcuffed and shackled, doubly secured by a long iron bar running through their leg-irons. But Pandora's box did not hold their ringleader, that damned Fletcher Christian, or eight others still at large somewhere in the South Pacific with the *Bounty*.

When the *Pandora* arrived at Tahiti months earlier, Christian and one faction of the quarrelsome mutineers had already sailed off on the *Bounty*, while the others were putting the finishing touches on a surprisingly well-built schooner. Edwards imprisoned the mutineers and put Lieutenant Oliver aboard the mutineer's schooner with a small crew to continue scouring the area for Christian. For three months the *Pandora* and the schooner chased one false lead after another in a hunt that had already cost five men lost in foul weather. Now, the schooner failing to make this last rendezvous, the taciturn Edwards dreaded further loss. His anger and resentment built with each passing day. Criminal dogs! He'd as soon hang those in hand and be done! But he had his orders: Bring the mutineers back for trial as an object lesson for British seamen who muttered discontent.

Edwards knew deep down that British seamen had legitimate grievances - impressment and late pay and short rations and capricious officers and arbitrary punishments - but such grievances were not sufficient cause to question authority!

The tall, solidly built captain removed his hat to wipe his face and thinning carrot-color hair. He was weary of these endless islands and treacherous, thieving savages. He wanted to work his way west, towards Djakarta, toward home. When he saw the approach of his paunchy, alcoholic surgeon he deliberately turned his back in pretended gaze over the taffrail.

"Excuse me, Captain."

Edwards didn't respond.

"Captain, I feel it's my duty to again inform you that the prisoners are in need of exercise. They've been ironed too long. Their limbs are losing strength. Some can hardly stand."

Cracking his knuckles behind his back, Edwards maintained his pretended gaze over the taffrail. "Are they getting full rations, Mr. Hamilton?"

"Yes, Captain, but...."

"Then they are getting everything they merit, Mr. Hamilton, and everything I am required to provide. They are criminals, Mr. Hamilton, or have you forgotten that, criminals who forcibly seized one of His Majesty's ships."

"Yes, Captain, but...."

Edwards wheeled so violently Hamilton retreated a step. Edwards glared at the surgeon, then raised his eyes to watch a lanky sailor pass down an emptied slop bucket through the iron hatch atop Pandora's box. As if speaking to a child Edwards said, "Those men know they will hang for their despicable deed, Mr. Hamilton, which is motivation enough for them to pursue any means of escape, however desperate. I will do nothing to facilitate them in their desperation, including the exercise of their limbs. If their limbs are not strong, they will be less capable of escape. That's all I have to say on the subject. All!" He lowered his eyes to fasten on Hamilton.

"Yes, Captain, but...."

Edwards' eyes turned to ice. Hamilton backed away, bobbed an awkward bow, and headed for the bottle in his cabin. "And Mr. Hamilton...."

"Yes, Captain?"

"I'll thank you not to inform me on this matter again. You have done your duty, most assiduously, a circumstance I will be pleased to put in my report." Hamilton bowed once more and retreated. The captain regarded his departing figure with amused contempt. Milksop!

~~~~~~~~~~

Once settled in Kupang, Mary B packed Quint's book in Little Bill's battered box with a troubled mind. Once they were free she hoped Will would change. She now saw why he might have wanted to leave her behind, why it might have been for her own safety, why he might have mentioned the banns to make it easier to bear his leaving her. She could appreciate too the stress he endured in pushing the crowded cutter toward Kupang, in keeping his leadership firm, the men in hand, in making decisions every day that, if wrong, could mean failure, even death. She could see all this, so she was still willing to try to forget the bad parts of the past.

But nothing changed. Even as success began to look possible (which even Will, damn his eyes! had to admit was due as much to her efforts as any) he stayed dismissive. He treated her with disdain, even as the mother of his own infant son. "My son!" he ranted. "These brats are yours!" He dissembled in public, but ignored her in private.

At first dismayed, Mary B soon grew bitter. How wrong to pretend her family's future was dashed on the shoals of New South Wales! She had no family! In truth, she had no husband! Yet she was bound by Emmanuel and by her troth for better, for worse, till death do us part. And the law said they were married.

But what of her fantasies of James. On Mary B's little island she'd carried his smile across the fire to her ablutions like a soothing unguent, had caressed herself with his almost palpable participation. But she was unable to turn to him now.

Although his smiles and kindness and cheerful play with the children nourished her spirits, her fantasies had to remain fantasies. Had she been more worldly she might have devised some clever sophistry, some devious plot and become involved with James again. He still liked her, she knew. But the real Mary B knew James too was pulled between his longing and some higher thing. They were alike in that, and she would not permit herself to be the first to do a wrong thing. Mary B felt bound by her vows as securely as if shackled, and she continued to suffer the consequences.

But more. In Kupang, Mary B grew guilty of her freedom, or more exactly, regretful of her flight. She lived afraid of being found out. She'd justified her escape by saying she'd suffered punishment enough, that she no longer deserved to be a prisoner, and that her rightful place was with her husband. But as the success of their escape became real and Will more remote, her invention failed. She was an escaped felon! Being married to a man who'd completed his own sentence made no difference at all. She'd deliberately broken the law - a big law. She was now a fugitive and would have to live the rest of her life looking over her shoulder.

Thus did Mary B live a lie in Kupang. A married woman - a woman cast aside. A victim of shipwreck - a dissembler. A free woman - a fugitive from the law. She carried her inner torment like a large, doughy child.

~~~~~~~~~

Timotheus Wanjon was troubled. Convinced the English refugees were not *Bounty* mutineers, he nonetheless found himself reconsidering the possibility they might be escapees from Port Jackson. Hints of something amiss were accumulating like slut's wool under a bed. Wanjon knew the reputation of the captain of the *Waaksamheyd*, Detmer Smit. And he knew that if one word described Smit it was "avaricious." Smit could very well have provided the English with their antique quadrant and little compass, which both looked Dutch-made. And with scars from fetters and flogging.... But the Governor always came back to this: To undertake a voyage of more than 3,000 miles in an open boat on an unfamiliar track was insane, especially with a woman and two small children!

The problem was, he couldn't prove or disprove the veracity of their story, and suspicion alone hardly justified offending a great power like England. What if they really were shipwrecks? On the other hand, by God! he wouldn't be duped! As the weeks passed his doubts nagged, but his dilemma remained unresolved.

The coronation of a new native king finally provided Wanjon with an excuse to distance himself from the escapees. Burdened with the need to provide endless entertainments in connection with the coronation, the Governor sent Mynheer Fruey to explain; with the demands of hosting so many native dignitaries His Excellency could no longer feed the shipwrecks from his own kitchen. He'd instead provide a slave for their needs. The convicts winked at each other. Just like lords!

The woman who appeared next day with Mynheer Fruey was small - barely five feet - with precise features and a tawny skin that showed the rough patina of outdoor labor. Her hair was glossy black, gathered atop her head by a large wood comb, and she wore the common native garb of a brownish-red sarong and a white, short-sleeved, open-front, jacket-like blouse. Shifting from foot to foot as Mynheer Fruey again explained the arrangement, she was clearly nervous, but she smiled a betel-stained smile when she saw Mary B with Emmanuel, for she carried a naked boy of about the same age on her own hip. A little girl peeped around her sarong. When several more slaves arrived with an assortment of pots and foodstuffs, the young woman proceeded to organize the cooking hut.

She was Rami Wek Fona, and to Mary B she was heaven-sent. That very afternoon Mary B mounted the short ladder to the cooking hut. She surprised Rami in her work, but the slave bowed a welcome, although with a tight smile. She wondered what the tall European woman wanted, and was nervous at the prospect of being unable to please. She was relieved to see Mary B's smile, however. She thought the woman's unfamiliar gray-color eyes looked sad.

Rami tried to determine what Mary B wanted by pointing first to a bucket of water on the table. Mary B shook her head. Rami pointed to other objects - a bag of rice, a garland of onions, a jar of oil, a pot, each time repeating a little phrase with a rising inflection, increasingly nervous with her failures, while Mary B's own smile grew as she shook her head and said, "No, no, not that, not that," until at last with a short laugh she took two strides to seize Rami's hands. "Just talk," she smiled down at the woman who'd appear in her dreams for the rest of her life. Then as she recognized the absurdity of her words, she nevertheless laughed again and repeated, "Just talk," knowing what she really meant was, "I'm lonely and I want a friend."

Had it come to this? that she was so lonely she sought out a slave for friendship?

In truth, she now felt so alien among Kupang's half-dozen European women she could no longer bear to stroll its short, narrow streets. After the early flurry of sympathetic attention she experienced as "a pitiful creature," after the torturous dinner at the Governor's table, and after a visit by Vrouw Fruey and the wife of the company Fiscal, Vrouw Ganser (during which they peered and gabbled and hissed like two fat, beribboned geese, departing at last but leaving behind a small Bible, in Dutch) - after all this, Mary B knew she couldn't dissemble. She couldn't hide her rough manners, her tongue-tied shyness, her ignorance, and knew she could never share the company of Kupang's white women, except perhaps as a mute, pitied attendant. A prisoner of a different sort, she felt as shut off in Kupang as on the *Dunkirk* or *Charlotte*, and desperately missed her women friends in Sydney.

~~~~~~~~~~

"What're you doing?" Mary B cried when she saw Will with writing materials and Quint's book. "Where'd you get that? You took it out of my box!" Will shrugged. "That's my book!" she said, "You don't need it anymore! You know where you are! That's Quint's book, Seedy's book! I promised her! You have no right!" Will knocked her hand away. "It's mine, Will! Seedy gave it to me! Not you! Me!"

When she again tried to seize the book, Will half rose and pushed her roughly, "Goddamn you, woman, keep away! Touch this book again and I'll wring your neck!" Mary B glared, panting with frustration. The man at the little plank table - her husband! - seemed a stranger, a skinny apparition. Unshaved for weeks, a tangled mass of gray-streaked hair shadowed his visage. His eyes were red-rimmed, baggy, haunted, as when he'd brooded about escape in Sydney. An atavistic shiver crept down Mary B's back. Will hunched over Quint's book like a cunning, bad-tempered dog with a bone.

"Goddamn you, Will Bryant! You're a terrible man!"

Will's face slowly cracked with a contemptuous grin. Mary B had never been more aware of his enmity. It reached out to envelop her like the sour odor of his body. Her hands clenched and unclenched. "You don't care! You don't care what happens to me! You'd just leave us! Whenever you felt! Even your own son! You're a terrible, awful man! I...I hate you!"

Will grinned back his contempt.

Mary B's anger seemed to rob her of further words. She swallowed repeatedly, her chest heaving. At last she blurted, "You won't get away with this!"

Still grinning his horrible grin, Will slowly, deliberately, and with obvious relish, raised his hand to make an obscene sign with his finger. Mary B spun from the room.

Why did Will expropriate Quint's book? Was he so indifferent to Mary B's pledge of safekeeping? Was it disdain? Was he satisfying some old grudge? perhaps Charlotte? Perhaps James? Or was it just bad judgment, like tweaking the nose of the revenue officer or selling his catch in Sydney? He could've asked Mary B, could've explained why he needed this ill-kept diary of smudged sketches, compass and quadrant readings and chicken-scratch calculations. But no, the imperious author at work, he filched Quint's book from Little Bill's box and bullied Mary B to pursue his secret project.

Actually, it was the admiration of the Dutch for his accomplishment that inspired Will to steal Quint's book. Feeling particularly embittered one day, he suddenly burst out laughing with a brilliant idea. Trapped in Kupang, he'd put his time to use by writing a narrative of his remarkable voyage! He'd add detail and comment and a few fanciful particulars and a lot of drama about raging storms and savages. And make his case for a free man fleeing a failing colony.

He spent much of that morning in reverie, gloating and fondling the worn cover of Quint's book, pursing and unpursing his lips in a habit picked up from Morton, imagining his narrative set in type, printed, stitched and bound, real in his hands. He imagined himself basking in praise, like Tench, whose narrative of the First Fleet caused quite a stir when several copies came back to Sydney by way of the Second Fleet. He daydreamed:

- Aye, Madam, 'twas a fearful moment, but the thought of England gave me strength!

- Savages? Sir, they cannot be tamed! They're upright beasts!

- Go back? For King and Country, Madam, for King and Country, 'pon my honor!

Rousing himself at last, he opened the book the wrong way and once again regarded Quint's picture with grudging admiration, and once again wished he himself could draw. He flipped the book to his own side and found the next blank page. Sharpening his pencil, breathing deeply to relax, he spent an hour in painstaking labor drawing and then tracing and shading the script of his title page in ink:

Rem^{arks} on a Voyage

from Sydney Cove, New South Wales

to Timor

by Wm Bryant

Done, he admired his work for a long time, imagining his celebrity status again, then yawned and closed the cover. Suddenly his head and limbs seemed heavy as lead. He rested his head on his arms and sank into dreamless sleep.

~~~~~~~~~~

After almost two months of freedom, it was anger and Quint's book and the arrival of Captain Dahlberg that made the convicts prisoners again. Much to Governor Wanjon's relief, the *Rembang* delivered Captain Dahlberg and his reinforcing company of soldiers just as hordes of natives were completing preparations for the coronation. For days chieftains and dignitaries had been gathering in Kupang, some by boat along the coast, some on tough, little horses from the highlands, still others borne by slaves on sedan-like litters, all with armed escorts. The nearby plain was spangled with night-fires of little tribal armies.

From past service both as an ally and an enemy of England, Captain Dahlberg had an excellent command of English, and at a convenient opportunity the Governor intended to use him to get better information about the shipwrecks. But opportunity knocked sooner than expected. One morning the Governor heard a commotion outside his reception room and looked up to see a watch of soldiers dragging in Will Bryant. Wanjon fixed the disheveled Bryant with a stern glare and sent for Dahlberg. This was the last straw!

While they awaited Dahlberg, a sergeant explained the watch had been summoned to break up another fight. When the sergeant arrived they found Will Bryant being restrained from reaching his wife, who knelt on the ground, dazed and bloodied. The sergeant chose not to go into detail about the mayhem of tussling, yelling, and swearing, of children bawling, dogs barking, and slaves yammering. He merely reported that after imposing order he came to understand that Bryant had beaten his wife because of the *Bible*.

Incredulous, the Governor asked, "*De Bijbel?*"

"*Ja, Excellentie, De Bijbel.*"

Wanjon sighed. Truly, these English had grown wearisome. He was tempted to string Bryant up by his thumbs right then, but the idea of a man beating his wife over the *Bible*.... With Captain Dahlberg's arrival, the Governor asked whether Bryant had beaten his wife according to a Biblical injunction. After conversing with Will, Dahlberg reported, No, it wasn't for that. Was it because she had failed in duties the Bible prescribed? No, not that. Some disagreement over an interpretation perhaps? No, it really had nothing to do with the *Bible*, Will said, why did the Governor keep asking about the damn *Bible*?

But wasn't their disturbance over the *Bible*?

Will mumbled. Dahlberg asked again. Will mumbled again. Dahlberg shrugged and said, "It wasn't the *Bible*, Excellency, it was just a book."

Governor Wanjon rolled his eyes, already regretting his curiosity. Then, thinking it might be a record of the family's accounts, he sent a look of faint amusement around the assembly. "*Een boek? een grootboek?*"

Dahlberg asked several questions to which Will mumbled replies. Dahlberg turned with a puzzled expression. "*Een blank boek, Excellentie.*"

A blank book? He beat his wife over a blank book?

Dahlberg spoke with Will for some time, then explained. "He says the book belongs to a dead friend. He wants to take it back to England for his friend's mother, but his wife got angry and hid it from him. That's why he beat her."

"*Waar es dit boek?*" the Governor demanded, his impatience now clear. Will said he didn't know, that his wife refused to tell him. Head bowed, Will stood mute.

Dahlberg said, "I don't trust him, Excellency."

The Governor mused, playing his lower lip with a finger, pleased to have Dahlberg as an ally in his own assessment of the Englishman. Judiciously, he asked the gathering, "He never mentioned a friend who died, did he?" Heads indicated No. Unless it happened before they were put in the cutter, someone suggested. The Governor folded his hands on an ample stomach and considered. Clearly they were getting nowhere. He leaned toward Dahlberg. "Get to the bottom of this. Find out who these people are. Don't be too rough."

Will broke before nightfall. Irritated with Will's obstinance in maintaining his story about the *Goodfellow*, Dahlberg took him to the barracks and ordered him stripped naked. The Dutchman mocked Will's protruding ribs, "You are thin, English. Are you worried?" He ordered Will's arms bound tightly behind and made him kneel. At Dahlberg's vague gesture, a soldier emptied the soupy contents of a slop bucket over Will's head. Liquefied shit ran into Will's eyes and ears and soaked his beard and found its way to his lips. Handkerchief to his nose, Dahlberg studied his prisoner in the spreading puddle of shit, then left. Two hours later, Will still refused to budge from his story. The Dutchman made him kneel on little stool, then left him under the watch of soldiers.

When Dahlberg returned Will's knees were throbbing.

"At some point you're going to tell me, English. In the end you'll tell me. Why not tell me now and save yourself this suffering." Will remained mute. Dahlberg ordered Will's ankles trussed to his wrists. The tangle of ropes at Will's back was then tied to another rope slung over a beam. Taking this rope himself, Dahlberg leaned back and drew Will's arms and ankles painfully up behind his back until his arched body swung free several inches above the stool. Will gasped. His shoulders and hips felt as if they were being pulled from their sockets. "Ah, no," Dahlberg said, "This is too cruel. My father learned this trick from you English, at St. Eustatius. Your Admiral Rodney had to have every last guilder, you see, and so you left my dear father hanging like this for six hours. After all these years he still feels the pain. But I won't do that to you, English. I'm a civilized man."

Dahlberg let go the rope and Will fell heavily, caroming off the stool onto the stone floor. The breath left his body. Struggling for air, he felt the warmth of blood in his nose. A soldier worked at the ropes, and in a few moments only his arms remained tied behind his back. Dahlberg made a tsk-tsk sound.

"I'm going to dinner now, English, and perhaps will take a nap. Unless you choose to tell me the truth, I'm afraid you'll have to kneel on the stool again." Will remained silent. "No?" Dahlberg sighed elaborately, and motioned for his prisoner to be knelt on the stool. He regarded Will with mild perplexity, then strode off.

By the time Dahlberg returned three hours later Will was unable to feel his legs. He'd fallen several times, but each time he wavered and lost his balance and pitched off he was repositioned to kneel again. He so dreaded remounting the innocuous little stool he wanted to cry.

Sucking on a toothpick, the Dutchman ordered Will unbound and dashed with a bucket of water. Will sprawled on the stone floor. He refused to massage his knees in front of Dahlberg. The Dutchman sat on the stool and regarded the inert body. "I think you will tell me now, won't you, English? I think now you will tell me the truth and spare yourself any more pain. Or maybe you like pain. Some men do. Do you like pain?"

Not sure why he wouldn't speak, Will remained unresponsive. It wasn't to save his own skin. He was safe. And not out of loyalty to the others. No matter what, they all knew in the end it was, "Look out for yourself and the Devil take the hindmost!" No, it was something else now, something with Dahlberg. Dahlberg should give him his due.

With an almost kindly gesture, Dahlberg lifted Will's hand to his knee. An involuntary groan escaped Will at the movement of his shoulder. Dahlberg spread Will's fingers and patted his hand. "No? Not yet?" He nodded to the soldiers and almost casually took the toothpick from his mouth, a sharpened splinter of bamboo. Dahlberg seized Will's middle finger in an iron grip and jammed the splinter deep under the nail. Will screamed again and again.

At last the pain abated to a searing throb. Will became aware of Dahlberg's gentle, persuasive voice, "I don't want to hurt you, English. I can see you are a strong man. Do you think I want to do this? No, I don't. No one wants to do this to another man, but you leave me no choice. You don't tell the truth. And see now where you've got yourself, English? You've got this little piece of wood stuck in your finger, this tittle of

God's vast creation, but it will cause more pain than you can bear. It's not fair that something so small can cause such pain, like a thorn in the lion's paw, no? The English lion's paw. Look, English, how easy...." and as if flicking a bit of lint Dahlberg flicked the protruding splinter with his finger. Will screamed.

Several times the convicts heard his screams. They imagined horrifying scenes of racks and wheels, of tongs and thumbscrews, of burning brands and hot irons, but when at last soldiers escorted Will back in the quick-falling tropical night they were surprised to see him moving under his own power in an odd, gingerly, stiff-legged shamble. He looked freshly bathed with his hair wet and slicked back. His teeth gleamed dully through his beard in a grimace that might have been mistaken for a brave smile. He held his left wrist.

Just as Captain Dahlberg predicted, Will told him everything, told him in a rush of words so rich with information Dahlberg could scarcely take it all in. And in the end Will thought his confession perhaps wasn't so damaging. After all, he kept insisting, he'd completed his sentence. Yes, he escaped from the penal colony, but that was only because the authorities refused to release him. In the eyes of the law he was a free man, free to go wherever he wished, even to Timor. Government no longer had a right to keep him prisoner. And if the others hadn't completed their sentences, well, that was their lookout. He didn't force them into the cutter, even his so-called wife. Here he explained how they weren't really married, not legally, but said she'd threatened to expose his plans if he didn't take her along.

"Yes," Will said when he finished his story, he'd lied to the Governor and was sorry about that, because His Excellency had been first-rate, but Will would happily work off his debts. The Governor could even have the cutter, none the worse for wear. The Governor might even come out ahead.

When Will finally stopped talking, Dahlberg still harbored doubts. He wondered, is the Englishman going to be hanged? Is that why he risked life itself in an impossible gamble for freedom? But the Dutch captain was unable to shake Will from this final version of his story.

The truth out, the worst over, the convicts were almost relieved. They awaited the Governor's decision.

Pleased finally to have the facts, the Governor was actually ready to accept Will's ardent claims. But then Cox came forward and maintained that he too had completed his sentence and deserved to be free. And then Bird popped up. Well, now who was to be believed! The Governor had heard enough. Let English authorities untangle this mess. He'd turn them over to the first English vessel that called. For the nonce, he ordered them under house arrest. Each day only two were allowed the freedom of the fortress. Because she was no threat to escape, Mary B was free to come and go.

As to the matter of the book.... After the Dutch retrieved Quint's book from its hiding place in a large jar of uncooked rice in the cooking hut, the Governor examined it carefully. He showed Quint's picture to the general admiration of his audience. When Dahlberg explained that the Englishman how wanted to use the book to write a narrative of his remarkable voyage, the Governor summoned Will. He could continue to write his narrative, but in the office of the Fiscal, where his book would be kept safe. Recognizing the value of Bryant's record of their remarkable voyage, Governor Wanjon was already resolved to keep Quint's book for his own.

~~~~~~~~~~

Mackesey rested wearily on his oar and watched Passmore try to coax the terrified cat into the boat. Its yellow fur salt-encrusted from sea-spray, the cat clung near the tip of the *Pandora*'s canted topgallant mast six feet above the water. "Come, puss, come, puss." Claws locked on the mast, staring without recognition at the outstretched hands and the rise and fall of the launch and the eight exhausted men at its oars, the cat refused to budge. She'd spent twenty-four hours clinging to this last piney hope for life. If Captain Edwards hadn't sent Passmore back to see what else he might salvage of the wreck of the *Pandora* the animal would have died of thirst.

Mackesey licked his cracked lips. Salt. He wanted to close his eyes against the glare of the morning sun. He wanted to sleep. A daydream born of exhaustion took him to a cool, shaded pool beneath a tumbling waterfall. He raised his face to the cataract and let the water pound his face and batter his lips and splash in his open mouth. He swallowed, and came back to the tableau of cat and sailing master: Numb Terror and Determined Rescue. Rescue, yes, but they might end up eating the scrawny animal. A reprieve for pussy, but not for long. Was it then rescue?

Mackesey shifted his attention from the drama of the cat to follow the ghostly form of the mast disappearing into the depths. Ninety feet below lay the tilted deck of the *Pandora*. And his chest and letters and little guitar and everything in the world he owned. And fresh water. Cask after cask of fresh water. And the bodies of thirty-five men, most of whom had exhausted themselves trying to save the frigate, only to be rewarded with death.

The *Pandora* had struck just after dark in stormy, running seas while trying to find a way through the Endeavor Straits. Edwards should have waited for daylight. A prudent man would have held back.

Mackesey would never forget the horrible, crunching shudder of the ship. Perhaps at shivering impact the thought flashed through everyone that all was lost, a fitting climax to a jinxed voyage.

With the *Pandora* leaking badly, every available hand took turns at the pumps and on bucket brigades that snaked down through the hatchways as Edwards tried to pack the staved bow. But relentless battering against the reefs opened the *Pandora*'s wounds wider. The water gained on them. Wind and waves finally bumped them over the reef, where Edwards dropped anchors to attempt repairs. Still hampered by darkness, he tried to work a sail under the ship as a patch as water continued to pour into the ship. He released the three hardiest *Bounty* mutineers from Pandora's box to take their turns at the pumps.

Abandoning the idea of a sail-patch that accomplished little, Edwards ordered the ship lightened. Men wrestled guns and carriages overboard. An ocean swell heaved the ship and sent a gun carriage careening across the deck to crush a hapless man against the bulwark. The first death. And still the *Pandora* settled deeper.

They fought all night, knowing that to abandon ship in the stormy darkness might mean death. They kept themselves going on ale and fear till the glimmer of dawn revealed the bow down, the portside awash to the scuppers, the deck tilted almost too steep to stand.

Still chained fast in Pandora's box, the remaining *Bounty* prisoners screamed for mercy. Edwards ordered spars cut away, then chicken coops and livestock stanchions and everything that would float was thrown overboard. A falling spar killed a second man. As dawn brightened, the *Pandora* heeled over nearly on her side. Finally released from their labors, men began jumping overboard. Unbidden, the ship's armorer worked frantically to free the long iron bar that slid through the leg-irons of the *Bounty* prisoners. He struggled to fit the key into the padlock as the *Pandora* groaned in her final misery. Succeeding at last, he scrambled to unbolt the iron hatch just as Edwards and the last of the crew leapt into the sea. Seven mutineers clambered out before the *Pandora* plunged bow-first beneath the waves so swiftly the last four men still trapped in Pandora's box were carried to the bottom.

The nightmare of the next half-hour would haunt Mackesey for the rest of his life. Earlier the ship's boats had been let down to hover near the endangered ship. Now they approached through storm-tossed waters, four boats for 137 survivors who'd worked frantically without rest for almost twelve hours, who'd gone without sleep for twenty-four. The men in the water cried for help. They clawed at the boats and lines and tried to clamber over those nearer rescue.

Boat crews hauled them aboard as quickly as possible, but some sank from sight even as hands reached to save them. The boats rode lower and lower in the spumy waves with their growing burden of survivors. Gradually the commotion died as those saved shrank into silence, and those who could no longer stay afloat disappeared. At last the only voices were those who called from the boats circling the stub of mast that jutted above the surface.

After a time the two most badly overloaded boats pulled away to make for a barren key some three miles distant, the only land in sight. The other two continued a fruitless search for another half-hour. Then they too gave up and to head for the safety of a patch of sand scarcely thirty yards long. There the survivors would regroup under the mounting sun. They'd count their losses, and plan their survival.

Mackesey heard a snuffling. Lame Johnny was shaking with sobs. Mackesey put a comforting hand on the boy's bare shoulder, already burning under the tropic sun. "It's all right, Johnny. He's resting now, in God's hands. They're all in God's hands." Lame Johnny's cousin was one of the thirty-five who'd found his grave in the depths. Like Lame Johnny, his cousin was a youthful landsman from Somerset. Like Mackesey, he'd been impressed into the King's service to make up for thousands of mariners serving the build-up of the Channel Fleet, a build-up fueled by the threat of yet another state of war with France.

Goddamn! Mackesey thought for the thousandth time, It's not right, these stupid wars! If he lived, if he ever reached home.... No more of this slavery! This damned slavery!

Passmore finally hooked a hand around the mast and grabbed the cat by its scruff. The animal screeched with terror and scrabbled wildly, raking the wood, then twisting to sink her desperate claws into Passmore's forearm. "Argh!" he cried, ripping the cat loose and flinging her to the floorboards. The cat scuttled under a thwart and crouched in fright.

His forearm bloodied, Sailing Master Passmore salvaged a length of copper lightning chain from the jutting mast, some of the rigging, and sawed off the top five feet of the mast. Although they rowed around for another hour and examined reefs awash in the low tide, they found nothing else of the *Pandora.*

For two days the survivors prepared for their next ordeal. Somewhere south the bleak northlands of New South Wales offered dubious advantage. Their only real hope was Kupang, 1,100 miles west, in boats fit to carry only half their number. Their provisions were pitiful. They possessed the useless iron of a few muskets and cartridges, bread enough for three ounces a day for two weeks, a little dried soup, a little wine. Inexplicably, when he ordered the boats away from the sinking *Pandora,* Edwards neglected to see them fully stocked. If they filled every conceivable container they could carry just four quarts of water per man...after they found water. But they had precious little. Finding water was their most urgent need.

They straightened the links of the lightning chain to fashion nails. Ripping up floorboards, they nailed them as stakes to the sides and stretched sailcloth around to keep waves from breaking into the overloaded craft. Other sailcloth they stretched for sunshades. Some they stitched into water bags. They caulked the compass boxes to hold water, even the carpenter's boots (which would hold more than a gallon each).

On September 1st, 1791 they pushed off from their sandy, barren prison to find water. As soon as they could hoist sails they relieved the crowding by putting oars across the gunwales, providing a platform for some to perch on. The largest vessel, an eight-oared launch, held 31 men, two smaller boats held 24 each, and the last 23. Plus the cat.

They sighted several islands that same day, and after retreating from an encounter with threatening natives, found fresh water. And next day set forth on their desperate venture to reach Kupang.

~~~~~~~~~~

When she hid Quint's book in a jar of rice it never crossed Mary B's mind that she was performing a symbolic act of separation, or that the book might prove the instrument of their actual separation. Rather, she hid Quint's book to assert her rights. Will couldn't have it both ways. He couldn't deny their marriage but play lord and master. Simple as that. And for that assertion she'd suffered another beating.

That was the last beating Mary B would endure from Will. Although unable to move his limbs without groaning, Will had to minister to himself. Mary B moved into the cooking hut with Charlotte and Emmanuel. There she shared sleeping mats and the split-bamboo floor with Rami and her two children.

"Good," Will thought. "Finally rid of her!" Later, when given leave to resume his narrative in the Fiscal's office, Will set to work with scarcely a thought of his one-time wife and child and the others. More than ever he believed his book represented his freedom.

To Mary B, possession of the book was now moot. The Governor had ignored - no - had never even bothered to explore her own claim to the book. As if she didn't exist! Now Will had Quint's book by the Governor's own hand, and she could do nothing. Will had what he wanted, and was rid of her too. He'd won the battle for the book and the war of the marriage. Mary B surrendered the battlefield to fight no more, and retreated to Rami's world.

Except through vague insights of instinct and empathy, Mary B would never understand the subtle status of this young woman of color. Although Kupang's slaves were usually brought from distant islands to make their return home more difficult, Rami was Timor-born. She wasn't really a concubine either, although such demands were part of her history, as evidenced by her two mixed-blood children. Neither was she a servant in the Western sense, with expectations of lifelong care and protection in exchange for faithful service. No, after her mother died, her debt-ridden father led Rami down to Kupang to fetch what she might bring. If he didn't exactly sell her at the slave market he nonetheless exchanged her for money with a Chinese merchant. Thereafter her fortunes followed the fortunes of her employers.

With domestic battles as common on Timor as on the sceptered isles, Rami certainly understood Mary B's flight from Will. She showed her empathy by inviting Mary B to join her on errands to the fish-sellers, to the twice-weekly market, to the river to wash clothes. Rami learned a few words of English and Mary B a few words of Atoni, Rami's language. Mostly each spoke her own language in a kind of intonational accompaniment to their real communication of gesture, expression, smiles, and frequent laughter. Although not of Kupang's principal clan, Rami had lived there for many years, and her acquaintance on the native side of the river was wide. Service in the Governor's household also lent her a measure of prestige, but the first time she and Mary B walked hand in hand to the market, word raced ahead. As if they'd never imagined such a sight, gawkers appeared from houses, shops, and gardens to watch the progress of the two women, babes on their hips, obviously enjoying each other's company. A white woman showing intimacy with a native, in public?

Before long Mary B was invited to witness a wedding, and then a birthing, each rich with ceremonial symbols she didn't understand. When her name was given to a baby girl, Mary B returned the honor (under Rami's guidance) with a symbolic gift of rice and a lock of her rich auburn hair, which pleased the parents greatly. They placed the hair with the baby's afterbirth and shriveled umbilical cord in a little bag, which they affixed to a ritual tree at the family shrine.

One day Rami wrapped sweetened sticky rice in a banana leaf and led Mary B and the children toward the nearby hills. The day was overcast and refreshingly cool, for the monsoon winds were reversing their track, a harbinger of the approaching rainy season. They walked for more than a hour at a leisurely pace, following a stream into the hills to reach an outcrop of rocks. Rami grew silent and watchful. She looked carefully behind, as if to make sure they weren't followed, then led Mary B around a large rock and pointed with solemnity. "Cave," Mary B said, naming the dark recess in the rocks. "Cob," Rami repeated. Then the young woman squatted to pantomime birth. After several more elaborate gestures, during which Rami repeated the birth pantomime and imitated someone emerging from the hole, she pointed to herself, to Kupang and the surrounding countryside. Only after another repetition of this elaborate mime did Mary B begin to understand that Rami was showing her the womb from which Rami's ancestors had emerged. She nodded her understanding, and embraced the young woman in thanks for showing her this private part of a strange new world.

The two became inseparable, Rami a mentor of things native, Mary B her eager student. After Mary B understood the symbolism of the cave birth, she intuited the significance of other symbols she'd observed during the birth ceremonies - round stones smooth as pregnant bellies, round little pots with vagina-like openings. In Rami's world too, Mary B now realized, women had their own symbols and secret wisdom.

On a more practical level, Rami taught Mary B how to cook rice in subtle variations of texture and taste, to plait grasses and palm leaves and viney materials into baskets, mats, hats, and fans. And to bathe. When Mary B first waded into the river with her long, fully cut European dress hiked up, she felt self-conscious and awkward. But after several twice-daily trips to the river she began to notice the sour, foul odor that Europeans carried and tried to smother with scents. Oh, she thought, how rotten I must have smelled!

One day she and Rami exchanged clothing. Nearly lost in the folds of Mary B's clothes, Rami so comically imitated the determined waddle of Vrouw Fruey that Mary B fell to her knees, helpless with laughter. And when Mary B stepped naked into Rami's sarong she found the freedom of the garment a surprise, and felt a little rush of sexual excitement. Bathing in Rami's sarong was much easier than in her own cumbersome clothing, and Mary B thereafter borrowed Rami's other sarong for their excursions to the river.

Rami also taught Mary B graceful female movements of native dance - slow, shuffling steps, gentle undulations of hips and abdomen, languorous motions of arms and hands, difficult movements of head and neck. At first awkward and clumsy, Mary B gradually approximated her friend's rhythms and movements, which possessed an economy of effort and grace unknown in the stiff, foot-stomping dances of Cornwall.

The incident that fixed the lifelong bond of the women occurred the afternoon Captain Edwards and the survivors of the *Pandora* reached Kupang. Mary B and Rami had set off with their children to gather a certain swamp grass used in making small, tightly woven baskets. They followed a path that ran near the sawpit around the back of the fortress, where they stopped briefly to chat with James, then walked along a stream for a mile or so to the edge of a saltwater marsh. There a clump of palm trees provided shade. They deposited the children on mats to play, and ranged nearby gathering the grass. They called back to the children every few minutes in reassurance. Mary B herself was not reassured. She feared snakes, and with each step she imagined stepping on a sleeping python or disturbing a cobra and finding herself face to face with the gaping jaws of an erect, weaving serpent. But it was a different set of gaping jaws that would bring her nightmares.

The women had gathered and tied several bundles and were starting back to the children when Rami dropped her burden with an awful scream and sprinted towards the palms. Heart pounding, Mary B dropped her own bundles and raced after, imagining a python wrapping its sinuous mottled form around Charlotte, around Emmanuel.

Rami stood statue-still in the dappled light. At first Mary B saw nothing amiss. The children continued their oblivious play. Laki, Rami's little girl, looked up and smiled. Mary B started to ask, "What's wro...?" Then also saw it. Also frozen in dappled sunlight, its jutting teeth lending a horrible, almost comic aspect to its endless mouth, only a short distance from the children, crouched the horny-ridged form of a massive crocodile. Rami must have seen it creeping up.

With another scream Rami snatched up a fallen palm frond. The crocodile swung with incredible speed, its mouth a-gape, hissing its warning. Its tail thwacked the trunk of a palm, shivering fronds far above and sending down a shower of debris. Now aware of the beast, Charlotte and Laki began to cry, which immediately set the two little boys squalling. And still Mary B stood frozen, her feet rooted, her mind blank except to register the fuzzy frantic motions of Rami harassing the enormous mouth with her thrusting frond.

Suddenly the crocodile charged in a blur. Rami leapt back, stumbled over a large rock, and for an instant lay helpless. Mary B would never forget her startled, frightened look. But the charge was a bluff. Rami sprang to her feet again as Mary B at last found herself moving towards the children. The crocodile swung again, hissing horribly, its tail flattening the grass like a heavy flail.

The women darted glances at each other and at their children. Bawling his protest at the commotion, Emmanuel began creeping toward his mother. Mary B screamed "Emmanuel, no!" His self-absorbed progress was bringing him closer to the crocodile. Her scream gave Emmanuel pause, but then he resumed his determined creep. Rami again began harassing the crocodile with her palm frond. Mary B had a brief thought that the frond was ridiculous. The crocodile lowered its body, and Mary B saw its claws grab the ground for purchase, its odd, elbowed legs ready to propel the massive body in attack. Emmanuel was a dozen feet from the crocodile. In an instant of insight, Mary B knew the crocodile had gauged the distance to her baby and was preparing to snatch her child and scuttle back to the safety of the water.

Later she would try to lift the rock a second time and find she could barely bring it to her waist, but now in one swift motion she stooped to grasp a rock bigger than Emmanuel, raise it high above her head, and thrust it with a grunt toward the hissing throat. She missed. Badly. The beast moved and the rock struck the giant reptile squarely on its left foreleg. She heard a popping crunch as the heavy missile collapsed the forequarter to the ground. Incongruously, the rock rebounded off the crocodile and rolled on to knock Emmanuel asprawl. His surprised scream of outrage was lost in the tremendous bellow of the crocodile as it lurched around, its broken foreleg useless. The giant reptile wigwagged awkwardly down the embankment with surprising speed. Boiling water, a furrow of underwater retreat, a water bird's startled cry, swamp weeds tossing, and in a few moments only the frightened yowling of the children and the raspy breaths of the women disturbed the tranquil scene.

~~~~~~~~~~

Although under house arrest with the other convicts, James Martin was permitted to continue as Mynheer Fruey's sawpit supervisor. He kept to his schedule of almost three months, rising just before dawn to tramp off to the sawpit, a rather insouciant tilt to his cap, whistling a little tune picked up from Cox. The fish-eye of the sentries at the gate sometimes gave his cheerfulness pause, but not for long, for James was a happy man. He was earning more than he'd ever earned in his life, two rix dollars a day, and lacking freedom to spend more or work less, he'd already saved seventy dollars. For the first time since his imprisonment five years earlier, he felt good about satisfying his simple wants with money he earned. He saw himself getting on in the world.

To enjoy his noon repast he'd climb to the tree-shaded shelter of a favorite little rise near the sawpit and there enjoy a view of the town and harbor below. Except for being too hot, he thought, Kupang is not such a bad place.

Had he been a different sort, James might have taken advantage of Mary B's situation and gone along with Will's arguments about banns and commissary records and proper legal domain. He might have tried to persuade Mary B with a "Why not? It's just you and me out here in the middle of nowhere." But James thought Will's legal gabble was self-serving nonsense. Mary B was Will's lawful wife whether Will liked it or not. If Mary B thought she was married, then she was married. James fantasized, of course. Some inchoate part of him knew he and Mary B were connected and would remain connected, not just by Charlotte, but by what they'd lived through. He knew they were mutually attracted, even if it was only in the half-life of imprisonment. But if at times he fantasized and if at times his longing tested his moral strength, James was developing a more confident sense of who he was. And of right and wrong. And he maintained a discrete distance.

When he spied Mary B and Rami returning from the saltwater marsh, their manner made him think something bad might have happened. They were tentative, as if picking their way on a path of sharp stones, and they carried their daughters on their shoulders and their sons on their hips. Where were their bundles?

James hurried down. Shaken, Mary B put the children down, fell into his arms, and burst into tears. Hardly able to follow her sobbing, breathless story, he led them to his resting-place and gave them water and what comfort he could.

Mary B needed several minutes to collect herself. She kept breaking into sobs. "So brave," James said, rubbing her back after she was at last able to choke out the story, "So brave."

"No! I was so frightened. I couldn't move. I couldn't think. All I saw was that horrible beast eating my children!" She began sobbing again and embraced Rami. The women clung to one another and sobbed.

The watchtower bell in the fortress began to clang. They looked toward the town. No column of black smoke announced a fire, no distant sail a ship. The *Rembang* lay listless in the glassy bay. A clutch of tiny figures ran out the fortress gate toward the waterfront. Soon two of them detached from the growing knot of people at the waterfront to walk rapidly in the direction of the sawpit. They soon broke into a run. "That's Cox and Bird!" James said at last. "They're at liberty today. What's going on?"

~~~~~~~~~~

Tench ruffled Ned's hair as he handed over his hanger and sword. Ned was no longer Sydney's "little Ned." In four years he'd stretched out, and when Tench ruffled the boy's hair he reached shoulder high. But scanty rations had kept Ned lean and wiry. Now he ducked from Tench's hand and assumed a mock-serious sparring stance. Tench chuckled and cuffed good-naturedly a couple times at Ned's head. The boy dodged with easy grace. "You're quick, Ned, you're quick." Tench unbuttoned his coat and dropped in a chair. He raised a booted leg. Ned knelt to pull.

"Do you know what the *Gorgon* means, Ned?"

Ned pulled the boot free. Tench raised his other leg and studied the boy's face for some reaction. Except for a protruding lower lip as he concentrated on the second boot, Ned's expression remained noncommittal. It was late September, 1791, and the *Gorgon* lay at anchor in Sydney Cove, newly arrived with the battalion's replacements.

"Means you'll be leaving, Cap'n."

"Aye, Ned, home to our motherland." Tench chuckled, then suddenly burst into song:

> *Oh, the roast beef of England!*
> *And old England's roast beef!*

Ned smiled and inspected the boots. "These need a good cleaning, Cap'n, and blackened up too. I'll do it tonight." Ned rose and fetched the Captain's bottle of canary. Fresh wine had arrived with the latest convict ships, the Third Fleet, along with a cornucopia of long-absent luxuries at exorbitant prices. Ned poured and handed a glass to Tench. The Marine swung his stockinged feet onto the table and tilted back. He sipped, smacked his lips, and nodded approval. "England, Ned, England! Do you remember? That pleasant, green land of fat sheep and contented cows, and London! My God, the people in London, Ned! It's the greatest city in the world! And newspapers! Newspapers everywhere, Ned, and music! aye, music all the time, and books! More books than you could ever read, and ladies dressed in silks and satins stepping from their chairs, ummm...."

A knock on the doorframe interrupted Tench. Ned admitted a young woman carrying the Marine's supper things wrapped in a cloth. Eyes averted, she curtsied to no one in particular, handed the bundle to Ned, and ducked out the door with another half-curtsy. Ned began to lay out Tench's supper. Tench shifted his feet an inch. "Ah, yes," he said, as if trying to find his place in a book, "England." He poured himself another glass, sipped, smacked his lips, and nodded approval again. Ned stood lost in thought. "Are you thinking of home, boy?"

Ned collected himself and realigned Tench's utensils. "A little, Cap'n. But I was thinking...every time I see her I wonder about Bryant and his gang and the way Cox just went off and left her and how he kept it a secret from her. And how she's so sad."

"You mean Sarah?" Tench asked, nodding toward the door.

"Aye. I just wonder how you can live with somebody and keep such a secret. You think she really didn't know Cox was going to leave her?"

Tench reddened slightly as he studied the thin film of canary sliding down the inside of his glass. "No, she didn't know. He couldn't tell her, not if they were going to succeed."

"You think they made it, Cap'n? Made it to freedom?"

Tench drained his canary and studied the play of light on the film of liquor remaining in the glass. "What's freedom, Ned? They got past the headlands, and that was free. But then what? Where would they be free?"

"Well, maybe a ship picked them up."

"There's no ships this way, lad, unless they come to Sydney. No, there was no ship." He looked indifferently at the covered dishes of his supper, then poured himself another glass. "They'd have to go north. But there's nothing north, not for thousands of miles."

"Well, what you think happened, Cap'n? You think they're all dead, even Mrs. Bryant and her babies?" He remembered the excitement her new baby brought to the *Charlotte*. When he

379

told his mother, not clear why the news stirred such a buzzing on the ship, she'd roused herself and said with a little smile, "Well, bless me," a scene that for some reason Ned still remembered vividly. He'd been on his best behavior to make amends for his misadventures in Rio.

"Dead?" Tench said. He sat with his glass resting on his groin. He stared absently at his new socks, another luxury arrived with the latest convicts, his old socks so darned he could scarcely tell where socks left off and repairs began.

Ned saw the familiar dark veil begin to fall over the captain. Tench would drink more canary and pick at his supper, or maybe forget eating altogether. Then he'd begin reciting - disjointed sentences, a phrase that rippled or rolled or had a special sound, or he'd sing a snatch of song or fumble in a book until he found some passage to mumble half-aloud. In the end he'd sink into gloomy withdrawal. And perhaps remember to wave Ned's dismissal. Or Ned would just slide away. Ned thought, this is what happens when Cap'n drinks canary alone. I must remember not to talk about dying. Ned started to say something, but Tench spoke. "Well, Ned, I hope not. I truly do. Maybe they got to Amboyna, or Kupang. Most improbable, but who knows?" Then, as if that would be the end, he added, "And then what?"

Ned waited a minute before he chirped, "Cap'n, shall I light a candle?" Tench looked surprised. "You said Latin this morning, Cap'n, tonight."

"Ah, yes," Tench said, brightening. He swung his feet down and humped his chair closer to the table. He peeked into a covered dish. The same. Three and a half years after leaving England they still ate salt meat. He wondered if the women remembered how to cook anything else. "Did you study your lesson today?"

"Yes, Cap'n," Ned said, retrieving a thin book from the shelf. Tench opened the book to its marked page.

"Ah, what was the word?"

"*Amore*, Cap'n."

"Ah, yes. Begin."

"*Amore. Amo. Amos. Amat.*"

"Good. Continue."

"*Amonus. Amotis. Amant.*"

"Good. Now use *amore* in a sentence."

~~~~~~~~~

Djakarta means "perfect victory," named for a battle fought long before Western eyes ever saw the strategic port at the gateway between the Indian Ocean and the China and the Java seas. Although the Portuguese arrived first, it was the Dutch who built a new Djakarta in their own image, a metropolis of 50,000 people, steepled churches, official buildings, sumptuous residences, warehouses, broad streets, and criss-crossed with those aqueous arteries that nourish the Dutch psyche: canals - miles and miles of canals. Outnumbered ten to one by natives, Europeans nonetheless controlled government, trade, factories, hospitals and resorts, Chinese and Arab middlemen, and thousands of slaves and servants and workmen whose labors kept the city going. Europeans lived in luxury rivaling the world's wealthiest potentates, but with the onset of the monsoon most of them fled the city.

The Dutch called their city "Batavia," a remembrance of the homeland, but both "Batavia" and "Djakarta" had come to death. Over the years the canals silted into cesspools and gardens of disease. Several thousand residents and visitors died of fever each year. Too many to bury. Bloated corpses - Malays and Javanese and Chinese and Indians, an occasional African, an occasional white - drifting slowly out to sea in the sluggish canals was a sight too common for comment.

Seamen dreaded this port of call. They considered themselves lucky to get away alive. Bligh lost three men here, even while they lived in the sanctuary of the huge maritime hospital. If they got away from Djakarta alive, mariners might still incubate the city's deadly diseases in their blood. Disability and death that followed from Djakarta might leave a homebound ship too short-handed to work. Separated from its convoy, more than one *Flying Dutchman* had drifted helpless in the vast Indian Ocean until a friendly ship happened along, or pirates.

It was mid-afternoon on November 7, 1791, and a soft, warm rain was falling when Captain Edward Edwards peered from under the awning of the skiff at the imposing bulk of the New Dutch Church, its broad cupola barely visible in the scudding mists. Beyond the church lay Djakarta's main square and his destination, the *Stadhuis*. The ferryman sculled the skiff towards the bank of the malodorous Ciliwung River, and

Edwards, in charity coat and charity shoes, a vinegar-damp handkerchief to his nose, prepared to step ashore.

He felt a burning in his chest and the familiar taste of bile. Edward Edwards, Post Captain of His Majesty's Navy, a supplicant to the Dutch!

Still, things had gone better than expected. Expressing sympathy at the Englishman's plight, the Governor-general told Edwards he'd recommend to the Council of Policy that the *Pandora* survivors be given every assistance. Meanwhile, Edwards had permission to sell the mutineer's schooner to put money in the pockets of the destitute crew.

Oh, what a relief when Edwards discovered the schooner and the eager face of Lieutenant Oliver! And what coincidence! Lieutenant Oliver had waited five weeks to rendezvous with Edwards, but at the wrong island! Finally, running short of supplies, Oliver made for Djakarta on a voyage that saw him fight off determined attacks by savages, run out of food and water, and finally limp into Semarang, some 300 miles short of Djakarta. But instead of finding relief, he and his crew were detained by the Dutch under the notion that *they* were the *Bounty* mutineers! Thus, when Edwards put in for water at Semarang, there were his missing men, prisoners all! Thank God for one stroke of good fortune!

Almost two months had passed since that inevitable afternoon when Cox and Bird rushed to the sawpit with breathless news. Boatloads of English Navy! Government! Authority! Certain of a life sentence if caught, Cox jogged into the hills, Bird tagging along. Two days later, hungry, disheveled, sullen, they were dragged back to the fortress by gleeful bounty hunters.

"I'm a free man, Captain!" Will asserted to Edwards when Governor Wanjon presented his principle prisoner. Will's eyes glittered. Under a dirty bandage his tortured finger throbbed. "I served my time. Government has no claim on me."

Edwards regarded him with indifference. "You're a thief and an absconder, Bryant. You stole that cutter in Port Jackson and you defrauded His Majesty's government here. You're a contemptible liar." He turned to address the Governor, then turned back to Will and extended his hand to accept an offering. "Where is your proof of discharge, your paper?"

Will groaned. "They won't give you one unless you stay, and that's not right!" He appealed to the Governor, "I haven't got one!" Governor Wanjon looked blank.

"So you stole a boat," Edwards said.

"That was my boat," Will insisted. "I bought it from the Dutchman on the *Waaksamheyd*, Mynheer Smit, and the compass and quadrant too, and the telescope and muskets!" The Governor squirmed at the translation of this information. He'd hoped the role of the *Waaksamheyd* in Bryant's story would not come up. If what Bryant said was true, the English would not be pleased.

Edwards bowed to the Governor in reassurance. "That is a Royal cutter, Bryant. You stole it. And then you lied to His Excellency here and brought discredit to His Majesty and the British people."

"But I was a free man! And I was trying to protect my wife because her term isn't up. But you can take her, I don't care!"

"Then you helped a felon to escape, and that's a felony. In fact, you helped all of them escape, and that's eight felonies. You'll probably be hanged."

"But it wasn't fair to keep me there!" Bryant cried. "I was free!" Surrounded by evasive eyes and indifferent faces, his voice helpless and defeated, Will repeated, "But I was free." Then, as if to reassure himself he repeated half aloud again, "I was free."

"Maybe, but we're not going to discuss it further. With His Excellency's permission," and here Edwards again bowed to the Governor, "I'll take you back to England. You can make your case there."

Governor Wanjon inclined his head in agreement.

Simple as that. Weeks of planning and headaches and sleepless nights, weeks of struggle against danger and overwhelming odds to reach freedom, all made worthless by the imperious decision of a shipless captain. His fate fixed, Will's protests were as futile as the struggles of an overturned turtle.

How sad, the Governor mused as he watched men from the *Pandora* lead Will away, After all this bother, I believe him now. But then his life comes to this. What made him struggle - a solitary sailor against the sea, a soldier against the redoubt? Why not just accept authority? Why not live in the new penal colony and try to make the best of it? Justice might have prevailed. He might have received his release. Then all would be well. But instead.... Was it foolish pride? Timotheus Wanjon sighed. Some never march in step. They never march in step and always get in trouble, and always pay a price. And sometimes the ultimate price.

Edwards wanted the mutineers from the *Bounty* back in irons, and the convicts too. Who knew where sympathies with their plight might lie? If mobs could free the prisoners of the Bastille, disgruntled seamen could free the prisoners of the Crown. He might be weeks rounding them up again.

When the Governor reported that Kupang didn't possess irons enough for so many felons, Edwards set the *Pandora*'s armorer to work. For simplicity the craftsman forged bilboes, a device developed by the Portuguese for the slave trade. The shackles slid on a short iron bar. A nine-inch bar meant a nine-inch step. To walk in bilboes was hard, to regain one's balance after a misstep impossible. Prisoners quickly learned the bilbo shuffle, a peculiar gait of tiny steps. They hated bilboes.

For lack of a prison the mutineers and convicts were confined in a warehouse basement. Although she was permitted to continue living with Rami, Mary B was put to work cooking for the *Pandora* men, whose shelter was the church and house vacated by the convicts. She had to report to the watch officer every morning and evening, but otherwise was permitted the freedom of the fortress and the native market. With money from James she was able to buy fruits and vegetables for the convicts, who now wondered why they'd lingered. Cox lamented, "We could be free. We could be on our way to America!"

"Ahhh, it was too easy here," Butcher said, "Like being on pension. Would you give up being on pension?"

"We should've gone north, along the coast, to the Portuguese."

"Aye, and been sold for slaves!"

"Well, we should've done something! What if a ship came? Say, bound for Valparaiso. We could've gone there."

Butcher considered the idea, an effort appropriately imagined as the clunking engagement of gigantic, creaking wooden gears. "Well, yes, I might've said Valparaiso."

Although time and again they retraced the twisting trail of real and imaginary events that found them once again in irons, long after everyone had sunk into their bleak private thoughts, Allie was likely to say, "Well, lads, at least we're going back to England."

Aye, they were. The new native king crowned, festivities done, Kupang quieted. Captain Edwards hastened his efforts. Charter terms for the *Rembang*, the ship that brought Captain Dahlberg to Kupang, were agreed to. Stores were laid aboard, farewell dinners eaten, and Will's infected finger amputated at the second joint. Angry red streaks spreading up Will's arm had prompted the *Pandora*'s surgeon to act, hoping to control the infection. "Infections are unpredictable," he told Mary B, "Usually we can get rid of them. I'll keep bleeding him, Mrs. Bryant. We'll wait and see."

It was an awkward moment. Mary B hadn't spoken to Will for several weeks. She'd made no attempt to see him when she carried food to the prisoners. Once, as she talked briefly with James through the wicket of the door, she heard Will call out her name in a voice of despair. James watched her shrink, and said he thought it was the infection talking.

"What does that mean?" she asked the surgeon, "Wait and see?" She avoided his eyes. Why should she care? Their estrangement was as complete as if they'd never known one another. Will had even tried to blame her to save his own skin! Why should she care one whit about a man who cared not one whit about her? What did she owe him?

Hamilton pursed his lips. He'd heard about Bryant's attempt to save himself by putting the blame on his wife. He remembered an old saying, My wife is my plague. Maybe it was simple as that. "Well," he said, smiling pleasantly, "Perhaps everything will be just fine. Say your prayers."

~~~~~~~~~~

Just before the *Rembang* sailed, Will tried to get Quint's book back. He had a hard time even gaining audience with Edwards, but succeeded when he sent word he had valuable information about the coast of New South Wales. As he stood weaving before his captor he could tell Edwards was interested. "It's all in my book, Captain. Coal! We chopped it right out of the riverbank. We found new harbors, rivers, information the government would want, that you.... Captain, I think Governor Wanjon wants to keep my book for himself, for the Dutch. That's why he made me leave it with the Fiscal every night." Bryant wiped his feverish face with a bandaged hand.

"Captain, I know you don't think much of me, but that book is all I've got. And I know you don't believe me, but my term is really up. I served my sentence. Government has no claim on me. And if all it comes down to is the damn cutter, well, maybe the book will make it right. See what I mean? My book for the cutter. Please, Captain, my book is all I got!"

Edwards studied the ill man. Bryant's eyes were feverish. He perspired profusely. Yes, Edwards could see possibilities, especially for the coal, and if not for Bryant then for himself. By the look of him Bryant might not live to see England anyway. "You have a point," he said, "I'll see what I can do."

But when he raised the subject with Mynheer Ganser he was informed that Bryant had written his account at the Governor's direction and at the Governor's expense and that the Governor considered the book Company property. Already indebted for so much, Edwards dropped the matter.

~~~~~~~~~

Wearing her blanket like a hooded cloak, Mary B protected Emmanuel from the soft, warm rain and looked across the slate-gray water. She was aboard the *Hoop*, a Dutch guardship housing the men from the *Pandora*. Although most of Edwards' prisoners had been sent ashore to the prison beneath the *Stadhuis* in Djakarta, several invalids remained aboard.

The heat and odor of close-packed bodies below stifled the air, and Mary B was on deck to find relief. Her eyes feasted on the familiar structures of Onrust, the fortified island that reminded her of Dock with its jumble of cribs and cranes, its warehouses and sawmills - a maritime workshop. In the opposite direction lay Djakarta, spread out like a vast, colorful collection of broken china, larger than any city Mary B had ever seen, almost too large to encompass after four years of isolation. Not since Rio had she seen such a sight. She was unable to reconcile this lovely city with its horror stories.

Today some of the *Pandora* crew would be boarding a Dutch ship, making ready to sail in an arrangement worked out with the Governor-general. *Pandora* men would help crew the Dutch vessel in exchange for passage back to Holland, with others to follow as ships become available. But Edwards had to purchase passage for the officers and prisoners. Whether officer or seaman or prisoner, all were eager to be quit of this city of death by whatever means.

The rain fell incessantly, sometimes heavily, sometimes softly, sometimes whipped into stinging spray by gusts of the westerly wind. But it fell relentlessly, hour after hour, day after day, breeding boredom and the bitter bile of loss, for when the *Rembang* sailed from Kupang it carried men from the *Pandora* who'd lost friends and relatives and everything in the world they owned; men from Sydney who faced banishment from their homeland for life; men from the *Bounty* who faced hanging; men who'd succumb to Djakarta's diseases and never see home again. A black melancholy had enveloped Mary B from Kupang to Djakarta.

Continued in her cooking chores, she was free to roam the *Rembang*, but she'd kept to herself, rebuffing with weariness the repeated advances of men from the *Pandora*. She felt utterly beaten down, lost in a swamp from which she'd never emerge. She missed the company of the other convicts and found herself thinking fondly of their adventurous voyage and of how stupid they were for not getting out of Kupang. She berated herself. "I should've said more, I should've done something instead of just sitting like a lump. That's all I've done all my life, sit like a lump while others run my life! Sit like a silly, giggling girl and say nothing!"

In her sadness she missed Seedy and Katie and Liz and so many others. And Rami too. How strange, she thought, to love someone I couldn't even talk to, someone so different. Yet the two had developed a bond as powerful as any Mary B ever knew. At their leave-taking Rami hugged Mary B's children with heart-rending urgency, had covered their faces with teary kisses, had keened her sorrow in the high, flat, staccato rhythms of her strange tongue. But she restrained herself with Mary B. She forced a tight smile and with averted eyes pressed a fold of cloth and a packet of tea into Mary B's hands, held her hands briefly, then fled. Only later did Mary B realize it was she and not Charlotte or Emmanuel who'd been the real object of Rami's affection and sorrow. Mary B felt a sudden pang of loneliness as she tenderly placed Rami farewell gifts in Little Bill's battered box, regretful she'd not been more demonstrative of her own love.

The rain fell relentlessly and bred dark stories of Djakarta. One told of a million people who died from a volcano eruption, another of the massacre of thousands of Chinese. Another told of daily executions of criminals in front of the *Stadhuis*, of how the Governor-general and his concubines would sit on the balcony where he'd let the women show by finger signals how the next victim would be dispatched - by sword, hanging, guillotine, burning, impalement, flaying. "Aye, they sit up there and drink and laugh and dine on fresh roasted hearts!" Oh, Djakarta was an evil place!

The rain fell relentlessly and bred sickness. Almost as soon as the *Rembang* dropped anchor, Djakarta's diseases began to fell the English, but the *Rembang* may have carried its typhus and malaria and dysentery in the orlop, where Edwards kept his prisoners in bilboes for the entire voyage, even when a fierce storm nearly sent the ship to the bottom. Plunging and rolling in typhoon-wild seas, remembering the nightmare of the *Pandora*, the *Bounty* mutineers screamed to be freed of their irons. But Edwards was obdurate, and kept them shackled.

Mary B stood on the foredeck in the soft rain and jiggled Emmanuel to quiet his fussing. She felt a premonitory uneasiness. Emmanuel's stool was runny, his appetite fallen off. He was hot with fever. The paunchy surgeon from the *Pandora* had pushed back the little fellow's eyelid, looked into his throat, thumped his chest, and presented her with a packet of powders. So far she saw no improvement.

She heard voices and movement near a hatchway and turned to watch several invalids being helped to the deck. She watched them shuffle to the ship's ladder, most of them assisted by companions. They were on their way to the maritime hospital, an institution situated on the very river that bred the city's notorious diseases.

Suddenly Mary B recognized a familiar figure supported on either side by *Pandora* men. He wore bilboes. His legs made feeble movements to assist his progress, but most of his weight was borne by his helpers. Glancing in her direction, he paused and turned slowly with his gaze fixed on Mary B. He swayed a little as he looked at her for a long moment, then slowly raised his bandaged arm in a hesitant, half-gesture of greeting. Mary B hesitated too before she shifted Emmanuel a little and half-raised her own hand. The men who supported Will looked from Mary B to their patient and turned their eyes away, as if the moment were too private for strangers. His bandaged arm still raised in that awkward signal of tentative greeting, Will kept his gaze fixed on Mary B until a sentry approached and motioned him toward the ladderway. He made an effort to shuffle under his own power to the ship's side, where he paused briefly to look back at her once again before he disappeared. She wondered later if she'd seen a small, rueful smile of farewell.

London
1791

"Can I hold 'im, Cap'n?"

Tench nodded to the convict groomsman to relinquish the bridle of the big russet gelding. Ned stroked the nickering muzzle. "He's a real beauty, isn't he, Cap'n?" Tench mounted, checked his baggage, then signaled with another nod for Ned to release his hold. The horse sidestepped, shook his head to settle the snaffle, blew, and quieted. "Yes, sir," Ned said, "A real beauty."

A small smile played at Tench's mouth. He removed his foot from the stirrup and held out a hand. "Come on up, Ned."

"God's eyes! you mean it, Cap'n?"

Ned made several attempts to get his foot into the stirrup as Tench slyly sidestepped the horse out of reach. Finally Tench quit fooling around and hauled the boy up, where he settled with a smile as wide as an open gate. Tench winked at the groomsman and kicked the horse into an easy canter. The boy's teeth rattled before he clamped his mouth shut.

Tench twisted a little from side to side. "You needn't hold so tight, Ned."

"Sorry, Cap'n."

The morning was cool, the sky bright blue, the sun warm on their shoulders. Light sparkled in the heavy dew, was caught in rose-gold fire by a few tiny windows, gleamed on thatch roofs, glinted on the wavelets of Sydney Cove. Tench breathed deeply, savoring the morning air, sweet with a familiar perfume of chimney-smoke, cooking, baking, distant brick-kilns, smoldering slash-piles. He turned out of the barracks area. "I'll carry you as far as the brick-works," he said.

With a wink from the Governor, Tench was making an excursion to survey the outlying communities. He had to prepare a report, but his real mission was to gather material for a fitting conclusion to his next book, which he planned to finish on his return voyage to England.

Tench slowed to an easy walk when he reached the new road, one of the Governor's latest projects. This high road to Parramatta would require several years of grueling convict labor but would end reliance on water travel to reach the inland settlements. Sydney now was a only stopover. Incoming convicts were processed and sent on, with almost 1,500 already settled in the area around Parramatta. Only convict servants and artificers remained in Sydney, where virtually all gardens and farms lay abandoned for better prospects inland.

As the big horse carried them steadily higher, the settlement shrank. Soon their eyes followed the faint outline of a cart track angling off through the scrub, the gallows track.

"Not many hang now, Cap'n."

"No, not many, Ned."

"Remember the big hanging, Cap'n?"

Tench was silent. How could he forget the ignominious end of six Marines, including Ned's father. "They burned the hanging tree then, didn't they, Cap'n?" Ned said this as a statement.

"Aye, they did, Ned."

"They sure were mad."

Tench remained silent. Making a bonfire of the hanging tree was not Major Ross's initiative, as the Governor suspected. Rather, destroying the ugly symbol was a point of honor that demanded satisfaction. As the fateful day drew to a close a pile of brushwood had begun to build around the tree. The pile grew with the offerings of Marines and wives and children until it reached as high as the branches from which the nooses had dangled. When the fire blazed to life after taps, silent figures ignored curfew to watch the flames roar into the night. They noted with quiet satisfaction the glow of the inferno reflecting from the distant windows of the Governor's house.

"The Governor was wrong, wasn't he, Cap'n?" Again the question like a statement.

"The law's the law, Ned, the same for all."

"But what about Porter, Cap'n? He didn't hang. The law wasn't the same for him."

"True, boy, but it's the law." They rode silently.

"Why, Cap'n? Why wasn't the law the same for Porter? He was the one they caught! Was it because he ratted?"

"It's the way the law works, Ned, if a man turns evidence."

Ned was quiet. Finally he said, "Well, just because it's the way the law works doesn't make it right."

Tench was ignorant of Ned's early fantasies of revenge, of the nights he'd lie sleepless with anger that his father was hanged, taken from him forever. Ned imagined himself in bloody scenes of swordplay to punish Porter and the Governor, pictured himself torturing the convict called Tartop and all who'd laughed at the death struggles of his father. Ned sorely missed his self-effacing father.

When they reached the top of the rise, Tench halted and looked back on Sydney. A thought suddenly struck him. "Tell me what you see down there, Ned."

"Cap'n?"

"What do you see down there?"

Uncertain what Tench was after, wondering if it might be another quiz, Ned studied the scene - the straggle of buildings, the half-dozen ships of the Third Fleet, the tiny figures. He began pointing. "Well, there's the Governor's house, and there's the wharf, and there's the storehouses, and there's the barracks, and the hospital, and Mr. White's house, and there's the battery, and...the *Gorgon*?" He named this last as if hoping he might identify the object Tench had in mind. Tench smiled and shook his head.

That morning Tench had awakened feeling wonderfully expansive - vigorous, confident, and glad to be going home. Seated atop this powerful horse on a hill overlooking the little settlement he'd helped create, he felt proprietary, as if he were the cynosure of all this budding colony and England and the *Gorgon* and this fine morning represented.

"Do you see a school down there, Ned?"

"A school, Cap'n?"

"Aye, Ned, a proper school, where a boy can learn his lessons - his Latin and mathematics and astronomy, his history and his poetry. Do you see such a school, Ned?"

Ned opened his mouth to speak, then wondered if some secret project had produced a school he'd be sentenced to attend. "No, Cap'n," he said warily, "There ain't no school like that in Sydney."

"*Isn't* any school, Ned, *isn't*. But you're correct, there isn't a school like that in Sydney. And that's why I want you to come back to England with me."

Ned sat stunned. England? The idea of England had never crossed his mind. Sydney was his home, these hills and coves, the Marines, the ships, his secret places and shortcuts. He barely remembered England. It was a crowded room in a barracks and busy markets. Go back to England?

Tench turned the horse and continued at a walking pace.

"But Mama, and Nell...."

"They'll be fine, Ned. I'll talk to Mrs. Thistlethwaite. You can come as my servant, just like now. You're quick, Ned, and school's the place for you." They rode in silence for another minute before Tench said, "I never told you, Ned, but I once got help from a man who wasn't my father, and that's something I'll always be grateful for, more grateful than he will ever know. And I...well, let's just say you're a lad I want to see do well."

If he hadn't felt so numb, Ned might have heard the catch in Tench's voice. For years the boy had worshipped Tench, had seen him larger than life, and had come to love him too. Tench was kind, and sometimes generous. But to leave his mother and sister and Sydney and everything familiar? Ned felt fear creep down his back.

They reached the brick-works where the new road angled west toward Parramatta. Tench halted his mount, helped Ned slide down, and repeated his instructions about chores. Hands in his pockets, his head down, the boy's bare feet scuffed the earth as he listened and repeated mechanically, "Yes, Cap'n, yes, Cap'n." Then in an uncharacteristic gesture, Tench extended his hand for a handshake of farewell. Ned looked into the Captain's eyes with tears, a confusion of admiration and disappointment, of love and trepidation.

"Will you be all right, Ned?"

Ned cleared his eyes with a swipe, nodded, and a couple minutes later returned Tench's wave as the Marine disappeared around a bend in the primitive road.

Tench stayed several days in Parramatta, systematically visiting little settlements of emancipated convicts and an occasional seaman who'd decided to make New South Wales his home. The settlement plan saw them living on allotments of 30 acres or more, depending on family size, but no one had more than two or three acres under cultivation, and all still relied on government handouts for sustenance. Soon men from Tench's battalion would be moving here also, for a surprising number had decided stay on after all.

Tench found some convicts doing well, some barely surviving, and some ready to give up. Those giving up would sell out to the more competent, who spoke optimistically of independence in the near future.

He found a handful of men and women from the *Charlotte*. In an area called Prospect Hill he visited Katie's religious mentor, Hannah Jackson, living on 70 acres with a husband and two quickly produced children. Nearby lived Mary Wickham and Fanny Anderson, also from the *Charlotte*. In a settlement called the Ponds he found Ann Smith, with one husband in England and another here, and Sick Liz Cole, who'd been sentenced in Exeter the same day as Mary B. Tench chatted amiably with them all, admired their babies and crops and crude homesteads, and after several days, quite satisfied with his mission, he headed back to Sydney. In five days the *Gorgon* would sail for England and he'd begin writing.

At his hut he found no sign of Ned.

~~~~~~~~~~

At first Mary B thought she was dreaming. She became aware of Charlotte's reassuring warmth curled next to her, but then wondered if that too was part of her dream. She raised a hand and touched her face. Yes, this was her hand and she felt fingers on her face. But were her fingers also a dream?

Bleary with febrile sleep, disoriented in the utter darkness, Mary B lay frowning, trying to bring back what had waked her, something out of place. She listened to the languorous creaking, the shush of displaced water, the periodic rumble of soft, manly laughter. If this was a ship, then she must still be aboard the *Horssen*. Yes, and if this was a ship she was awake. Suddenly she felt better when she realized she still had Allie with her. Just the three of them now.

As her throat tightened with sadness she wanted to lose herself again in sleep, but her memory wouldn't rest and began bringing forth images of the past few weeks:

- Emmanuel dead in his sleep aboard the *Hoop*, carried away by fever, his shroud disappearing in gray, rain-hissing water, a little bundle of body, a sad trifle of shattered hope that vanished in brief bubbles.

- The surgeon's expression just three weeks later when he brought word of Will's death. Will's bandaged hand half-lifted in farewell was the last she ever saw of him. He had succumbed to Djakarta's notorious fevers.

Inwardly Mary B groaned. She regretted she hadn't seen him once more. Perhaps with Emmanuel's death he might have wanted to set things right, even if on his deathbed. She sometimes still wondered if she should have tried one more time with Will. Or was that feeling only her regret of all that had gone so bad? Too late now.

"Did he say anything?" she'd asked as she searched the surgeon's eyes. She remembered clutching Charlotte with fierce protectiveness. She wanted the surgeon to tell her Will had expressed sorrow, admitted to his selfishness, perhaps said he'd been wrong to treat her so badly.

Hamilton shook his head. With apology in his voice he explained, "He was delirious, you see, out of his head. He said your name, and a lot of names. He talked of a little man, but it was raving. I'm sorry...."

Mary B shivered and kissed the sweaty warmth of Charlotte's forehead. Emmanuel dead, Will dead, Morton dead, Bird dead, Cox drowned - add them all to Djakarta's death toll. And where were James and Lilley and Butcher? Were they all dead? Was it only her and Charlotte and Allie?

There it was again! What had waked her, Cox's song! But Cox was dead! Or was he plucked from the sea while she slept? Mary B shifted her arm, kissed Charlotte's sleeping forehead again, and groped her way to the ladder. Through the hatchway she saw dim stars scattered in the tangle of rigging. She heard the experimental strumming of a guitar, words half-sung, the murmur of men's voices.

"All right now, Mac, let's have it from the beginning. You've got it now, don't you?" The guitar thrummed a couple of times and Mary B heard an unintelligible reply. "All right now, boys, pipe down. Mac's going to give us his new song."

There was a hush and a couple more notes on the guitar. Mary B eased herself up until she could peer over the coaming. The air felt moist and balmy and a rising quarter-moon cast the scene in soft silver. A half-dozen men from the *Pandora* reclined at ease, their faces turned toward the one called Mackesey. He sat with his back against a bight, his legs sprawling, his face down-turned to a small guitar in his lap.

"Well, I don't know...."

Voices of protest, of encouragement. A few more experimental thrums, a chuckle, and Mackesey sat straighter. His voice was gentle, a husky baritone, and when Mary B heard the song Cox had sung so often in his clear tenor voice she felt she was hearing a different song. Cox was plaintive and lamenting. Mackesey was resigned, with a chuckling appreciation of the singer's fate.

My father sailed the salty sea
To see the world around, oh
Around, oh, around, oh
To see the world around
 My mother cried, Don't you go too
To see the world around, oh
Around, oh, around, oh
To see the world around
 But I did go as he did go
To see the world around, oh
Around, oh, around, oh
To see the world around
 Now I am drowned in the salty sea
And the world it goes around, oh
Around, oh, around, oh
And the world it goes around

Mackesey finished to silence, then a few words of quiet praise as thoughtful pipes knocked against horny palms. When men stirred and Mary B began to withdraw, the ship's bell struck, faces turned, and she found herself looking directly into Mackesey's eyes. She shrank beneath the coaming and retreated to her coign, but not before she caught for an extra moment the gentle, reassuring smile the Irish seaman sent across the moonlit deck to her, a smile he would share many times in the coming weeks as he extended Mary B a hand of friendship.

~~~~~~~~~

Yes, Cox was dead, they said. Jumped in his shackles into the treacherous currents off Onrust. Along with James, Lilley, and Butcher, he'd been roused in the pre-dawn darkness to help man the *Hoornway*'s capstan, and only moments later, looking about and not seeing him, James surmised that the incongruous soft splash was Cox going over the side. Not a word of farewell, not a hint of his plan, if he had a plan. Not missed for a half hour, he either drowned or got away, for by then the *Hoornway* was battling the currents of the strait and would not turn back.

When Captain Edwards learned of the missing man he ordered the prisoners double-ironed for the duration of the voyage. He was obsessed with keeping his prisoners secure, but then, what else did he have to show for the loss of his ship and so many lives?

~~~~~~~~~~

"Do you really think she wanted us, Cap'n, wanted to keep us?"

"No question, Ned, no question." Ned twisted on the thwart of the scoot to watch the pyramidal mass of Devil's Peak loom higher and higher. After several days in Cape Town, Tench had taken Ned to Anna Labonne's. Standing in the street regarding the small stone house, Tench said, "I just wanted you to see, so you'd know that part of her life, that she wanted to take you back to England. She was a strong-willed woman, Ned, but she only wanted the best for you." Ned's memories of his mother abandoning him on the *Charlotte* were dim - his hot tears and the hoots and cruel laughter of others.

Tench thought the old Huguenot would be long dead, but as he and Ned talked in the street before her house, the top half of the front door creaked open and a white-palmed hand beckoned. He was surprised at the old Hottentot's recognition, but how would Tench have known of the hours she'd spent describing the romantic reconciliation of four years earlier, painting for Anna Labonne's sightless eyes every nuance she could conjure of the handsome British officer come to woo back his wife, of how the distraught woman brightened so in anticipation of the captain's return. After her Hottentot companion had exhausted her store of real and fanciful details, Anna Labonne would invariably sigh, "*J'espère seulement qu'ils soient heureux. Oui, qu'ils soient heureux.*" And her companion would nod.

No, although they groaned with every movement of their shrunken bones and withered flesh, Anna Labonne and her ancient companion still creaked about.

Ned couldn't stop staring at Madame Labonne's milky, sightless eyes, at her tiny, wizened, yet somehow elegant face. He'd never imagined anyone so old. He stood transfixed as Tench leaned close to shout his inquiries to the old woman, and felt he could listen for hours to the soft, deliberate cadence of her replies. At one point Tench turned with a little smile. "She thinks you're my son."

At the end of their brief visit Ned approached for her blessing. She surprised him by fumbling on a side-table for a small, worn book she pressed into his hands. "*Pour le bon fils*," she said. Tench protested, but Anna Labonne insisted: "*Oui, oui, mon Capitaine, parce qui'l est un bon garçon. Je le connais.*" She found Ned's face and traced his features with soft fingers that smelled of flowers. Her fingertips lingered near his eyes, wonderfully comforting. Ned thought, this old woman is the gentlest thing I have ever known.

"*Comme sa mère*," she said, "*Comme sa mère.*" She groped for a hand and held it between her own. "*Ecoutez! Ecoutez l'instruction d'un père. Cultivez la sagesse. Oui, cultivez l'intelligence au prix de tout ton avoir.*"

"*Oui*," Ned answered in a small voice, the only French he knew.

"What did she say when she was talking to me?" he asked later.

Tench ruffled Ned's hair, "She said you should mind me and go to school."

At Ned's feet in the scoot lay a clutch of parcels - clothes and shoes that were gifts from Tench. He held onto the old woman's gift with its crumbling leather cover and a French title Tench had explained but Ned couldn't remember. He basked in this unexpected addition to his riches, puzzled by the old woman's generosity.

"Well, Ned," Tench explained, "Things of the earth don't matter much when you're ready to die."

But how could a boy of thirteen who still pocketed odd-colored rocks and hoarded the rare farthing understand Anna Labonne's systematic divestment of her earthly possessions, this last chore of one who has outlived every human pleasure, every human connection? And how could such a boy have understood Tench's sudden flush at the old woman's final words?

She called out her admonition once more in a surprisingly strong voice as the captain closed the door, "*Ecoutez l'instruction d'un père!*"

Ned made a surreptitious swipe at his eyes and turned his attention from Devil's Peak to the sight of the *Gorgon* a half-mile ahead. "So she really wanted to take us back to England, Cap'n?" he asked again.

"No question, Ned, that's what she wanted all along, even in Sydney."

In absent-minded betrayal of his tumbling thoughts, Ned's thumbs unconsciously scrubbed the cover of his book. Suddenly aware of a change in texture, he looked down to see he'd effaced part of the timeworn gilt lettering.

s Prover

de la

nte Bib

With a guilty, sidelong glance, he stuffed the book into his shirt and focused his attention on the *Gorgon*, still ignorant of the bitterness Tench caused by his decision to carry Ned away to England.

When Tench had returned from his excursion to Parramatta and found no sign of Ned, he sought out Hester Thistlethwaite. Flushed with anger, she scolded him. "Well, I hope you're satisfied, because you've killed her off at last. You couldn't have it any tidier, could you? Kill off the mother and make off with the boy."

Speechless, ignorant of the reason for Hester's attack, Tench shook his head in denial.

"Well, of course you did," she berated him, "You might as well have put the rocks in her pockets yourself. Did you think she was such a vegetable? Poor woman, she knew something of what went on. Good heavens, you should have heard her awful groaning! She wasn't deaf, you know. She understood things. Surprising things. She understood the boy's torment.

"He doesn't want to go, you know. Did you ever think of that? This is his home, Sydney, you arrogant man, you! Here! Sydney! With his poor mother and sister! How could you just yank him away? Did you ever think of that?

"You know how he loves you, how he worships the ground you walk on! Well, he was in torment, utter torment! And he's run away, did you know that?"

No, Tench didn't know, for he knew nothing of events that followed after he dropped Ned off at the brick-kilns. Now he learned Ned had gone straight to Hester to break the news that Tench was taking him back to England. That night Malvina filled her apron pockets with rocks and disappeared into the waters of Sydney Cove.

Two days passed before her partially eaten body bobbed to the surface near the town wharf. Then Ned came up missing. Speculation ran high that the boy had also done himself in. As a crew dragged the water near the wharf, shore side observers commented how "it" ran in families.

Ned was not drowned, however, but was hiding in the woods near Bennelong's point, terrified that his tale of going back to England with Tench had driven his mother to suicide. By the time Tench returned to Sydney, Bennelong himself had tracked the boy down. And Hester Thistlethwaite had made up her mind. Tench would take the boy to England as planned, and she would raise Nell.

Amidst the bustle of his departure, Tench found himself pondering Hester's accusations. Admittedly she did give him pause, but by the time he left Sydney he'd convinced himself that he could hardly have been the instrument of Malvina's suicide. The woman was deranged! Unpredictable! Any number of things could have sent her over the edge!

Nevertheless, Tench was sobered, and he sailed from Sydney with a sense of responsibility for Ned's future far more serious than his impulsive idea that pristine morning on the hill. For her part, Hester would remain ignorant for years of Nell's feelings of guilt, how Nell believed the shame of her violation by Henry Wright had somehow caused her mother's derangement. And ultimately her suicide.

The *Gorgon* towered above the scoot, a 44-gun frigate that next day would begin the final leg of the voyage up the coast of Africa to England. Tench craved English civilization, and Cape Town was but an appetizer. As he followed Ned's awkward progress up the ship's ladder his spirits felt rejuvenated, and he found amusement in the boy's determination to carry all his own possessions.

As soon as Tench stepped on deck and saw Lieutenant Creswell's sleek, knowing smile, he knew something was afoot.

"John?" Tench greeted. Creswell bowed, drew himself up, and declaimed in an odd, high-pitched, stentorian voice,

What though the field be lost?
All is not lost; the unconquerable will,
The study of revenge, immortal hate,
And courage never to submit or yield.

Tench smiled. "*Vain wisdom all, and false philosophy,*" he quoted back. "And how long have you labored at that? I thought you hated Milton more than Latin."

Creswell's self-satisfied smile plumped his cheeks. "Ah, but I have news, Wat, and torturous Milton is its theme."

"Revenge and hate? Unconquerable will?"

"And courage never to submit or yield," Creswell prompted.

Tench thought in good-natured puzzlement. His face brightened. "Ahah! You were rebuffed again by the beautiful Juffrouw Van Schaaldier." Creswell laughed and slapped his knee. He shook his head.

"All right," Tench said, "I yield. What news?"

Creswell laid a confiding hand on his shoulder. "Bryant's gang, Wat, the ones who escaped. They were put aboard from some Dutch ships, and so are the *Bounty* mutineers."

~~~~~~~~~~

Stork nodded a greeting to the sentry and offered snuff. The Marine guard took it gratefully. Two levels below main deck, the light was dim, the air malodorous and oppressive. The sentry inhaled the spiced tobacco deeply and sneezed, then nodded his thanks. Stork said, "I know some of those people down there, from Sydney. Can I see them?" The sentry looked him up and down. Knees bent, his head craned awkwardly in the low headroom, Stork wore just a shirt and breeches. He carried a small string bag of garden vegetables. The sentry turned the bag with an indifferent paw. "Be quick," he said.

Voices silenced as Stork's bare feet appeared on the ladder. He felt he was descending into a dark pool of invisible, suspicious eyes. At the bottom of the ladder he waited for his own eyes to adjust.

"We got no money," a voice mocked. Stork heard a rattle of shifting shackles. Outstretched bare feet and the dull gleam of bilboes ringed the small space where Stork stood. "I'm not selling," he said, "I'm looking for the men from Sydney." Silence.

"What you want?" a voice called from the darkness.

Stork raised the string bag.

"By God," another voice said, "If that ain't...!"

Stork heard a groan of effort, and after a long moment a bearded, sallow-faced figure did the bilbo-shuffle from the darkness. Stork was taken aback. It took a moment for him to recognize the skeletal visage of James Martin.

"You're Pottle, right?" Stork nodded and raised his bag. James looked at it. "For us?" There was a long silence before James asked, "Why?"

The question seemed to defeat Stork. His shoulders sagged, his eyes searched around his feet, his Adam's apple worked desperately. He suddenly looked over his shoulder and caught sight of Quint jerking his hand above his head and sticking out his bleeding, blackened tongue and rolling his eyes in a horrible, comic caricature of being hanged. Stork looked as if reason suddenly abandoned him. James touched him on the shoulder and almost fell when Stork recoiled. "Easy," James breathed, as if gentling a horse, "Easy now. Easy." His quiet voice seemed to reassure Stork. Taking a huge breath, Stork again proffered the bag.

"Never mind why," James said, and half-laughed as he turned the bag to examine its contents. "Look here, boys, potatoes, carrots, onions!" He fished out a carrot, took a bite with his side teeth, and tossed the bag into the dim recess. "Share it out, Allie, and make sure you get something soft for yourself!" He chewed as he studied the carrot for a moment, then swallowed, then had to swallow several times to get down the rest of the badly chewed chunks. "Whew," he said, when he at last looked up at Stork with an expression of dog-like appreciation, "Not used to it." Stork's eyes teared.

James bit off another chunk of carrot. "Pottle, right?" he said around his chewing as he extended his hand. Stork nodded as they shook hands. "So, shipmates again, eh?" James said with an odd smile. Their handshake lingered. Then James laughed a short laugh that turned into a racking cough, sending shreds of carrot onto Pottle's shirt.

~~~~~~~~~~

Tench laid aside his quill and turned Mary B's hand so the whitened end of her foreshortened little finger caught the candlelight. "I remember this. Odd, but you probably lost that fingertip within a hundred miles of here. Doesn't that seem odd, that we'd be together again like this, in this same place, four years later, after so much has happened? It's like predestination."

Mary B didn't answer. Neither did she withdraw her hand, which Tench continued to fondle and turn in the light. He sat on an upturned box so he could write at his little cabin table, while she sat in his chair with a glass of his canary. She'd just finished recounting the story of their escape, their voyage up the coast of New South Wales, their time in Kupang, in Djakarta, and on the Dutch ships. Tench seemed impressed, nodding thoughtfully as he scratched away, even exclaiming in admiration when she told how Will outwitted the fearsome war canoes.

Mary B had begun to feel a rare sense of well-being. Although Charlotte lay sick below, being here with Tench, sipping wine and telling her story to someone familiar, she felt as if she'd stepped through a door into a different world and entered a dream life. Forgotten were her griminess and the embarrassment of her threadbare, soiled dress, faded to a hint of blue. She found herself admiring the gleam of candlelight on Tench's fair hair and the starchy whiteness of his shirt. She marveled at the smooth, rapid movement of his pen so effortlessly converting her words into something permanent. "Thank you for helping," she said at last as she withdrew her hand.

Tench waved off her thanks. "I'm sorry I can't do more. Edwards, you know, and the wives."

Mary B nodded. The wives and children of returning Marines were packed below with the baggage and accoutrements of the returning battalion. Their jealousy and resentment over the special attention paid to Mary B were plain to see. Not confined with the other prisoners, she and Charlotte lived in a small space on the second deck, a stifling berth, but far better than the hold where the rest of the prisoners were kept shackled.

At first sight of Mary B, Tench had been shocked at her appearance - thin, pale, haggard, worried. "I've been sick," she explained, "and my little girl too." Despite the silent disapproval of Captain Edwards, Tench asked the *Gorgon*'s surgeon to prescribe daily wine and extra deck time for Mary B and her daughter. For the many kindnesses of this familiar Marine, Mary B felt a deep obligation.

Tench drained the last of his glass, poured another, and topped off hers. "This wine isn't very good," he said, "but we'll do better in Santa Cruz."

Mary B nodded politely and sipped. "It tastes good. Thank you. You're very kind." Reluctantly, she thought it might be time to leave.

Tench left his box to sprawl on his narrow bed, in the process slopping a little wine. He'd already finished most of a bottle. Now he sighted Mary B over the brim of his glass as if aiming a pistol. "I admire you, Mrs. Bryant, I truly do. No, I mean that. You're a remarkable woman, you know that? What you've done?"

With downcast eyes Mary B shook her head in slow negation, absently turning the novelty of the stemmed glass between her thumb and fingers. Flattered by his solicitude, his attention and praise, she flushed, then flushed more deeply from guilt over her lengthening absence from Charlotte. Several days ago the little girl had taken a turn for the worse, becoming feverish and diarrheic and coughing a frightening, croupy cough. The surgeon said it was another case of pernicious grippe, which was making the rounds of the children, but Mary B thought the illness more serious. Although uneasy about leaving her sick child, at Tench's summons she'd left Charlotte under Pottle's care.

"In fact, Mrs. Bryant, I'm a little jealous. You know why? Because when we get to England you're going to attract a lot of attention. Aye, more than anyone, even the *Bounty* men. Aye, because you're a woman, see, and what you've done is remarkable!"

He drained his glass and suddenly sat up. With a decisive motion he took Mary B's glass from her, put it on the table, and seized her hands. "A remarkable woman," he said, searching her eyes. He stroked the tops of her hands with his thumbs. Mary B dropped her look. Tench raised one of her hands toward the light and peered closely at the texture of her skin. "Capable hands," he murmured. "Jim Martin told me how you kept them all going." He raised her hand to his lips. "You know, I never trusted your husband. None of us did. I didn't like him either. I never thought he was good enough for you." He kissed each of her hands in turn. "I dream about you sometimes, do you know that? Do you ever dream about me?" He turned a hand to kiss her palm.

404

There! Mary B thought, her mind awhirl, and Captain Tench a gentleman! Does he really admire me? Does he really dream about me? Am I really remarkable? Tench was now telling her how happy she must make a man, how lucky a man would be to have such a wife, how much she must miss having a man to comfort her. He kissed her hands continuously, turning them over to kiss her palms, her wrists, looking up from under his brows with a small smile of conspiratorial appeal, nipping the end of her foreshortened finger with his lips. "How sad and lonely," he murmured, "To lose so much, how brave you are to bear so much alone, how admirable."

Mary B watched his lips move in the soft light, now praising, now beseeching, now nipping. "He wants me," she suddenly realized. "Why? I'm a hag. I'm unwell, and I can't keep clean." She suddenly remembered the tawny gleam of Rami in the clean, warm river. They had washed each other.

Tench was tugging at her, trying to draw her closer. "What does he see in me, really?" she wondered. "His words are just words, like Will's. He'll say anything to get what he wants. Flatter me and tell me lies. He thinks he can have me for words and wine! Such a small mouth he has, a suckling mouth full of lies."

She thought of Mackesey's wide grin and his bright blue eyes, of his manly gentleness. He'd shown her so much kindness during the long voyage from Djakarta, and had never asked for a thing.

Tench was still busy with her hands. She realized she missed Mackesey. Such an interesting man. And his talk! Oh, his talk was like a song! She wondered if she might ever see him again. The *Gorgon* would be weeks at sea before the *Pandora* men left Cape Town. She hoped she would.

Tench was sucking her fingers. "Stop," she said, abruptly withdrawing her hands. Tench almost fell off his bed at her sudden retreat. "I must go, Captain Tench. My little girl is sick." Behind her back she wiped her hands on her dress. Tench squinted up. He looked rather drunk.

"I have so much of Sydney to tell you," he said, reaching for her. She twisted away from his outstretched hand. "You know I went to see Hannah Jackson and the others?"

Mary B shook her head.

"Aye, I saw them all. Do you know Hannah has another little one? a boy I think it was."

Mary B backed away from him. "I must go, Captain. Another time I'd like to hear your news, I truly would, but I must see to Charlotte." She hesitated. "Do you know anything about Seedy, or Katie?" Tench furrowed his brow in thought, then shook his head. Mary B turned to leave.

"I was a prisoner too," Tench appealed, "Did you know that?" He made another clumsy reach for her. One eye over her shoulder, Mary B pressed the door. "Aye, Prometheus bound. A prisoner of war, Mrs. Bryant. Goddamned Americans!"

Mary B felt for the bolt, then suddenly faced him, "Were you kept in bilboes, Captain?"

Tench appeared dumbfounded at the sharpness of her question. He dropped his eyes. He stood and busied himself for a few moments, putting things back in the little box that held his writing material, slid it under his bed with a foot, and rearranged his chair. He plopped himself heavily at his table, legs sprawling, and dripped the last of the bottle into what remained in Mary B's glass. "No, no bilboes. No chains." He picked up her glass and studied it. As Mary B's hand moved the bolt he said, "Edwards was wrong, Mrs. Bryant, very wrong. He shouldn't treat prisoners like that." He moved his glass to admire the color of the liquor against the candlelight.

"It wasn't just Edwards," Mary B said quietly, "You all took us from everything we knew and sent us into a wild place. And you were going to keep us there forever. We struggled and starved and...."

"We all struggled and we all starved, Mrs. Bryant!"

Silence. She avoided his eyes. "Yes," she said quietly, "You were fair about that."

After a long silence, Tench also spoke quietly, "No, no chains, Mrs. Bryant. But I wore humiliation." He remembered the moment of both shame and relief when his vessel - the *Mermaid* - struck her colors. "For an officer to be captured, even a junior officer, is humiliating, Mrs. Bryant. Capture is defeat. Defeat is humiliating. I carried my humiliation like a weight of chains." His voice was sad, reflective. "Do you know what that means, Mrs. Bryant, humiliation?"

She shook her head. "It means shame. Surely you know what shame is."

She winced at the jab. She'd begun to feel a bond, but now she resented his sudden smug superiority. Words! He was all words!

"Then you know you don't need chains to be shamed, don't you?" he continued.

She nodded again, her mouth clamped in resentment. She felt put in her place by his easy words! Oh, why didn't she have the strength of words?

"So we both know what it is to be shamed." He downed her wine with a large gulp, choked a little, and waved vaguely at the door. "I'm a gentleman, Mrs. Bryant. You're free to go. You obviously do not care to return my favors." He wiped his mouth with a sleeve. "But tell me, I really want to know, and after all this time I still don't know, so I'd appreciate your telling me. You'll tell me the truth, won't you?"

Wary, she barely nodded. "Well, what I want to know is, are you a real felon? I mean, did you really lead a life of crime? Were you a dealer, or did you steal or counterfeit, or were you a strolling woman? I'd like to know what you did to deserve your punishment. I don't care, understand, but I'd like to hear it from your own lips."

Tears brimming, Mary B blinked in disbelief. After all your words, she thought, after all we've gone through together, after the miracle of coming together again....

When she answered her voice was choked. "I could have you in me. I could have you in me like a husband. But you'd just take your pleasure and nothing would be different. You'd still think of me as something small, something to play with, something to use. You don't even see me." She had to fumble with the bolt before she could flee.

Next day Tench didn't really apologize for his behavior the evening before, but Mary B thought him quite genuine when he confessed he'd had too much to drink and hadn't meant to embarrass her. For days he continued such a show of correct behavior and such concern for her sinking child that Mary B found herself unable to carry a grudge. Despite her disappointment with Tench's boozy manipulation and seduction, despite her sad realization that a vast gulf would forever separate them, she still found things in Tench to admire. On the voyage out he'd chosen her to receive the benefit of his domestic needs and had again interceded for her on the voyage home. Besides, she asked herself after yet another little kindness by Tench, am I so innocent? I married a man I didn't love because I was afraid. I used him too.

~~~~~~~~~~

Charlotte died off the coast of Africa in early May, the last of several children to succumb in the steamy doldrums. Relief from the tropic heat impossible, her little body racked with a relentless fever, she died quietly in her mother's arms. Had she survived to see El Pico she might have recovered, for the temperate latitudes would revive the *Gorgon's* human cargo. The nearness of home would lighten hearts and the anticipation of reunion would nurture hope. But only death would cool Charlotte's brow.

When Tench offered his condolences, Mary B asked if James Martin could attend the burial. The Marine looked surprised.

"Jim Martin?"

"Well, James - James was her father. Didn't you know?" Tench shook his head. "Aye, he was always good to her." Her voice caught and she began to cry. "I didn't think I could trust Will on the escape."

Tench wore a dumbfounded expression. So that was why Martin joined Bryant's gang!

Mary B recovered from her tears and asked, "Captain, would you say the words?" A little surprised, Tench collected himself and bowed. That evening Tench brooded in his cabin. How often things become clear only later, he thought, after we feel safe telling our secrets, making our confessions, acknowledging our passions. How wondrously the past changes. We think we know this, so we believe that; then we learn something different, and we change what we believe. Each time the past changes we think we have the real truth, but we should probably feel even more uncertain. The past is like the sea, constantly changing with wind and light. It's the same sea, but it looks different; it's the same past, but it looks different.

Only a few gathered next morning in the waist of the ship - James, Allie, Stork, Tench, Ned, and a small fringe of Marines who remembered the baby born on the *Charlotte*. Stitched in old sailcloth and weighted with a ballast stone, Charlotte's remains rested on a gangway balanced by two sailors. Tench made his finger a bookmark as he surveyed those gathered. He cleared his throat.

"When the disciples asked Jesus, `Who is the greatest in the kingdom of heaven?' Jesus called a little child to him. And he said, 'Unless you become like this little child, you shall not enter into the Kingdom of Heaven.' What Jesus meant was that those who are as humble as children are the greatest in heaven, and have a special place there. Charlotte is already on her way to her special place in heaven."

Tench addressed Mary B. "Jesus also said those who take care of children are special, and that He will also welcome them into heaven so they can be with their children again."

He paused to survey the gathering, a little surprised they seemed to be listening. "If any of you have ever been close to death you know there comes a moment when you want to measure your life, to answer the question, 'How worthwhile was my living? Did I make a difference?' But how do we measure the life of a child like Charlotte, plucked from our midst so soon in her life? Little Charlotte endured all our troubles in Sydney. She survived much, and until the end of her struggles she did not complain, trusting that her mother always did the best she could for her. We were fortunate to have known this little life. She represents the best in all of us."

James sagged when he suddenly realized that his connection with Mary B was gone. Stork put an arm around his shoulder.

Tench took a big breath. "How do we measure the life of this child? Here is how: we take away with us today what she gave us, a child's love and trust and the pleasure of seeing her grow; we take the laughter and joy she brought and the hope that she offered for the future. These are the things she had to give, the only things. She leaves no riches but our memories of her, my friends, but these riches are measure enough for Charlotte, or for any of us."

Ned thought of his box of possessions. When Charlotte played with his rocks and shells she usually tried to make off with her favorites.

Tench opened his book. *In sure and certain hope of the resurrection to eternal life through our Lord Jesus Christ, we commit the body of this child, Charlotte Spence Bryant, to the sea. The Lord bless her and keep her, the Lord make his face shine upon her and be gracious unto her, the Lord lift up his countenance upon her, and give her peace, both now and evermore.*

Tench nodded, the sailors raised the gangway, the bundle began to slide, then stopped, caught on something unseen. They raised the platform higher, then higher, and higher still until the gangway seemed to stand almost on end. Horror-stricken, Mary B put out a hand and had begun to take a step when the bundle let go with a swoosh and splashed into the glassy sea. With a cry, Mary B stumbled to the side to see her first-born child and best companion disappear in a brief rush of pale, greenish, evanescent bubbles.

A day or two later when Tench spoke to Mary B, he asked, "You said her full name was Charlotte Spence Bryant. Who was the Spence?"

"My grandmother," Mary B answered with a catch in her voice and her eyes suddenly brimming. "When I was young I promised her I'd keep her family name alive." She turned away from Tench and he saw her shoulders shake with quiet sobs.

~~~~~~~~~~

He swung on his crutch toward the infirmary and thought it odd that the idiot woman wasn't in her familiar place. "Katie!" he called, "Katie Prior! Come on out!" A basket of bandages in hand, Katie peeked out the infirmary door into the chilly June morning. She took a long moment to recognize the bearded figure of her Cornwall countryman, the convict hero of the burning *Sirius*, John Arscott.

"John, I saw your ship was back. What happened? Are you hurt?"

The hirsute sailor laughed. "Nah! The surgeon fed a couple of my toes to the fish! The rest of me is sound, though, and with news you'll want to hear!"

"Oh, John, how did it happen?" Katie asked, limping down the steps, "Are you in pain? Come sit down. How was the voyage? Has a proper surgeon tended you? Shall I get somebody? How did it happen?"

Arscott's laugh was rich and full as he suddenly scooped Katie up by the waist in his free arm, basket and all, kissed her loudly on the cheek, and set her down again before she could protest. "Questions, questions! No wonder a man goes to sea! Listen, Katie, here's a question - whatever happened to your friend Mary B, eh? Would hearing that news be worth a kiss?" Hand to her mouth, her eyes suddenly huge with fear, Katie stood stock-still. "Nay!" Arscott said, "Not bad news, girl. She's alive, aye, and the babies too, alive and on their way to England! We heard in Calcutta."

"England?"

"Aye, England. The news came from Djakarta. They made it to Kupang, all of them, all safe."

"What?"

"Aye, to Kupang, then Djakarta."

"Safe?"

"Aye, every one, but they got caught again in Kupang and now they're on their way to England with the *Bounty* pirates."

"Engla..." Katie began to say. Her eyelids fluttered, the basket slipped, her knees buckled. John Arscott caught her. "Here now," he said to his unconscious burden, "I always said you had gimp! But when I come to claim you for a wife you faint away before I get the words out! Let's not have the future Mrs. Arscott trying to slip away like that!"

~~~~~~~~~~

Some 1,500 years before Mary B was born, Romans sailed up the River Thames to subdue and occupy a small, cluttered trading center. They laid out streets and sewers, erected barracks and administrative buildings, built storehouses and wharves, and gradually replaced a wooden palisade with an imposing wall of stone, defended by several fortified gates. They kept the local name: London. After a score of generations the Romans retreated from England, but for centuries afterward London's paved streets and the wall with its fortified gates were reminders of their conquest.

Locals called one of these gates, "Newgate," which a succession of additions and modifications established as one of the principal keeps of London, a labyrinth of dungeons, passageways, and strongrooms designed not only to hold the enemy without, but prisoners within - whether for debt or crimes or political intrigue. The great fire of 1666 which destroyed most of London gnawed at Newgate, but the ancient stronghold was well built, and repairs kept it going for another century, when neighbors complaining of the stench and disease finally saw it razed. A new prison rose on the same site, however, and although London had long since sprawled beyond its Roman walls and the new prison was no longer a gatehouse, it too was called "Newgate." To this prison Mary B and the four remaining escaped convicts were committed in early July, 1792.

They were a weak, emaciated lot. After disembarking the Marines at Portsmouth, the *Gorgon* proceeded to Purfleet on the Thames, several miles below London, where the convicts were transferred to civil authorities. After months of sea-motion, the jolting, pounding progress of the prison-wagon from

Purfleet was torturous, yet the commonplace sights of the countryside were opulent to Mary B's hungry eyes - comfortable manors, verdant cropland more productive than any she'd ever seen. The well-fed people, the fat sheep and cattle, the plump, squawking fowls sent scurrying - the comparative wealth of England's everyday life filled her with awe and brought tears to her eyes. How beautiful! she thought, I don't care if I am going back to jail, I'm glad to be home.

Rising through the smudge of thousands of smoking chimneys, the distant spires of the world's greatest city came into view, and then the famous tumescent dome of St Paul's. In a while they rumbled through a clutter of wharves and warehouses and mills and manufactories, shacks and shanties and fetid ditches, suddenly emerging to behold a panorama of buildings stretching north and south along the Thames as far as they could see. The convicts sat mute with wonder. "Well, lads," Allie finally shouted above the din of the wheels, "Was it worth it?"

"Good God!" was all Mary B could breathe. Mouth agape, she craned and twisted, her eyes feasting on each startling new image - there a building that stretched forever! See that gilded carriage pulled by glossy, beplumed black horses? See those blackamoors in livery? See the crossroads over there, filled with men and women in endless variety bearing an endless variety of burdens - baskets and bags and crates and buckets and firkins and trays and boxes! Look! Over there! A woman in finery getting into a carriage. And there's another getting into a chair. And over there, a woman in rags, and up there, three women looking down from a window. Look, a ragamuffin being chased by a fat man! And look, an old blind woman crossing the street! See that train of attendants bearing parcels? See the hurry, hurry, everywhere?

Few of the myriad faces turned to gawk at the convicts in their cumbersome prison as it bumped and jerked to an alley off Bow Street. There they were taken to the quiet rear of the Public Office and locked in cramped holding rooms - warm, moist, and redolent with the odors of human and animal waste, unwashed bodies, vomit, sooty coal smoke, rotting food, a vanished bouquet of London summer air.

London vibrated with activity, with commerce, with ideas, teemed like a corrupt and paradoxical human beehive. Home for three-quarters of a million people, one of every ten in the Kingdom, London boasted dozens of theaters, dozens of fiercely partisan newspapers and magazines, hundreds of clubs and societies. It knew the extremes of wealth and poverty, privilege and oppression, piety and depravity. The most populous city in the western world, the seat of British government, London was the heart of a nation becoming the world's greatest power.

But London could cup its ear and hear the clamor of a blood-soaked revolution working its way through the venerable institutions of France. Upstart American ideas had invaded France. Was England next?

Shortly after the convicts spent that first night in the Bow Street Public Office, James Boswell held up a newspaper in tentative greeting to Mary B's four convict companions and nodded to his old acquaintance, the doddering keeper of Newgate, Richard Akerman. When Akerman's attendant placed the stool he carried for the old man's comfort, the jailer sat and briefly rubbed his arthritic knees. He spoke with a wheezy old man's voice, his words deliberate, his cadence measured. "This, gentlemen, is Mr. Boswell, James Boswell, Laird of Auchinleck, distinguished local barrister, author of a fine book, the world's greatest biography. And he would like to speak with you."

Boswell smiled at Akerman's little joke about his book and turned to the convicts with a good-natured bow. "I read about you in the *London Chronicle*," he said, holding the newspaper aloft again, "And I've been to see Mr. Bond at the Public Office. He was very taken with you. He said he seldom saw anyone who'd suffered so and who so deserved mercy." He paused to regard each of them in turn. "And I must say, he seems right about the suffering. Was it really that bad?"

The prisoners looked sidelong and shuffled in their chains. Finally, his voice quavery, James Martin spoke. "We're convicts, sir, not animals. We were starving in Sydney and afraid of the savages. Captain Edwards treated us like slaves. No better than the *Bounty* men. We almost drowned. If it wasn't for the *Gorgon* we'd all be dead. Dead as doornails, sir, like all the others. In Sydney, I mean, because we were starving there, and afraid of the savages. God's witness for that, sir."

His speech had not come out well. For months the convicts had debated which reasons for their escape would gain the most sympathy. But when a time came to speak, as now, sense seemed to fly from their heads. They sounded foolish, like stupid boys, and James and other three now hung their heads at still another failure.

But James Martin's disjointed speech seemed not to bother Boswell. He nodded. "That's what Mrs. Bryant said too. She asked me to look in on you." After a long pause he said brightly, "Well, would you like to hear what the newspaper said about you a couple weeks ago? The headline? *'Wonderful Escape from Botany Bay?'* Would you like to hear?" The response of Butcher, Lilley, Allie, and James was enthusiastic gratitude. Their spirits perked up. Boswell positioned himself against a pillar as the four seated themselves on the plank floor like school children. Butcher nudged Lilley and slid him a smirk. All right! Sending deferential looks towards Akerman, other curious convicts rattled closer to eavesdrop, and within a minute or two Boswell had an audience of a dozen Newgate felons.

Boswell read, *On Saturday, James Martin, John Butcher, William Allen, Nathaniel Lilley, and Mary Bryant, were brought by several of Sir Sampson Wright's officers, from on board the Gorgon frigate, to this office. They are all that survive of eleven persons who escaped from the settlement at Sydney Cove. This escape was, perhaps, the most hazardous and wonderful effort ever made by nine persons (for two were infants) to regain their liberty, which they declare they should not have ventured on but from the dread of starving, and the certainty that if they didn't survive the period for which they were transported they should never again see their native country....* Boswell took some time to finish the long story, during which the convicts were unable to restrain themselves from asking each other with looks, "What?" or "Who said that?" The story ended with a brief description of each of them and their crimes.

By the time Boswell finished, the four escaped convicts were in a funk. The sum of their lives was presented in those last brief sentences - impetuous, stupid thinking that caught them up in the great gears of law and knocked them about until they were sometimes unable to say for certain whether they were men or animals.

Boswell looked from one to the other. He suddenly felt at a loss. Their pale, ravaged countenances wore such hopeless dejection and sadness his heart went out to them. He hadn't even intended to see these men, but had come to Newgate on a whim, moved by the newspaper story, not sure what he was seeking, but curious about the woman the newspaper called an "Amazon," a woman of resolution "*hardly to be paralleled.*"

She certainly hadn't looked like an Amazon. Indeed, when Akerman accompanied him up the malodorous stairs and passageways to the women's ward on the second floor, Boswell's first glimpse of Mary B had moved him to murmur, "My God, she's an angel!" She stared out the barred window of the inner courtyard at a slice of St Paul's looming dome, the pale afternoon light casting an aura around her slim figure and the half-profile of her face.

Cruel illusion! She was not slim, but thin, and her hair was stringy and unkempt, her face sallow, her eyes sunken, and she sagged under an unseen burden. She was barefoot, and her threadbare, faded blue dress hung listless on a bony frame. When she turned at the sound of her name he saw such a frightened, woebegone expression he was taken aback. She's not pretty, Boswell thought, not pretty at all. But despite her distraught appearance, he recognized an appealing frankness in her level gaze, and he liked her open brow. Clean her up and she might be quite presentable.

"Mrs. Bryant?" he repeated meaninglessly, suddenly distracted by a fantasy of tupping this abject woman in a splashy bathtub.

She shook her head. "Mary Broad." Her voice was husky with weariness. "My name is Mary Broad. I'm not married."

Boswell bowed low, a gesture that so surprised Mary B she retreated a step. "Ah, yes," he said, "The newspaper said you were convicted under a different name. But I'm honored to make your acquaintance under any name. My own name is Boswell, James Boswell. Perhaps you've heard of me...."

At first suspicious, Mary B gradually warmed to her curious visitor. A conversationalist who would not be deterred, Boswell tried to link their lives. He told her how much he would miss his own little Betsy should God take her away, how much he missed his dead wife. When he asked about her family in Cornwall, Mary B began to cry, so he told of his own conflicts with his deceased father. They ended up talking for nearly an hour, with both frequently wiping their eyes. Boswell was struck by the chain of events that had sent this abject young woman to the end of the world, by her courage and determination to survive, by the successive blows of misfortune that had deprived her of everything. "This little box," she said with a short, sardonic laugh, nudging Little Bill's battered box with a foot, "This little box is all I have left."

When Boswell rose at last from their visit, his face shone with purpose. Old Akerman sighed quietly. Even before Boswell spoke, the jailer knew his plump friend was about to attack another windmill. "Bozzy" did so periodically, embracing questionable causes with such lusty vigor his friends sometimes avoided him. "I'll speak to the authorities," Boswell now told Mary B. "I know Henry Dundas. He's an old school friend. I'll get you out. You don't belong here. At the least you should be out on rules. Don't you think she should be out on rules?" he asked Akerman, "She's no felon!"

Akerman raised his bushy, white eyebrows as he squirmed on his stool, uncomfortable with the spot Boswell was putting him in. "Rules" gave the imprisoned daytime freedom of the neighborhood.

At Akerman's obvious discomfort Boswell said, "Well, never mind, I'll see what I can do. You have my word I'll get you out, Mrs. Bryant, I mean Mrs. Broad, I mean Miss Broad. Is there anything you need now?"

She shook her head and thanked him, her hands clasped as in prayer. As Boswell turned to leave she said with a rush, "My husband said our marriage wasn't legal, Mr. Boswell, that we were never married. He said the banns weren't read and that the minister wasn't legal because he was a Methodist. He didn't want to take me with him, Mr. Boswell. He would've left me there all alone with my children. No good could come of that, Mr. Boswell. I'd have been a married woman with two little children and all alone in that hard place, harder than you can think. I had to escape, Mr. Boswell, I had to escape, even if it was with my husband."

Boswell blinked at this last statement, then seized Mary B's clasped hands and bent to kiss them. "You are a remarkable woman," he said fervently.

It was then that Mary B had asked him to look in on her four companions. "We'd all sooner die than go back," she said.

"Listen," Boswell said now to the four men, putting energy in his voice, "It's not so bad. I think what you've done is remarkable. The coffee houses are buzzing, and no one wants to see you sent back. I have friends, I'll speak to them."

Their expressions downcast, they remained silent until Allie raised his eyes as high as Boswell's paunch-plumped, velvet weskit. "Begging your Honor's pardon," he said, his toothless words lippy, "But why should you care? We're nobodies, just small fry. We got no money. Some gentlemen at the Bailey passed the hat, but that's gone." Allie swept the others with a glance and drew himself up a little straighter. "Me myself, I served my Sovereign in two wars. Aye, on the *Ramillies* under Moutray himself, and..." then he faltered and his voice cracked and when he resumed it was to say, "But I'm nobody now, just a convict. We're all nobodies. We're all just convicts." Abject heads nodded meek agreement.

"But I don't care," Boswell said, "I don't care if you've got no money or if you think you're nobodies. I've got friends and I'm a barrister. I can't promise you anything, but I'll see what I can do. And no fees. You've suffered enough."

The convicts exchanged glances heavy with disbelief. Finally, James said, "Aw," and got to his feet. Then they all rose and Boswell pressed each man's hand in turn. He sent for wine for himself and Akerman and ale for the men and soon was talking easily, clapping the men on their backs, invited to feel the scabs and scars from wounds and bilboes. When he mentioned the name Henry Dundas again, this time as an old school chum, he drew blank looks. "He's Home Secretary," Boswell explained, "He can release you from your sentences. He can get you pardons." Faces brightened. Through this voluble stranger they might at least be allowed to complete their sentences in England. In the best of all possible worlds they might even see freedom. Giddy with enthusiasm, Boswell suggested that they prepare a little pamphlet for others awaiting transportation to Sydney. "Tell them what it's like, what they'll need. You could get real money for your efforts." As they continued talking, hope built on hope, and when finally Boswell left with the promise that he would return next day to gather more information regarding their particulars he left four smiling men with hearts filled with joy for the first time in a year.

As he followed the unsteady footsteps of Akerman, Boswell's spirits were oddly elevated. He'd come to Newgate on a whim to view an Amazon, and now found himself somehow committed to the cause of five escaped felons. His voice had a jocular, self-deprecating quality when he passed into Akerman's quarters.

"Dick, Dick, Dick," he addressed the old jailer, "Why do I do it? Why do I always get myself into these things?"

Forty-odd years keeper of Newgate, witness to countless instances of morbid curiosity transformed into compassion, the venerable Akerman shook his head and replied in his asthmatic measures. "Because, Mr. Boswell, in addition to your probative wit, your political passion, and your literary achievements, you are at bottom, a man of heart." Boswell smiled, pleased at the description, and reached for his purse to get money for his new charges. Akerman put out a palsied hand. "Unfortunately," the old man continued, patting Boswell on the arm, "This city is a hard place, in a hard time, to be a man of heart. Go home to your children, James, and I'll see to the welfare, of your new clients."

Thus did James Boswell become champion of the Botany Bay five.

But how could the convicts know that despite Boswell's old Scottish title and respectable estate, their champion was a depressive alcoholic and a debauheé; that his career as a barrister was a joke, that he was a barrister without briefs, a man whose unabashed seeking after recognition and preferment had pared his friends to a last faithful few, and these few did not include Henry Dundas. True, by virtue of his recent biography of "Dictionary" Johnson he'd recently been elected to office in the Royal Academy, but his biography had also offended many, and had probably closed more influential doors than it opened. After all, how can you trust a book-writing blabbermouth - and a boozer to boot? In short, Boswell possessed dubious entré to London's power structure.

So Boswell, now 52 years old, his legal career a shambles, juggled debts and worried constantly about his five motherless children. Adrift in the London he loved, he agonized over his thwarted ambitions and flitted from one fantasy to the next - to be ambassador to Corsica, to serve in China, to write a biography of Sir Joshua Reynolds. In his private brooding he fantasized acquiring a wife with a fortune, a sinecure in government, and to be well thought of, even loved. Perhaps in the sickly, penniless Botany Bay Five he saw that he might at least realize this last fantasy.

True to his word, Boswell returned next day to garner more information. For the next several weeks he tracked down records, wrote letters to families and friends, gathered character references, delivered petitions to the anterooms of busy officials, and every few days saw to the welfare of his clients. But to no avail. Despite Boswell's petitions and pestering, the Botany Bay five languished in Newgate.

They did not lack for visitors. Boswell brought John Courtenay and William Wilburforce, Members of Parliament, to hear the convicts' first-hand accounts of life in the penal colony. Jeremy Bentham, the philosopher and penologist, came several times and talked about a writing project. Friends and relatives of a boatload of women convicts bound for Sydney on the *Bellona* approached to learn the real story of their chimerical destination.

But London was a vast, busy hive, and Mary B and her friends fast-fading flowers. Newspapers that brought them into momentary view moved on to other matters, and as the novelty of their story wore thin and August heat cooked up the annual noxious stench of open sewers and rotting garbage, London society began its summer flight to the fresh air of the country, and Boswell's efforts wavered.

He left for Cornwall on a long-planned trip to visit an old friend. "I'll be at Penrhyn," he told Mary B, "Not far from Fowey. Do you want me to look up your people?"

She seemed to shrink. She shuddered and shook her head. At her reaction Boswell felt a powerful empathy for her loneliness and isolation. He resolved to try to effect reconciliation between her and her family. He smiled gently, "I understand."

Before departing for Cornwall he wrote a hasty appeal to Henry Dundas for assurance that "*nothing harsh shall be done to the unfortunate adventurers from New South Wales, for whom I interest myself, and whose extraordinary case surely will not found a precedent.*" The reference to precedent concerned a bit of gossip that John Courtenay picked up at a club, that Dundas would not release the prisoners before the expiration of their sentences, because doing so would set a precedent of rewarding escape, no matter how extraordinary the case. When Courtney reported this tidbit, Boswell slapped the table and declaimed, "As Mr. Burke said, '*Privilegium non transit in exemplum*!' an exception does not create a precedent!" Courtenay patted him gently on the arm. "Bozzy, I believe that principle applies only to civil law."

~~~~~~~~~

The raised voices of a half dozen men gathered around a corner table in the Dog and Bear caught the attention of a tall, thin man. Wearing an ill-fitting coat, he sat alone, not so much nursing his ale as holding the mug for company. Occasionally his lips moved in silent speech, and once he turned as if someone had spoken his name. "What?" he said aloud, then sheepishly buried his nose in his alepot. If he heard his name it wasn't spoken by anyone there, for he was a stranger in this Southwerk public house.

He eased his frame around to survey the noisy scene at the corner table, trying to gauge whether the growing argument would turn violent. Some of the men wore sailor dress and looked tough. One was an older boy. Another in town clothes sat slightly apart, plucking idly at a little guitar. The tall, thin man had been conscious of the vaguely familiar tune for some time, but was unable to place it.

"And I tell you it ain't right!" one of the sailors rebutted, slapping his hand on the newspaper, "Not with Bligh not here! A man's got a right to face his accuser, that's the law!"

"But Bligh ain't the accuser, Dick! The law's the accuser! They broke the law! 'Tis nothing to do with Bligh!"

"'Tis so, Jack! Bligh was why they broke the law! You heard what they said about his temper and the way he got after Christian. Sometimes laws have to be broken!"

"Easy, Dick," the man with the little guitar said quietly. He'd noticed the tall, thin man studying them. Spies were everywhere. The table fell silent as the others turned, their sun-dark faces suspicious and challenging. When the tall man saw them all staring, he drew his feet under him, took a big breath, and approached their table. His head nearly scraped the low ceiling beams.

"I came up in the *Gorgon* from the Cape," he said. "'With the *Bounty* men." He still held his alepot as if warming his hands. His jumping Adam's apple betrayed his nervousness. "I got to know some of them. My name's Pottle."

For several days newspapers had carried lengthy accounts of the court martial taking place in Portsmouth. As the proceedings drew to a close, opinions clashed in taverns and clubs all over the south of England. Those who worried over the excesses of republican France and feared the mutter of sailors in His Majesty's fleet would just as soon see all ten *Bounty* mutineers hanged out of hand, a clear message to rabble that sought to usurp authority. And those who'd experienced the harsh rule and abuses of the quarterdeck, who'd climbed the shrouds and bent the lines, thought a fair trial as unlikely as the promises of a press-master.

The men at the corner table were the last of the *Pandora* crew, finally back in England after serving their Dutch ship to its home port of Ostend. Impressed more than two years earlier, these landsmen had suffered shipwreck, hunger, thirst, exposure, and saw nearly a third of their shipmates lost to drowning and disease in a circumnavigation of the globe. Now after days of run-around and delay, they'd finally received their back pay and the abrupt thanks of their government. But another scrape with the law, an unguarded word of discontent, a nap in the wrong place at the wrong time, and they could land right back aboard ship against their will. If there were advantages to a career in the British navy, these weren't men who'd affirm them.

In a few minutes Pottle was drinking his ale among the *Pandora* men and sharing his opinions of the trial. "I doubt they'll all hang," he said.

"And why's that, Mr. Pottle?"

"Because there's officers - young Heywood and Stewart. They're just boys really. They won't hang young officers, not with France getting hot."

The men were silent. They puffed their pipes, scratched their jaws, picked their scabs, and after a long moment the one called Dick said, "Well, that's a deep thought, Mr. Pottle. That thought bears some minding." Heads nodded sagely.

When conversation about the trial came to a lull, the man with the little guitar asked, "The prisoners from Botany Bay, they were on the *Gorgon* too?"

"Aye," Stork said, "Right with the *Bounty* men. They're all here in London now, in Newgate. I see them now and then."

"You do?" the man with the little guitar asked, making a good-natured question sound with a pluck of a guitar string. Stork nodded. "And the woman - her friends called her Mary B - do you see her too?"

Now it was Stork who looked surprised. "You mean Mrs. Bryant. Do you know her? She lost her little girl, you know."

"Lost her little girl?"

Pottle nodded. "Off the coast of Africa."

"I'm sorry to hear that. And do you see her too?"

Pottle nodded again, then asked, "Do you know her?"

"No," the man said, bending to pluck a few notes of the familiar little melody that escaped Stork's recollection, "not really."

The men at the table exchanged amused glances. Then one informed Stork that the woman had sailed aboard their Dutch ship from Djakarta to Cape Town. Winking at his mates, he said, "Well, now, Mac, will you be going to call? We thought she had a kindly look for you." The table stirred with anticipation of manly teasing.

"You thought so, eh?" Mac didn't raise his eyes, but proceeded to pluck out the opening notes of "A-roving."

"Well, indeed I did, indeed I did! Didn't you, Dick, didn't you, Jack? Why even Lame Johnny here. Didn't we all?" Amused faces nodded. Mackesey plucked a few more notes, then thrummed a finale and stood. He tousled Jack's hair and Jack ducked away and laughed. Stork was surprised to see that "Mac" and he were of almost equal height. Still smiling, Mackesey said, "Well, then, if that's what you think, maybe I'll be off to see her, eh? My name is Mackesey, Mr. Pottle, but the intrepid call me 'Mac.' Would you care to join my perambulation? Perhaps we shall sit upon the ground and tell old tales."

~~~~~~~~~~

William Bligh wiped his neck and face and hands. Shit! He was getting sick. Clammy sweat, throat feeling raw, spasms of shivering - the first telltale signs of the fever. It was mid-afternoon in Kupang, and he lay propped beneath netting in a second floor bedroom of Governor Wanjon's residence. Outside he heard the occasional clip-clop of a horse, the squeak of a wagon wheel, the staccato rhythms of native speech, an occasional gabble of Dutch. He ventured a hand from beneath the netting to remove the cloth covering a pitcher and tumbler, poured a measure of tepid water, and sucked it down.

He lay back and mopped his neck and face again. Sick or no, he'd be undeterred this time, by God! Yes, sir! Those London fancy pants would choke on 'Breadfruit Bligh!' So far everything looked good on his second attempt to bring breadfruit trees to the West Indies.

Bligh picked up Bryant's book. The Governor smiled when he told Bligh about Bryant's wife hiding it in a jar of rice.

Strange how truth is sometimes revealed, Bligh thought. When he'd met up with Captain Hunter at Cape Town on the *Waaksamheyd*, Hunter could barely contain his anger. "God Almighty, Mr. Bligh, that idiot Smit insisted that to make for Kupang was too dangerous. So we piddle-paddled through the Spice Islands for more than six months! To reach Djakarta, Mr. Bligh! Six Goddamned months for a three- month voyage! The man's an idiot, Mr. Bligh, an idiot!"

Aye, a wily idiot, Bligh thought, for now Bligh knew why Smit had piddle-paddled. Not knowing how soon the escaped convicts might reach their destination, Smit had kept Hunter and the crew of the *Sirius* clear of Kupang! Although Governor Wanjon would never admit to it, he knew as well as Bligh that Smit had aided and abetted the escaping convicts, and far beyond just selling them a few instruments and supplies. Bligh thought, Smit must have an axe to grind with somebody of authority in these Dutch East Indies.

Bryant's book lay unopened on Bligh's knees as his thoughts turned to the news he learned from the Governor about the *Bounty* mutineers. So Edwards had been unsuccessful in tracking down Christian. Did that mean the *Bounty* was perhaps lost? That the rest of the mutineers were dead? Oh, how Bligh wished them dead! All of them! Dead and burning in Hell! What wreckage that scoundrel Christian caused for his petty grievances! Two ships and dozens of lives!

But a guilty thought nagged Bligh. He should have seen the trouble with Christian coming! He'd never trusted the man's damp, messy hands. He should have paid more heed to his gut feeling that Christian was a burning fuse. But how was he to know it would be the Tahiti women who'd ignite mutiny, those fresh, young *wahines*, those tawny lovelies of bright smiles and firm, willing flesh? How was he to know they'd be the powder that blew up his mission? Oh, that Fletcher Christian were dead! Dead! Dead!

Bligh opened Bryant's book and found himself staring in surprise and puzzlement. "Peter," "Quint," "His," "Book," he read. Bligh turned the page and read the nearly unintelligible words that described Peter Quint's life and love and impending death. What the...? He read the text again. Was it truth or jest? He returned to the picture and brought the book nearer, drawn to the intricacies of the panels. He first recognized the unique bulk of Table Rock at Cape Town and then the mouse-like shape of the kangaroo. Then he saw that the lighthouse was the Eddystone in Plymouth Sound. In another panel he identified the skyline of Portsmouth. Now each panel revealed its meaning - the cone of El Pico, the Sugar Loaf at Rio. And that must be Governor Arthur Phillip on the stump! Amazing! This single page depicted the entire voyage of the First Fleet and the early history of Botany Bay, a pictorial history on a page. And the second page was a history of a different sort. Who was this Peter Quint? The Marine in the center? He gave an amused snort when he noticed the grinning pig. "Incredible pig!" he mused.

Bligh lost interest and found Bryant's log on the flip side of Quint's work. He chuckled appreciatively when he saw a correct form for a ship's log - the hour of the day, the knots, the bottom in fathoms, the direction of the wind. He scanned the esoteric numbers and notations as a musician might read music, tracing in his mind's eye the circumstances of the little boat. When he came to Bryant's title page he exclaimed aloud. "Ahah! Now we'll see what the Governor's been hoarding."

~~~~~~~~~~

A chilly November wind spattered plump drops of rain against the soot-blackened windows of Newgate. A few women still visited quietly around a guttering rag-light, the incoherent sibilance of their voices a sussurant lullaby. The ward was peaceful. Snug beneath a blanket, warmed by Boswell's wine, Mary B's eyes were heavy, but her mind swarmed with images and words from the exhausting day.

First had come Mackesey and the woman. Oh, Mackesey! What a blessing! So different from anyone she'd ever known! So kind, so gentle, so caring.

On the *Horssen* she'd been suspicious at first, wondering what he wanted, why he never hinted at an underlying purpose. When finally she screwed up courage enough to ask, he looked puzzled, then hurt. "Why, Mary B!" he said, for he'd adopted her women's name from hearing her stories, "You're a remarkable woman, and you're stealing my heart!"

He now appeared frequently (if unpredictably) at Newgate, sometimes with a sweet, sometimes with a bottle of soup or a bit of stew. Once he brought her an orange and wondered at her sudden tears.

When she expressed concern about the cost of his generosity, he said he performed between plays in a little theater and did well by coins tossed on stage. "Besides," he said cryptically, "how much is a minute worth to a drowning man?"

Sometimes he came with Mr. Pottle, who seemed ever more distracted. Although Mackesey might stay only briefly, he more often lingered as if he had nothing better to do; he'd strum his little guitar and sing quaint songs to entertain the women on the ward. Once the ward bully tried to sit on his lap and fondle him. Mary B was ready to fly at her, but somehow Mackesey turned the incident into a funny diversion. Big Bertha ended up sitting on Mary B's lap, sucking her thumb and mugging rejection to the belly-laughter of everyone, including the prison matron.

Sometimes he read aloud. Mary B had come to love his Irish lilt. He read books that told of castles and ghosts and young ladies who swooned, of doughty knights and women of virtue. Sometimes he read about serious matters, "idea books" he called them. Women with sewing would stitch while they listened to his reading, and always express surprise at how fast the time flew. But best of all Mackesey liked verses, which Mary B had come to love best of all too. Even if she had difficulty getting the full sense of what he read, the rhythm of his voice soothed her; the sounds of his words eased her mind as lapping wavelets had soothed her troubled mind in Sydney.

One day Mackesey was scarcely able to contain his excitement. He carried a thin paperbound book he said was real poetry by a real man. He read several of the poems, sometimes stumbling over the words, but laughing often. Mary B especially liked the poem about the plowman who talked to a little mouse turned up by his plow. On another day Mackesey brought Mary B a chapbook. "Here," he said, "this can help you get your letters." Each page was divided into twenty squares. In each square was a picture of an object. The name of the object was printed below with the first letter exaggerated - **A**pple, **B**ear, **C**ow, **D**og, **E**ar, **F**ox, **G**oat. Tucked in the bodice of her threadbare dress Mary B now carried a paper on which Mackesey had helped her copy the initials of her name. "Practice how to make these letters and you won't have to write your name with an X," he said.

She felt proud to be learning. No one had ever helped her learn before, and sometimes abed at night she'd secretly trace the shape of her initials and other letters on the blanket with her finger. On still another day Mackesey was so happy he danced a jig as he excitedly held up a little magazine. "I'm in here, Mary B! I sold a poem! I sold a poem in London!" But when he read his poem aloud she heard sorrow and anger. His poem told of the sinking of the *Pandora*.

Today, however, Mackesey had come with a pretty woman and kissed Mary B's dubious hand with feigned reserve before giving her a hug that sent her heart soaring and made her warm all over. Turning to his companion, Mackesey said, "And this is the famous Mary B!"

The visitor was a gentlewoman, but when she put out her hand like a man Mary B couldn't help smiling back. She only caught the first name - Mary - because the last name was too long and too hard. Then Mackesey asked Mary B to tell Mary about Botany Bay, especially about the lot of the women. Mary B felt a surprising flush of jealousy and anger. She wanted only Mackesey to hear her stories. She realized she wanted to reserve her thoughts and feelings for him alone.

But she acquiesced, and during the next two hours she came to like Mackesey's friend, a woman who seemed to know exactly what Mary B meant when she described the drudgery and hardship of the colony - how women were given the most lowly work and the most witless tasks, how unmarried women were made housekeepers for huts full of men and had to fight them off.

When Mary B told of their starvation and of the dreadful death toll on the Second Fleet and of the deaths of so many of her shipmates and of her own two children, this other Mary could only shake her head with sadness.

"Isn't this a story?" Mackesey exclaimed sadly when Mary B seemed unwilling to continue, "Don't you think Mr. Johnson should take this story on?"

The woman thought. "Oh, I don't know, Mr. Mackesey. There's so much going on with the trouble in France. He's frightened of what might happen if he keeps stirring the pot, and such a story as this would surely stir the pot!"

Mackesey explained to Mary B that a man named James Johnson, the fine old fellow who bought his poem, had published the first part of *Rights of Man*, an important "idea book." But when the second part of *Rights of Man* appeared the courts issued a warrant for the author's arrest. Thomas Paine was wanted for sedition.

"These are hard times, Mary B, what with the riots and massacres in France. And for once it's not just the poor people getting killed. It's nobles and priests! Did you know the people in France have the King under arrest? Bigwigs here are scared, and when they're scared they're dangerous."

Later, when Mary B seemed unwilling to talk more about Botany Bay, the other Mary asked, "And how did you come to be sent away to Botany Bay?"

Mary B's eyes filled. She swallowed several times as she mindlessly tortured an old handkerchief. A flurry of thoughts about sin and guilt and doubt left her wordless. This is what sets me apart, she thought, what will always set me apart! She was unable to answer, except to say in a small voice, "I was trying to help a friend, and things went bad."

Suddenly she resented the woman's prying and resented Mackesey for bringing someone to gawk and cluck like a mother hen. Mary B kept her eyes on the floor, knowing she was being rude. After a long, awkward silence, the other Mary began to gather her things, saying she'd had stayed too long and was sorry to have been so taxing. But the damage was done. When she and Mackesey stood to depart, again there was Mackesey's hug, again the woman's man-like handshake, but this time hurried, as if a wind were blowing up rain and visitors were anxious to be home.

After they left, Mary B felt drained. Who was this other Mary? And what was she to Mackesey? Dare she even ask? What if he never came back! "What right do *I* have to him?" she asked aloud. Faces in the ward turned at her sudden outburst. She bit her lip and crawled into the bed she shared with two others. She pulled her blanket over her head.

She was in prison, and he was out there! "I'm in here, and she's out there!" she announced aloud as tears came. Before Djakarta she'd hardly ever cried. Now she was constantly surprised at her quick tears. "I should be thankful he comes at all," she told herself. But it was no good. Thankful for what? She'd told her story a hundred times, and nothing changed. Boswell was off in Cornwall, she never saw James or Allie or the others, even old Mr. Akerman never came around anymore. And now Mackesey was leaving her for another woman....

Perhaps her visitors had passed on the stairs, for it was only minutes later when the matron called, "Mary B, Mr. Boswell is here to see you."

Had he not been pre-occupied Boswell might have noticed the gleam of Mary B's vestige tears, but he managed a smile that was more a grimace as he awkwardly shifted a paper-wrapped parcel to jerk a bow. In the dim light Mary B saw his distracted look. Boswell began to fish for a handkerchief, but was momentarily confounded by the parcel, which he suddenly thrust towards Mary B. "Here, this is for you. A gown, from my Euphemia. I'm sorry I've been remiss, but I was, you know, I went to Cornwall, and I, I, I've been rather...indisposed."

Mary B curtsied her thanks as Boswell wiped his palms, then his face. She'd not seen him for almost three months, and found herself elated by his appearance.

Boswell made a show of smiling again, then suddenly blurted, "Akerman's dying. I just came from him. He's an old friend, you see." He wiped his face again, belched quietly, and blew his nose. He found a half-shilling for the matron. "Come, I want to go to the public room. I want wine. Oh, God, do I want wine!"

Outside Newgate, the other Mary had seated herself in a chair and presented her hand to Mackesey. "I hope you'll reassure your friend that...well, you'll reassure her, won't you?"

Mackesey bent his lips to the gloved hand. "Most assuredly, Madame" he said in a deep, affected voice.

Mary Wollstonecraft laughed. "Oh, you goose! You have the capacity to make everyone look ridiculous, as if they weren't ridiculous enough already." Mackesey winked and stood back. Mary Wollstonecraft looked serious. "Mr. Mackesey, I've thoroughly enjoyed making your acquaintance, and I shall miss talking with you a great deal. You know that, don't you? You are such a complete man! But I've put myself in a ridiculous position and I simply cannot stay here longer. But of course you know that, don't you? This whole evil city knows that."

Mackesey inclined his head in polite acknowledgement. Perhaps not the whole city, but plenty knew this spirited polemicist for women's rights had fallen in love with a married man. Unable to bear the torture of his Platonic returns, she was fleeing to revolutionary France.

Mackesey inclined his head again. "I shall miss you greatly too, my friend, but I'm afraid we'll both be leaving." He glanced around, then leaned close to the window of her chair. "I've got to see what's going on in Ireland. I thought I'd never go back, but it's my home, where my people are, and I must see for myself."

Mary Wollstonecraft reached for Mackesey's fingers to grip them and say, "Oh, of course, Mr. Mackesey, of course." She looked away for an instant, then leaned near and spoke earnestly, "We've so much to do, haven't we, so many wrongs to right? I sometimes think I'll go to my grave a scold, an old crone shaking my finger at the world, croaking out 'And another thing....'"

Mackesey joined in laughing at her self-mockery as she released her grip to admonish the world with her finger. Then her voice became confiding again. "But it can be dangerous, especially for you, Mr. Mackesey, such a free spirit. Do take care, won't you? Who knows, perhaps one day we'll persuade the French to liberate Ireland also."

She signaled to the bearers and was hoisted from the cobblestones. "And I hope you'll not leave that fine woman alone in the world. There's something special about her, although I'm afraid she's a little jealous. But tell her I'm gone, out of her way for good, won't you? And tell her that I admire her greatly and that I think she's a remarkable woman and that I wish her well." And with those words Mary Wollstonecraft was borne away.

In the women's ward the figures at the table were now shadowy, whispering, spirit voices.

How strange, Boswell's visit, Mary B mused sleepily in her bed. He'd gulped a glass of wine, poured another, and drunk half that away before he wiped face and hands again. "I hate death," he announced with a sigh. "A fine old man, Akkie, bound for the grave. Like us all." Chin sunk in his fleshy neck, Boswell stared vacantly at his outstretched legs. As if addressing his feet he began to ramble about family and friends and acquaintances who'd died in the past few years, names that meant nothing to Mary B. When he talked about his wife he began to cry, but soon made an effort to recover his composure.

He told her he'd stopped by to see her four friends and found them all in a state of dejection. "It's the *Bounty* mutineers," he said, "Did you know any of the *Bounty* men? Any of those hanged?"

Mary B shook her head. "Not to know them," she said. Mackesey had told her of the outcome - four convicted, three

429

hanged, one appealing, the rest acquitted, including the young officers. Boswell said, "The boys say Captain Edwards treated the *Bounty* men like dogs." Suddenly in the hubbub of the public room Mary B was aboard the typhoon-battered *Rembang*. She heard the horrible screams of Edwards' prisoners trapped below. She mashed her sweaty handkerchief in her lap. How long would these terrifying memories keep flooding back? "They suffered together," she said quietly, "It makes a bond, to suffer together, to come through it." She saw Charlotte disappearing in a blur of greenish bubbles.

Boswell suddenly roused himself and fixed her with an imploring look. "You don't believe in Hell, do you, Mary B? Eternal punishment? Damnation forever? Unendurable pain, forever and ever, burning? You don't believe in that, do you?" He didn't wait for her answer before he continued. "You know, only a year or two ago they burned a woman right here in Newgate. That's what they do to the women. They strangle them, then burn them. The punishment of Hell!" He gave a short, barking laugh. "Because hanging isn't seemly, see? Because it's not seemly to watch a woman dance at the end of a rope. Not seemly!" He gulped his wine and poured another.

He told her he used to attend executions to write about them for the newspapers and had once watched fifteen men hanged at the same time. "I told myself it was for the money. Me, destined to be Laird of Auchinleck, writing about those final, cruel, grotesque moments for money! Imagine me prying at those last thoughts, snatching at those last words, looking for some insight that might...that might bridge that instant between life and death.

"The thing is, they knew they'd be dead in just minutes, see, and I hoped they could see into the hereafter, that they could tell me about what...what it's like to be dead!" He took a long pull of wine from the bottle. "For the money? Pah! Not the money, and not morbid curiosity either! It was fear, Mary B! Fear for my own death!"

He expelled a huge sigh, as if he'd just rid his conscience of a secret sin. He sank back and stayed quiet for a long moment.

Then he sat upright as if to change the subject. "Cornwall reminded me of Scotland." He winked. "The worst parts." He loosed his braying laugh.

Then, more seriously. "It's so rough, so wild, so bleak. For some reason it made me think of you. It made me think of you a lot. God, I was desperate to get back!" He smiled a sad smile. "My friend Sam used to say, if you're tired of London, you're tired of life!" He suddenly seized the bottle and guzzled wine so violently some spilled down his front. He wiped his mouth and chin with a sleeve. "Temple thinks I drink too much. Christ, I felt his eyes on the back of my neck the whole time. But it's his wife. She doesn't like me."

He leaned close and spoke in confidence. "You know, Mary B, I saw the assize house in Exeter, Rougemont. Aye, I went in to stand in the very dock where you stood! It was evening, see, and I went in and stood there and imagined I was you all those years ago, a young woman from Cornwall, all alone, frightened." He didn't tell her of his visit next evening, when he returned with a young streetwalker for sex in the docket.

After a long pause he said, "And now you're the girl from Botany Bay! Aye, the girl from Botany Bay. Remarkable!"

Boswell had finished that bottle, ordered another, and finished that one too, all the while rambling more and more disjointedly of his blasted hopes, his detractors and rivals, his inadequacies as a father. When he talked once again about his absence from home when his wife died he began bawling. Mary B reached across the table to comfort him. He looked at her with such drunken sorrow and loneliness her heart went out to him.

"Mus' go," he mumbled, his words so thickened by wine they were scarcely intelligible, "You're no' th' one." He made an effort to stand, failed, tried again, failed again, and gave a snuffle of resignation. "M'prisoned b' wine," he said, and sank back as if prepared to remain the night. Mary B went around the table and with some effort got her arms around him and hauled him to his feet and supported him to the door. As they lurched across the room he mumbled, "Mar' B."

"Aye, Mr. Boswell."

"Mar' B," he mumbled. They'd reached the door. Boswell fumbled with his hat. His wig askew, his eyes glazed, he swayed ominously, then leaned into her so heavily she had to brace one foot behind to push him straight again. He dropped his stick. Mary B held him upright against a wall with a hand on his paunch as she bent to retrieve it. "Mar' B."

"Aye, Mr. Boswell." She settled his hat on his head, rather crookedly.

"Mar' B," he mumbled again as she tried to arrange it better.

"Yes, Mr. Boswell," she said emphatically, laughing a little. James Boswell had become a funny drunk.

"Mar' B," Boswell said thickly, drawing up his swaying frame as straight as he could, mustering his resources, "You are 'markable woman. `markable. An' I am prifa... prifa... prifalich know you. Know 'at, Mar' B? eh, know 'at? Prifalich! But I ha'e... wee reques'" He raised a hand to make a sign of small with his finger and thumb, then swayed backwards until he bumped against the doorway. After an effort to straighten himself he gave up and remained slumped against his backrest. "Reques', Mar' B."

"Yes, Mr. Boswell, tell me what it is."

"Mar' B...man...honor, min' you." He drew a huge breath and exhaled the wine-soaked words, "Jus' one, long, ver' long, kiss, eh? Wet!" And he ridiculously cocked an eyebrow.

Disappointed, Mary B affected a soft laugh. She put her fingers to her lips, then pressed them gently to Boswell's own.

"Thank you, Mar' B. Jus' I thought. `markable woman." His feet began to slide. He slowly lost height, gathered momentum, and as Mary B stepped aside, slid full length onto the floor and began to snore. Even as the turnkeys hauled him to the street and put him in a chair for home he kept snoring.

Then shortly after the turnkeys carried Boswell out, Mackesey came again, his voice insistent outside the wicket. He was permitted to stay only long enough to present Mary B with a bouquet of cranesbill, to exchange a few words of reassurance, and to enfold her in his arms for his first kiss upon her lips.

Oh, Mary B thought, her emotions soaring, his arms are holding me, his arms are enfolding me. He's holding me, holding me, holding me safe.

Mary B slept.

~~~~~~~~~

Arthur Phillip sat wearily back in his chair, fingers interlaced on his desk. He stared vacantly at a scramble of papers and considered John Arscott's request. The strapping sailor wore an uncomfortable half-smile. Arscott's sweat-slick hands nervously twisted a cane behind his back. The Governor looks old, he thought, old and tired and shriveled up. Blue-green bags burdened the Governor's eyes.

Arthur Phillip spoke. "You realize of course if you return to England you'll be subject to arrest if you're found out?"

"Yes, Excellency."

"But you have absolutely no intention of returning?"

"No, Excellency."

Arthur Phillip thought, *How many have said that?* To Arscott he said, "But what if your wife finds out and decides to come out here?" Even as he spoke, Arthur Phillip realized the absurdity of his question. Another year of hardship and hunger was drawing to an end. Almost 500 deaths in the past year, one in ten of the colony. What wife in her right mind would come to New South Wales to join a convict husband, even an emancipated convict husband?

For Arthur Phillip's benefit, John Arscott appeared to consider the question. Of course he and Katie had not only turned that prospect over, but also a dozen other possibilities that might through some twisting thread of unhappy circumstance jeopardize their future together. "Begging Your Excellency's pardon, but even if my wife knew where I was she wouldn't come, least of all to a place like this...begging Your Excellency's pardon."

An appreciative smile played about Arthur Phillip's mouth. He nodded. "And your intended, does she know the full circumstances of your, ah, situation?"

"Aye, Excellency. And we love each other."

Arthur Phillip lowered his gaze, and absently picked up his little hourglass. "To my beloved Arthur, on the occasion of his fortieth birthday." Fourteen years. "*Omnia Tempus Revelat.*" Five years since he'd seen her. Surely she was dead by now.

Arscott shifted his weight. His sore foot was beginning to hurt again and he wished he could sit down. As if reading his thoughts, Arthur Phillip asked, "What about your foot? Will you be able to stand the work? Farming's not watch and watch, you know, it's all the time."

Arscott straightened. "I'll be fine, Excellency, and Katie's a damn good worker. She's got gimp, and after we're settled better I'll hire a man or two." He hesitated, "But it's the land, Excellency."

"The land?"

Arscott nodded. "l want everything good and legal, so our young ones will have something, an inheritance, all legal, see?"

It was Arthur Phillip's turn to nod. Yes, he thought, this is the kind of man we want here, a man ready to start a family, to build for the future. If I had a thousand convicts like Arscott I wouldn't care if they all left wives in England and took new wives here. "All right, Mr. Arscott," he said, "You have my permission to marry."

Arscott beamed.

"And I suppose the point of all this is to put it in writing, eh?" Still beaming, his eyes tearing with relief, Arscott nodded. Arthur Phillip bent to scratch away, then called in his secretary to witness his signature. As he handed the paper to Arscott, he said, "Good luck to you, John Arscott, and don't forget the banns." He shook Arscott's hand. Twice before the governor had shaken John Arscott's hand, once when he returned from Norfolk after saving the *Sirius*, the second time when Arthur Phillip presented him with his emancipation paper.

As Arscott reached the door, Arthur Phillip asked, "Mr. Arscott, about the *Atlantic*; what do you think of her?"

Arscott, who at that moment might have said the Devil himself was not such a bad sort, wore an irrepressible grin. "She's a good ship, Governor. She'll get you back to England in a fair way."

~~~~~~~~~~

The morning was overcast, the light gray, and the streets strangely empty. Mary Wollstonecraft heard a funereal drumbeat, the clatter of hooves, the harsh, grating rumble of iron on cobblestones. Through her apartment window she watched an open carriage pass by bearing Louis XVI from prison to his trial. Surrounded by a mounted guard, the King looked pale and frightened. Behind closed windows across the street other faces watched the passing entourage, silent and impassive.

That night, shaken by the scene, she wrote to James Johnson, her London publisher, "*I want to see something alive. Death in so many frightful shapes has taken hold of my fancy. I am going to bed, and for the first time in my life, I cannot put out the candle.*"

She found herself thinking of that other Mary imprisoned across the Channel, the woman they called Mary B.

~~~~~~~~~~

Mary B was watching the dim light of guttering sconces cast shadowy obscurations of lonely, frightened women peering from the courtyard windows of Newgate. London nights had fattened on December days until she treasured every hour of winter's mist-shrouded half-light. She dreaded the inevitable darkness that descended to paint the high windows with nighttime despair. Just a year ago today - Emmanuel. And then Will. And then Morton, Bird, and Cox.

Had Mr. Boswell given up? Although his reappearance was brief, a flurry of activity brought hope to her heart once again. But a few days after his drunken spectacle he'd returned in a state of discouragement to tell her that hope for early release seemed futile. Henry Dundas was adamant; he would not reward escape. Mary B must serve out her sentence. The best Dundas would promise was that she would not be returned to Botany Bay.

"The ingrate!" Boswell fumed, trying to rouse his sense of outrage, "After all I've done for him! I'll not give up! You don't deserve to be in prison, and I'll not give up."

Poor Boswell! How could he bring himself to tell Mary B he was unable even to gain audience with Dundas?

With Akerman dead, Boswell's visits had become infrequent, and as Mary B brooded by the window she mentally counted - three weeks since Boswell last came.

But it was Mackesey's absence that most hurt. Oh, that was hard! While Boswell brought brief happiness and hope, Mackesey brought warmth and reassurance. Mary B liked Boswell. She was flattered by his attention and encouraged by his enthusiasm, but she knew he'd first come because she was a curiosity. Mackesey had come because he cared. Boswell thought her important because she'd done a remarkable thing - she'd survived terrible ordeals. Mackesey valued her for who she was.

"I'm not sure when I'll be back," the tall sailor had told her. She'd felt her face flush with resentment. Not sure? Did that a mean a month, six months, a year, perhaps never? Was it another woman, that other Mary? For all Mackesey's good points, Mary B found his secretiveness and unpredictable coming and going annoying. Why couldn't he say when he'd be back?

She'd dreaded the hour of his departure, and when she said she didn't want him to go he tried to be reassuring. "I don't want to leave you, Mary B. Lord, I do not! But I must see what's going on for myself. Maybe it's nothing, but I need to be there. Maybe there's something I can do so I can go home again someday, to a free Ireland. Do you understand? But I'll come back, Mary B, no matter what, I'll come back to you. You have my word." He kissed her long and lovingly on the lips.

Despite fears that his absence would lead to another crushing disappointment in her life, Mary B had tried to put on a brave front. She'd joined in with affected good spirits when Mackesey led the women in songs for Christmas. He sang several new songs from a little Scots magazine, and caused Mary B to blush when he sent her a smile and sang,

My Mary's asleep by thy murmuring stream

Flow gently, sweet Afton, disturb not her dream.

As the inevitable time for his departure drew near, he'd turned to a page in the little magazine and said, "Now here's a poem to remember me by. It's by that real man poet, the Scot, but it's my tune, so it's Irish...and we'll sing it in an English prison. So what else is new?" They'd all laughed. As they learned the words and joined in chorus after chorus, some became teary-eyed at the sentiments of friendship and separation.

For auld lang syne, my dear,

For auld lang syne

We'll take a cup of kindness yet

For auld lang syne

With their final embrace, Mary B whispered in his ear, "Oh, Mackesey, Mackesey, please come back to me. Please come back to me." Then he was gone.

"Oh, Mackesey, Mackesey, please come back to me," she mouthed to the bleak December night.

Fowey
1793

Following his drunken soliloquy and Mary B's fingerborne kiss, Boswell became more and more caught up in the turbulent times. He'd arrive at Newgate with agitated energy and breathy gossip from his coffeehouse haunts. Louis XVI beheaded, the guillotine swarmed by rabble snatching at bloody souvenirs. While Ireland rumbled with discontent and Scots fomented sedition, England feared war with the French regicides. Country oafs catechized Tom Paine's *Rights of Man* and Irish exiles solicited French arms. Throughout the country, England's loyalist militia drilled in preparation for trouble.

But Boswell's political tidings mattered little to Mary B. What did Paine's *Rights of Man* mean to her? Those were men's rights! What did mobs mean to her? Would they storm Newgate and set her free? She tried to be patient with his perpetual hopes, his appeals for faith, for calm, for forbearance. "I saw Lord So-and-so," he might say, "and he was much affected by your story," or "I heard Lady Such-and-such is interested in your case," or "They talked of you at whist last night." But nothing changed, and if she had to serve out her sentence...well, at least her sentence was nearing its end.

She'd learned to pass time by living in her head. Inspired by a sliver of St. Paul's dome she could see from a window of the inner court, she saw herself among cool, stone columns and polished pews. She pictured a quiet corner where she sat and looked up to the very top. But after a time her imagination would fail. What was it really like? None of the women on the ward knew. "Not the likes of us," they said. Oddly, she felt reluctant to ask Boswell, as if her curiosity about such a grand object wasn't fitting. So she imagined what she could - smooth, cool columns, polished pews, a vaulted ceiling impossibly high.

With such games of the mind she passed the hours, for without Mackesey her chapbook lacked interest. Thus would one endless, dreary, drizzly day of sitting and shuffling and mindless blather drag to an end, expire into oblivious sleep, only to be revived by the gray birth of yet another dawn and another day, and another, and another.

In their impatience to be free, her four friends seized on chimerical schemes. When Allie heard the Channel Fleet was again beefing up for France, he sent word to the Admiralty Office that he would happily perform his patriotic duty in the King's navy. Several days passed. He amended his offer and said he'd serve at half pay. Still he heard nothing, and he marked his 57th birthday in Newgate, waiting for word that never came.

A convict forger wrote a letter for Butcher, who thought he'd never heard anything so grand: "*May it please your Honor, It ill becomes a person in the low sphere I move in to address a person of your exalted character, nor should I have presumed to have taken that liberty but for the following reason....*"

As soon as the letter left Newgate, Butcher haunted the door awaiting the Home Secretary's response, mouthing the forger's promises: how Butcher would return to Sydney to bring "*indifferent lands to perfection*," would select and plant crops that would thrive, and would be willing return "*on proper terms*," even though he'd "*suffered a great deal in going and coming*." Sure of his eloquent appeal, Butcher imagined Henry Dundas slapping his desk and saying, "This is the man we need! Bring him to me!" Butcher waited two weeks, as hope in those magical words leaked from his heart like water from a sprung bucket. He gave up waiting by the door. In the end, he heard nothing.

Compared to the other three, Nat Lilley's situation was enviable. Boswell had managed to track down relatives who pursued respectable lives in London, and with their help Lilley began to make fishnets for a Fleet Street merchant. After a time his earnings helped support his wife and three children, who came to live in London with his sister in the old city. And thanks to a single, hurried conjugal union in Newgate, Emma Lilley now carried their fourth child. Safe in prison from the pursuit of bad ways, gainfully employed, visited regularly by his family, Lilley felt an odd equanimity, and was prepared to accept whatever personal fate authority might decree. If he had to go back to New South Wales, well, he'd take his wife and children under a new policy that permitted spouses. On the other hand, if he had to serve out his sentence in Newgate...well, that was fine too. With work for his hands, with a once-a-week visit from his wife, with an occasional glimpse of his children and a few coins in his pocket, Newgate wasn't such a bad life.

438

James Martin struggled with the project suggested by Jeremy Bentham, who puffed a pet proposal in prison design he called the "Panopticon," a forerunner of the walled penitentiary. Believing his Panopticon would make convict exile to such horror-lands as New South Wales unnecessary, he sought to discredit Botany Bay. As an account of desperate escape from the starving colony might prove useful to his purpose, Bentham offered the four men five guineas for a publishable account of their story. Five guineas! Generous! Until they learned that every Grub Street hack wanted half for helping them.

James decided to write the piece himself. He scrounged the ward for writing materials. To write a dozen words was a forbidding task, but through the darkest days of winter he struggled with false starts. Then, after scratching out a few tortured paragraphs, he fell ill. Sweating, shivering, drifting in and out of delirium, he dreamed of struggling up an endless ladder with too many bricks, of digging a hole in which the dirt ever slid back. When Pottle came to visit on an afternoon of lucidity, he spent an hour taking quavery dictation, but the effort obviously exhausted James. Pottle refused to continue. "Get well first," he said. James next induced Lilley's brother-in-law to help. But he quit too when James suffered a fit of coughing. "It's not worth the pain," he said.

As James remained an intermittent prisoner of fever, hovering between illness and recovery, his project languished. He offered Bentham what he'd written for a guinea. No response. Still sickly, James slipped into deep despair. In the end, the crinkled half-dozen sheets of paper begged on the ward with such enthusiasm lay folded under his pallet, abandoned.

~~~~~~~~~~

Of course Mary B was wrong. She was not immune from the effect of events outside Newgate. As winter winds tugged loose the last stubborn leaves, those events swirled in to touch her too. So far, England had maintained a strict, official neutrality regarding the turmoil in France. But France seemed out of control, and the beheading of Louis XVI was proof that a bloody-mouthed monster was aborning. Now republican France had chased the Prussians across the Rhine, routed the Austrians from Belgium, annexed Savoy, and deliberately broken a long-standing commercial treaty with England. Increasingly threatened by this Jacobin adventurism, the British decided to act, and after much bluster England invoked a state of war in February, 1793. Soon after, the war and publication of his second book brought Captain Tench to London, and to Newgate.

Boswell had written Tench for confirmation of several points regarding Mary B's story and asked if Tench could provide a few words "*in favor of her character.*" He'd read Tench's book with great interest, of course, ("*a commendable work*") and invited Tench to call on him when in London.

Boswell's letter flattered Tench, for the captain was not long back in England before he learned that Boswell's biography of the eminent Samuel Johnson was a popular subject and a unique achievement. Tench wrote back affirming that Boswell's points were essentially correct and that Mary B was a woman of good character, "*indeed, in her own way, a remarkable woman.*" Tench also informed Boswell that he'd read Boswell's Hebrides work en route to Botany Bay with great interest and looked forward to reading his famous biography of Johnson. And as he soon needed to be in London, Tench would be pleased to provide whatever support he could in the cause of Mary B. In truth, Tench smelled patronage.

Due in Portsmouth to receive his new assignment with the Channel Fleet, Tench hurried to London to deliver his manuscript to the publisher, but when he called on Boswell he learned he'd gone to Auchinleck. Disappointed, Tench left his card at Boswell's residence and made a quick trip to Newgate.

Mary B thought Tench looked grand, and a big smile never left her face during his brief visit. His face freshened by the winter air, his long sidehair a little windblown, he looked invigorated and confident, and the ward women gasped when he bowed to kiss Mary B's hand with a natural grace. "I was so surprised when I heard from Mr. Boswell. I knew you were going to be famous, but I never imagined someone like Mr. Boswell would become your champion. Is there anything I can do for you?"

Mary B told of Boswell's efforts and frustrations and his difficulties with Dundas. "I know he tries to keep a happy face, but it's all delays and run-arounds. And it seems such a waste to sit in prison. What good am I here? At least in Sydney I did something useful."

Tench smiled. "Would you like to go back, Mary B? I can probably arrange it."

After feeling momentary shock at the question, Mary B realized Tench was joking, but she kept a straight face as she replied, "Of course I'll go back."

Now it was Tench's turn to look shocked.

"I'll go when you go, Captain, just say when." They both laughed as Tench nodded appreciatively. After their laughter settled, Mary B said, "I'll say no more, sir. I'll get out when I get out, but that will be the happiest day of my life."

Tench proffered a small package. "I wonder if you'd give this to Mr. Boswell for me? I didn't want to just leave it at his house. I thought you might want to explain." Mary B looked blank. "Smell," Tench said, holding the package to her nose, "Sweet-tea leaves, from Sydney." Mary B held the package in both hands and inhaled deeply, her tears starting. For a moment Tench thought she might collapse. He put a hand on her shoulder. She stared at the parcel, her thoughts swarming with the faces of friends who'd shared this make-do tea. "I know," Tench said with genuine sympathy as he patted her shoulder, "Powerful stuff."

From Newgate, Tench hurried to the Home Office. Before leaving Sydney he'd renewed his acquaintance with a captain in the replacement detachment, Nicholas Nepean, whose brother served as Under Secretary to Henry Dundas. Maybe I can move the mountain, Tench thought.

Evan Nepean shoved papers aside and lifted the top of a chased container to offer Tench snuff. The naturally upturned corners of Evan Nepean's mouth gave him the same appearance of a perpetual little smile as his brother, a family trait. For several minutes Tench related anecdotes about Evan's brother Nicholas and the penal colony, while the Under Secretary nodded and seemed to smile his innocuous little smile. Finally, drawing a breath, Tench asked about the status of the escaped convicts. Nepean stopped nodding and cocked an eyebrow. His mouth twitched. "Has Boswell enlisted you, too?" He picked up a small Indian dagger he used as a paperweight.

"Only as a character reference. No, she was on my ship on the way out and I came back with her from the Cape on the *Gorgon*. She's a good woman, and I'd like to give her a hand."

"How's she holding up?"

"She's got gimp. She'll make it...if she has hope of getting out." Immediately he regretted his lame after-thought.

Nepean swung slightly away and poised the dagger between the fingertips of his two hands, entertained by the play of light on the polished metal. "That Boswell," he said half aloud, "What a pop-gun!" He shook his head slowly at some private thought, then abruptly laughed, and turned back to Tench. "Mr. Boswell can count himself fortunate if Henry Dundas doesn't do violent harm to him before that poor woman's time is up, Captain. The man has pestered and picked and bothered and bedeviled this office to no end! He's had everybody and his uncle interfering, my friend, and he reached the point long ago where he's doing her no good. But the fool doesn't know when to quit! Jesus! I dread the sight of him!"

Nepean's outburst took Tench aback, and the captain's expression betrayed his concern that he'd made a mistake in coming. Nepean raised a reassuring hand and stretched his little smile a bit, "No, no, no, Tench, it's all right. She'll get out when her time's up, but not a day sooner. Boswell knows that too, but he's like a spoiled child and won't be satisfied until he gets his way. What I didn't tell him but will tell you is that the Secretary may even keep her locked up a little extra. And not just to vex Boswell," he quickly added when he saw Tench's reaction. "There's nothing wrong with having her pay for her part in the escape." He turned again to gaze at the Thames. "She's not blameless, you know. She was sent out to Botany Bay on her own account and came back on her own account. She could have chosen not to escape. But that's not even the point."

Nepean stood, signaling the interview at an end. "Henry Dundas is provoked, Captain. Bozzy says Henry owes him, but, well, Bozzy's going to get his nose tweaked." Nepean extended a hand, his eyes a-twinkle. "Thanks for coming by. It was good to meet you."

"Jesus!" Tench murmured as he replayed the conversation. He hunched deeper in his greatcoat as the jerking coach rattled through early darkness toward the lights of Portsmouth. Tweak a nose! Mary B is the instrument by which the Home Secretary is going to tweak James Boswell's nose! What crap! She sits in prison because he wants to tweak a nose!

Later he wondered, are we going to war because some perfumed fops in fancy offices want to tweak a French nose? Jesus, am I being sent in harm's way to tweak a nose?

Mackesey struggled up again from a long way down, dragging his incredibly heavy body toward a pinpoint of light he could barely make out. He was terrified the lion would catch him again and pin him with its indifferent paw, a paw so heavy and painful it smothered his breath. Behind he heard the deep throaty panting. He tried to crawl faster, fearing if he failed to reach the light this time the lion would drag him back down the abyss and he'd lack strength to free himself again. When he suddenly felt the monster close behind, he gave up. He was too tired, too tired to live. He closed his eyes and felt the beast nose his body and begin licking his face with its rough tongue. The realization that he need not struggle further was a relief. It would be all right to die. Everlasting sleep would be wonderful.

Then his eyes must have been open, because he saw Mary B sitting in sunlight just a little way ahead. She smiled and beckoned. Mackesey realized then he'd been deep in a cave. Mary B stood and walked into an intense light that swallowed her, but before she disappeared she turned to beckon once more. Mackesey tried to stand. He put a hand on the rough stone. His hand slipped and he felt the stone abrade his flesh.

At first he didn't see the girl, but after his vision cleared he recognized the crude stonework of a doorway and saw her sitting in its light. She was perhaps eight or nine years old and dressed in a rag of a dress and a dirty shoulder-wrap. Her dark brown hair was straggly, her face grimy. She sucked her thumb as she regarded him with large, impassive eyes. When Mackesey tried to rise, a blinding light stabbed his brain and he fell back, exhausted. He wondered if the awful pain in his head was what kept dragging him back into the abyss.

When he awakened again an old woman was bent over him croaking an indistinct, monotonous litany. She took up a wooden bowl and chanted briefly, then drank and held the liquid in her mouth while she raised his head, She put her lips to his. He tasted something warm and bitter. With great effort he raised his hand to see its ache and discovered a blood-encrusted rag. He slept.

~~~~~~~~~~

"Thorn? What kind of name is that, Scotch?"

"I don't know, sir, it was my father's."

The master of the *Sugar Cane* rolled his eyes and draped an arm over the back of his chair. The boy was fair-sized and wiry and promised agility in the rigging. Thinly spattered with adolescent pimples, his face had an outdoors look. He seemed bright enough, but a furtive shadow in his green eyes hinted at something hidden. "You're not bound, are you, boy, not running away? It'll go hard if you are." The master had to ask, but he'd take the lad anyway; so far his voyage had been just one headache after another. After delays in Deptford and Portsmouth, he not only was short-handed, but the two dozen New South Wales Corps recruits he carried promised as much difficulty as his convict cargo. Indifferent to their mission, more concerned with their own unhappy lot than their duties, lacking an officer to maintain discipline, four had deserted in his ship's boat, three claimed they were too ill to rise, and one quarrelsome brute now lived in irons. Unsure of his authority over these so-called soldiers, the master of the *Sugar Cane*, Thomas Musgrave, couldn't even find a bigwig who cared. Everything was war, war, war with France, and the problems of a chartered transport carrying Irish convicts to Sydney didn't much matter. Now, to top all, no sooner did he drop anchor in the Catwater than a third of his crew were impressed for the Navy's men-o-war. All over the Kingdom men were being rounded up to feed the war.

The youth's voice cracked and he blushed, hoping the damnable squeak wouldn't betray his lie. No, he was not *actually* a bound apprentice, because servants weren't bound. But he was running away, and only through luck had he heard of the *Sugar Cane*'s dilemma and tracked down Musgrave at the Running Fox.

Musgrave paid no heed to the break in the youth's voice. "You know we'll be gone near two years, boy?" The youth nodded. "What?"

"Aye, sir!"

"And there's no turning back! You can't get out and walk home!" Again the youth nodded. "What?"

"Aye, sir!"

Captain Musgrave gave him a long look. "All right, Ned Thorn, I'll give you a try. You might make a foretopman someday." Musgrave wagged a finger, "But you'll answer to your mates if you don't pull your weight. There's no man's pay for a boy's work." Musgrave opened a ledger and scrawled an entry. "Ten shillings a month and the usual stoppages. Can you write your name?" The youth nodded. "Damn you, boy, I didn't hear you!"

"Aye, sir!" the youth shouted. And with the proffered quill Ned Thornapple signed on to the *Sugar Cane* as Ned Thorn. "Good," the master said when he saw the youth could write his name, "Need money for necessaries?"

"Aye, sir!"

Captain Musgrave smiled a little smile and made another notation in the ledger. Ned signed, and Musgrave counted out a month's pay. "Lame Johnny!"

A youth somewhat Ned's senior detached himself from watching a nearby game of dominos and came to attention before Captain Musgrave. "Aye, sir," he said, flicking a sidelong glance at Ned.

"Lame Johnny: Ned Thorn. Stick close and see he gets to the New Quay by three bells. Mr. Denwright will take you out. We'll be catching the ebb." Ned grinned and touched two fingers to his brow.

Since stepping off the *Gorgon* almost a year earlier, Ned had shot up a good four inches. His shoulders had broadened and his chest filled out. Sometimes abed at night his joints ached, which Tench said was due to weedy growth. "Like a weed, Ned, you grow an inch overnight."

With the development of his man's body Ned had come to think more of manly things. Enough of books and learning! Sister stuff! Almost sixteen, he wanted adventure! An ensign's rating and shipboard life.

But Tench was adamant. A thousand times he preached, "First school, then career."

But Tench had become dull, always writing or revising or gossiping with his brother officers about regulations and rules and the customs of the service and promotions and commissions and brevets and backpay and halfpay and messpay and postings, and who got what and how and why and on and on. All quite boring. Then too, Tench was courting a Navy surgeon's daughter, and the way he mooned over Anna Maria Savage was unlike any behavior Ned had ever seen, not at all manly. "Unsullied, Ned," Tench would say as he doted on her miniature, "an unsullied woman!" But with the state of war and a brevet Major commission, Tench had a new posting, and Ned wasn't going to Portsmouth as his servant. Ned was destined to remain in service to another officer in Dock, and in school.

Damn it! Ned thought, if I can't go to war too I'll take my life in my own hands. I'll run away, ship out, and see the world! And when he learned about the plight of the *Sugar Cane* he thought, what better adventure than to sail back to Sydney? He'd see the sea, and see his sister, the only family in the world he knew. He missed his sister.

From the Running Fox, the youth called Lame Johnny and Ned headed up the hill toward the Hoe Gate. After sidelong study, Ned asked, "How come he calls you Lame Johnny? You don't look lame."

The older boy snorted. "Oh, that was Tahiti."

"Tahiti!"

Lame Johnny's chest swelled at Ned's exclamation. "Aye, Tahiti. You know where that is?"

Ned shrugged. Vaguely. Marines and convicts had talked constantly about the women of Tahiti, the *wahines*. This boy scarcely older has been to Tahiti?

"Well, I was younger then," Lame Johnny said, "Younger than you even. Anyway, I'm a really good runner, see. I could beat you easy, I bet. Well, some of the boys set up a race, me against a native named Oto. It was supposed to be all fun, but things got out of hand. Pretty soon the boys were betting buttons and nails and anything they could get their hands on, and the natives were betting chickens and pigs and cloth and even their women." He was thoughtful, then shrugged. "I could've beat him. I was ahead till I got this cramp in my leg. Damnedest cramp I ever had! Jesus, my leg felt like a piece of iron...hot iron!" A long pause. "Anyway, this Oto beat me. But I wouldn't quit even after he already won! I never quit! So I kind of skip-hopped to the end. Now I'm Lame Johnny." They walked in silence. "They weren't really mad, my shipmates, but after that I was always good for a laugh, especially when some booby started skip-hopping around with his tongue hanging out like they said I did." He snorted and shook his head. "Anyway, that don't matter. Topside I'm a match for any man."

Despite his denial, Ned had a feeling that Lame Johnny didn't like his name. "Well, they shouldn't call you a name for something you couldn't help!"

Lame Johnny shrugged. "That don't matter."

"Well, I won't call you Lame Johnny unless you want me to. You'll just be Johnny to me, John if you like." Ned studied the cobbles in their path with a serious mien. Lame Johnny glanced over to see if Ned was teasing.

"All right, call me Johnny, I don't care." A long pause. "I had a shipmate once that never called me Lame Johnny either. Saved my life too."

"Really? How?"

"Shipwreck." Lame Johnny took a deep breath to relieve a sudden tightening in his throat. His words were strained, as if he had to work to squeeze them out. "I was so scared. My cousin and me said we'd stick together, but it was dark and we couldn't find the boats and we kept calling to each other. But the waves...." He swallowed repeatedly before he could continue. "We couldn't keep together. And then...and then I never heard or saw him again." He said this last in a very small voice. He took several big breaths. "Then Mackesey was there and he kept me going."

Awed, Ned asked, "Where was this? Was he the only one that...you know?"

"Nah," Lame Johnny said, suddenly affecting gruffness. "Plenty others, and plenty I knew too. Thirty-five, they said. But where wouldn't mean anything to you." Lame Johnny remained silent for a moment, then said, "New South Wales."

Ned stopped and stared. Lame Johnny fumbled his words at the expression on Ned's face as he explained, "The Endeavor Straits, when the *Pandora* went down."

~~~~~~~~~~

The sound of horses coming fast over the hill, the warning barking of a dog, the squawk and flutter of chickens, the stamp and splash of hooves in the mud outside the hut. Through the low doorway the glimpse of a stirruped boot. The squeak of leather, sucking footsteps, a musket clicked to half cock, a rough voice, "All right, you Goddamned rebel, come on out, we know you're in there." The old woman raised a hand to indicate Mackesey should stay.

A bent-over crone emerged from the stone hut dug into the side of a hillock. Gray, matted hair strayed from beneath a tattered shawl she clutched beneath her chin. She squinted with her good eye at the mounted militiamen, her other eye milky and opaque. "Jesus!" the sergeant said, shielding his face, "Turn away, you Goddamned witch! You'll put us in Hell with your evil eye!" The men laughed. The sergeant led his horse a few steps closer to the doorway. "Come on out, Mackesey, we know you're in there."

A girl of perhaps eight appeared in the doorway and stepped to the woman's side, cold mud squishing between her bare toes. Her dress too was dirty and ragged, and she clutched a shawl and sucked the thumb of her free hand as she stared at the impatient horseman in blue. "Come on out, Mackesey, we're not after your bed-partners here." The soldiers laughed again, and one stepped forward to prod the girl with his musket. Seeing movement in the dim interior, the sergeant said, "Back!"

His eyes averted from the brightness, Mackesey groped through the low, narrow doorway. Despite the overcast sky, too much light pierced his eyes. He put up a hand to shield his sight and turned half away. With a deliberate move almost too swift to follow, the soldier took a step and clubbed him in the back with the butt of his musket. Mackesey collapsed to his knees. A wave of nausea threatened. When he put out his bandaged hand to steady himself against the wall the soldier smacked it with his musket. Through a roaring noise inside his head Mackesey heard the soldiers laugh before the mud jumped up to his face.

The woman and her granddaughter watched the four horsemen ascend the hill to meet the road to Dungarvan. They heard the sergeant's voice and the answering laughter of the soldiers. Hands tied behind his back, roped by his neck, Mackesey stumbled after. He concentrated on not falling. He knew it was seven miles.

~~~~~~~~~~

To Mary B's relief, March 20, 1793 dragged to an end with no one coming to set her free. This was the day that completed her seven-year sentence! But now what? What would she do out on the street? For seven years she'd looked forward to the day when she'd be free. But with freedom imminent she was suddenly frozen, completely at a loss. She imagined herself on the steps of Newgate, turned out to shift for herself. Should she go left or right? She'd have no more reason to go one way than the other.

Except for Boswell, she was friendless in London. But Boswell was off to Scotland again, leaving her with a pound note for necessaries. But, good Lord! what was a pound in this alien city, this anthill of sharpers and thieves! What good was money? She needed a safe place and someone to help her live on her own. She felt stupid and helpless, angry with Boswell for being gone, angry with Mackesey for abandoning her, at herself for feeling so stupid and helpless.

There was Mr. Pottle, of course, but Mr. Pottle was Mr. Pottle, and he'd taken up the Bible, finding a dozen occasions a day to point his finger heavenward: "*Judge not, that ye be judged, for with what judgment ye judge, ye shall be judged....*" New inmates mocked the gangling evangelist, but the old hands soon set them straight. With a letter of introduction from John Kirby, Akerman's successor as keeper of Newgate, Pottle collected food in the neighborhood for the prisoners, for the tupence a day the government provided made for poor fare. Many depended on the charity of bakers and merchants for enough to eat. Pottle also earned a bit running errands for some of the inmates, for debtors in hock and celebrity felons never seemed to lack for gold. With a pallet under the main stairwell and with regular food and work to do, Pottle seemed content. He regularly talked with Quint, and if anyone seemed interested he could point him out peeping around a corner or grinning up from beneath a table. Pottle's touch of craziness made Mary B uncomfortable.

Her great, secret hope was Mackesey. He'd given his word that he'd come back to her, and she ardently wanted him to keep his promise. She wanted him to be dependable and solid, an anchor in her life. Her feelings for Mackesey were more than friendship, but she wouldn't let herself say she loved him. She'd already made that mistake once, and what had come of it? After all, prison life wasn't real life, and certainly not a place to trust one's heart. If Mackesey came back to her, well, that would be the first thing. And when she got out of prison, well, that would be another thing. And then if something came of both those things, then that was something else again.

But while she could tell herself all this and pretend life was very tidy, at bottom she held faint faith in good things happening. She lived with the trepidation of an uncertain future. Oh, she wanted Mackesey to hold her, to enfold her. "Don't leave me, Mackesey!" she prayed, "Not you too."

~~~~~~~~~~

Early evening, and two slim figures perched on the upper yard of the *Sugar Cane*'s foremast, unconcerned that their bare feet dangled fifty feet above the calm waters of Cork Harbor. They saw the warm glow of the setting sun on the hills of Great

Island, on the clutter of Cove of Cork a half mile distant. Ned Thornapple glanced over his shoulder and motioned for Lame Johnny to look. Several miles distant, sunlight glinted on a prominent structure beetling over a hill the sun had painted a rosy hue. Lame Johnny nodded absently and continued his reasoning, "The thing is, he saved my life, Ned. I'd be dead if it wasn't for him. I can't just let him be carried off. I know he's not a criminal."

Ned nodded. He'd never forget Lame Johnny's shock when the youth had recognized his friend Mackesey from the *Pandora* being hustled aboard the *Sugar Cane* with the convicts. Tears started in Lame Johnny's eyes, stirring Ned's memories of his own tears at seeing his weeping father being escorted in the company of thieves to the hanging tree. Oh, how Ned fantasized during those awful days before the hanging! Somehow he'd bamboozle the sentry, somehow pick the locks, and somehow sneak his father to the safety of his secret cave near Bennelong's Point.

"I've got to do something, Ned, I've got to! He saved my life!"

Ned nodded, and bent to spit. He watched his foamy gob sail down and strike the water and begin its almost imperceptible drift toward the Irish sea. Ned already knew he'd help Lame Johnny, that his help had to do with their friendship, and doing what was right. And maybe adventure. And because he liked Mackesey's kind eyes and gentle smile. But he still didn't understand what Mackesey had done wrong.

"Mac!" Lame Johnny blurted to the haggard shadow of the man he'd last seen in London. Befuddled, linked to four others by a chain, Mackesey brightened at the sight of Lame Johnny, then quickly signed silence. Although not scheduled for transportation in the *Sugar Cane*, the sheriff brought them down from Waterford because, he argued, his jail was full, the prisoners insurrectionists, and the Lord Lieutenant most insistent on ridding the countryside of such troublemakers. The *Boddington* had already taken on extra convicts; surely the *Sugar Cane* could too! A half-plug of tobacco persuaded a sentry to grant the two youths access to Mackesey.

They found the convicts in the below-decks surprisingly cheerful. After baths, new clothing, and better food than most had eaten in months, they lolled in their confinement with an odd, lighthearted acceptance. From the other side of a bulkhead came the laughter of women. Refurbished, disinfected, and freshly whitewashed, the *Sugar Cane* was a paradise compared to Ireland's fever-infested jails and dungeons, and especially the cesspool on Spike Island where the convicts had been assembled for transport.

Their heads close in conversation, Lame Johnny talked with Mackesey while Ned inspected the deck beams with great care. They whispered for several minutes before Lame Johnny asked in a voice almost normal, "You mean the Defenders?"

"Shhh!"

Mackesey scowled at the convicts near them. Satisfying himself no one eavesdropped, he whispered urgently, "You want to join me in chains? Don't ever use that word, even in jest! Don't trust anybody! There are men here who'd sell you out for a pinch of snuff!" The two had whispered a few minutes more before the boys left for their perch on the upper yard.

"It's worse here, Ned, because the Catholics can't vote, see, and they got to pay for the English church. And when he came to Dublin they stopped him because he had a *Rights of Man*."

"But I still don't see what he did, what he did wrong! What was so bad?"

"Well, there's these Defenders, see? And they protect the Catholics."

"From what?"

"From...well, I don't know exactly, but I guess anybody that does bad to Catholics. Anyway, all the Catholics rent their land, see, because the law used to say they couldn't own land. Anyway, now the law says they can, but if they work and work to get the land better and get money to buy it, then landlords raise their rents because the land's better. So they can't ever get ahead! The Catholics can't ever buy the land they work. So the Defenders help."

"But what do they do?" Ned's frustration was growing.

Lame Johnny shrugged. "Well, I guess they...mostly get Catholics to stick together."

"Is Mackesey a Defender?"

"No, but in Dublin he joined a traveling company, you know, actors and musicians and such, and he told me some of them were Defenders and he just got more and more sucked in. He said all he did was write the oath, you know, because he can write, but up around Wexford some Defenders were in a tavern and the militia came. They arrest Defenders whenever they have meetings, see. Anyway, there was a terrible fight and Mac got hurt. He got away, but the law was after him. The Defenders kept moving him around, but somebody snitched."

The boys were silent while Ned digested this story. He spit again and watched his breeze-blown gob. If Mackesey was a criminal, then maybe the law was wrong. How could they send a man to Botany Bay for writing words? What harm could words do? Ned asked, "Is he a good sailor?"

His *non sequitur* drew a blank look. Lame Johnny nodded vaguely, "Aye, a good sailor."

"Well, I've got an idea," Ned said, "But the thing is, we have to bide our time." He lowered his voice. "Listen, now is when they expect them to try to get away, when we're still close to land. But if we bide our time and let things settle down...." Lame Johnny nodded, still puzzled. Ned laughed, amused that Lame Johnny didn't grasp the simplicity of his plan. "Look, we're shorthanded, right? and Captain Musgrave will look for help among the convicts. After that, we just have to find the right time and the right place to get him free."

~~~~~~~~~~

"Very interesting," Boswell said, working his lips and tongue, "and you say it grows wild?"

"Yes, sir. I don't know what we would've done without it. The convicts, I mean." Tench was irritated with his nervousness, especially after his disappointment in his first encounter with the Laird of Auchinleck - blowzy-faced, red-eyed, rumpled, hung-over, and repeatedly yawning.

"Yes, yes, I suppose," Boswell said, taking another sip. He set aside cup and saucer to scratch under his gown with introspective pleasure. He'd insisted on serving a cup of Tench's own "Botany Bay tea" in honor of the captain's visit.

Their eyes met briefly and they exchanged forced smiles. "Well, I'm so glad to finally meet you, Captain - pardon me, Major! Your uniform looks splendid! Is it new?"

Tench nodded, conscious of their sartorial gulf. Boswell's faded dressing gown was badly frayed.

Now Boswell yawned again. "Pardon me. Too late a night, I'm afraid."

Tench nodded politely.

Boswell made a weak effort to sit straighter. "I much enjoyed your book, Major Tench." The men exchanged compliments and brief anecdotes about their books. Boswell was finishing up the second edition of his *Life of Johnson* and Tench was reading the proofs for his sequel, a work he was calling *Complete Account of the Settlement at Port Jackson.*

"Well, Major," Boswell said, shifting to the matter at hand, "Mary B talks of you a great deal. She says she's much beholden to you."

On his recent visit to Newgate, Tench was shocked by Mary B's haggard appearance. She said she wasn't sleeping well, that she'd been having recurring nightmares about her daughter and a crocodile. She'd begun to cry. "She was such a little nothing at the end, Captain. Without life she was such a little nothing when we put her into the sea. Yet she was so big alive. It was just a year ago, you know."

"She's a courageous woman, Mr. Boswell, very sensible. When I came to know her better I wondered how she came to be...well, what she came to be...I mean, sent to Botany Bay, and all."

"Yes, yes, well put," Boswell said as he absently began to scratch the back of his hand. Tench noted Boswell's dirty fingernails. "That, of course, is a question we can all ask, isn't it? How we came to be what we came to be. I was firstborn at Auchinleck, you know, and what I came to be I came to as soon as I came to be." He brayed at his own joke. "Too bad," he continued, braying so hard his eyes teared and he made a mental note to use his joke again, "Too bad I didn't come to be in Kent!"

Tench laughed politely and held up a hand. "Oh, but Mr. Boswell, it's you she talks of with such admiration, of how good you are, how determined. I was surprised to find her still in prison, actually. I'd expected her gone back to Cornwall, especially after all your efforts."

"Yes, yes, well...efforts," Boswell said, his words and gaze trailing away. He continued scratching with a distracted, pawing motion. He looked at Tench with an expression of bafflement. "Her time's up, you know...has been for almost two months. She should've been released in March. I thought she'd be included in the latest pardons. I don't understand...." He looked thoughtful for a moment, then said, "I hear Governor

453

Phillip is due back on the *Atlantic*. I thought I'd see him. Do you think that might do some good? Did he know her?"

Tench shrugged. "I suppose. He certainly knew her husband, for all the trouble he was."

Boswell considered this, then took a big breath and expelled it in a huge sigh. "I expect this all has to do with Dundas and me," he said. "Two Scots, stubborn, contesting the figure of a woman, supine." He brayed again, then sighed. "I much regret she's caught in the middle, though." Then, as if the fact were relevant, he said, "I've arranged shilling lodgings for her nearby, with Madame Duval." He looked down to see he'd broken the skin of his hand with his scratching. He licked his fingertips and began to rub the sensitive area.

Tench glanced at the mantle clock. "Would you mind if I stopped by the Home Office again? I served with Nepean's brother, you see. It might help."

"Oh, yes, yes, of course," Boswell said, "Please do! Whatever we can do to get the poor woman out of prison must be done. Justice now serves her most cruelly!"

At the door Boswell again began scratching his hand. He smiled wearily. "Major, I've defended many criminals over the years, not always successfully I'm afraid, but I was often struck by their basic decency. Some didn't seem to be very bad people. Sneaky, yes, and lazy, and probably not very bright, or very truthful. And not dependable of course, and not...well, but still and all, basically decent. And the same with the people from Botany Bay. They seem decent. And I can't help feeling that if given a proper chance they'd lead good, decent lives." He gazed into the middle distance. "And now, of course, we're sending all other sorts to Botany Bay, not criminals at all, but political men, dissenters." His voice trailed away. "I'm of two minds on that." Tench shifted his feet. Boswell collected himself and sighed, "Well, I expect there'll be trouble with that before we're done. Anyway, Major, I'd like your opinion. You lived with these convicts. You're a man of judgment. You know Botany Bay. What I'd like to know is, am I right? Are they decent people? Can convicts be the building blocks of our new colony?"

Tench noted a blooming drop of blood on the back of Boswell's hand. He gathered himself to answer, then caught himself when a scene from Sydney suddenly came to mind. He'd completed his farewell excursion on Arthur Phillip's big gelding and was returning to the settlement. He'd come upon a natives watching something going on in a patch of pasture, giggling like children as they took turns at a wildly indecent pantomime of bestial intercourse. They pointed out a convict half hidden by brush in the pasture. Tartop. On his knees, pants down, holding captive a confused ewe by its fleecy flanks, Tartop pumped vigorously.

Tench smiled his own weary smile. "I don't know, Mr. Boswell. In the end, I'm afraid I just don't know."

A short time later Tench sat in Evan Nepean's office as the Under Secretary sifted a mound of paper. "Let's see," Nepean said through the appearance of a perpetual little smile, "I do seem to remember...." He paused to study a paper. "Well, my goodness," he said, his expression turning wry, "So that's where you've been hiding!" He put the paper carefully aside, then sighed and clutched up a double handful, "Paper, Tench, paper! We're drowning in paper!" He began sifting again. "And Boswell isn't helping! Look, here's one from Boswell, and here's another! And look at this, would you!"

With an expression of amazement Nepean held up a letter by a corner, "Can you believe this, Tench? He even wrote to Henry's new wife! Henry Dundas' wife, Tench? His bride of two weeks? What is there about that convict woman that's got Boswell baying up every tree? Does he want her for a mistress?"

Flinging Boswell's letter to the floor in a fit of pique, Nepean began sifting again. "Petitions, Tench, petitions! Everybody wants something! A job, a contract, a promotion, sell a project, buy a pardon! We're drowning in paper!" Suddenly he ceased his sifting with an "Ahhh!" and leaned to open a drawer. "Yes, here we are." He retrieved a document and glanced over it, saying "Hmmm, hmmm, yes. Mary Bryant alias Mary Broad. Hmmm, yes, yes." His eyes engaged Tench. "Well, I'd say you're in luck, Tench. Of course it still needs Henry's signature, but I could get that tomorrow." Tench waited, his hands folded atop the pommel of his sword. "Yes," Nepean said, perusing the paper again, "Yes, this might be what you're looking for." He put the paper aside and leaned back and placed the tips of his fingers together. He stretched his perpetual little smile. Tench thought of a cat that had just cornered a mouse. "Well," Nepean said, "What're your plans, Major, off to the war?"

It was nearly dusk when Tench knocked at the door of Newgate's keeper, the luxury of warm, springtime air lost to his pique. How many weeks more would that damned pardon have sat in Nepean's desk-drawer, a symbolic tweak of Boswell's nose? Did they want the man to beg? Did the law mean anything to these men of power?

Kirby wasn't in, but Tench told a clerk he'd deliver the release next day and asked that Mary B's accounts be readied, that Boswell would be looking after them. The clerk nodded agreeably and gave Tench writing materials to dash off a note telling Boswell the good news. Tench said that he could be reached at the Turk's Head Inn should Boswell have further need of his services. Dispatching the note, Tench cautioned the clerk, "I think it would be better if she didn't know about this yet. No need to get her worked up." The clerk nodded agreeably again. Tench headed back to his room, suddenly exhausted.

It was almost midnight when he was awakened by insistent rapping. Groggy, he stood in his nightshirt to read a note from Boswell by light of the porter's candle. Boswell congratulated him, thanked him for his efforts, and asked Tench to escort Mary B to Madame Duval's on Little Titchfield Street. Boswell would have a carriage for Tench at the Turk's Head at nine in the morning, unless Boswell heard otherwise. Tench scribbled an affirmative reply and crawled sleepily back into bed. He was just dropping off when he had an irritating thought. Before he could see Mary B released, he'd have to cough up her jail fees.

~~~~~~~~~~

Alert for French privateers and warships, the *Sugar Cane* plodded south on the now familiar route to Sydney - the Canaries, the Cape Verde Isles, Rio, Cape Town. Perhaps not Cape Town. French forces were rumored en route, and Captain Musgrave would be cautious. Meanwhile, Mackesey and a half dozen convicts drafted to work the shorthanded vessel became *de facto* members of Musgrave's catch-as-catch-can crew. Although his hand still hurt and he still got dizzy spells, Mackesey disguised his problems and was soon a popular figure in the yards. In the below-decks the Irish convicts remained so well behaved and cooperative their example soon settled even the most disgruntled among the New South Wales Corps recruits. By the time the ship made Santa Cruz, the *Sugar Cane* had become an easy-going community.

For three months Ned kept silent about his past, even with Lame Johnny, concealing an adolescent fear that somehow the long parental arm of Tench would snake out to snatch him back to Plymouth. But by early June, with the *Sugar Cane* bound for Rio, Ned's fear finally gave way to pride.

Crossing the Line was a grave matter among the crew. King Neptune demanded proper respect from those who trespassed his boundary for the first time. Heads need be shaved, aye, and tattoos cut, aye, and a good scrubbing with a dry brush, aye, or at least a good ducking. They recounted the horrors of past initiations with relish. Remember how this one cowered like a frightened lamb? How that one went mad? Aye, and then there was the gift, a week's rum or a shilling apiece for a man's yard mates. Youngest in the yards, Ned was their favorite target for this constant prattle.

Oh, what a surprise when Ned one day informed his mates that he too was a veteran of the seas, that this would be his third crossing of the Line and his second voyage to New South Wales! To Lame Johnny's open-mouthed astonishment, Ned backed up his claims with detailed descriptions of Rio and Cape Town. Amazement over-topped the crew's disappointment over a lost initiation, and Ned became much sought after for information. Playing the sage, he was fond of beginning his discourse by saying, "Now, if you gents was smart," then tell them stories of the great hunger, of opportunities for bringing in shoes and tea and tobacco and good soap and ribbon, almost anything. Some asked about prospects for farming, but most wanted to hear about the naked native women, the bouncy kangaroos that carried purses, or the odd little bear that lived in trees and feared no enemies.

One day Mackesey pulled him aside. "Did you happen to know a convict woman? They called her Mary B."

~~~~~~~~~~

Wearing a smile of anticipation, humming tunelessly, Mary B brushed her hair by the window of her second-floor room on Little Titchfield and looked down on the now-familiar street. Two years ago she'd awakened in Kupang, free but not free. Today she was truly free. Among her scant belongings in Little Bill's battered box was a boon that gratified Boswell - a full pardon, with all of her civil rights and privileges restored.

Other mornings had not been so happy, for after a few days of euphoric rushing about, Mary B's life had shrunk to loneliness in her shilling-a-day lodgings. In prison she'd looked forward to being useful on the outside. Now outside, freedom found her immobilized.

At first she rose each morning determined to take charge of her life, to brave these alien streets, but then she'd find herself puttering and fussing and procrastinating until the welcome intrusion of a visitor, or tea, or dinner, or the threat of a shower made venturing out untimely or unwise. She spent whole days watching the street below. Little Titchfield hardly looked threatening - three-story houses cheek-by-jowl, noisy children at play, pedestrians of all descriptions, street vendors, boisterous boys, the coming and going of sedan chairs, an occasional cart or carriage. Twice she got as far as the front step, only to be assaulted by such frightful noise and confusion that she fled back to her room. Rather than growing, her confidence was leaking away. Mackesey's chapbook lay unused. The prospect of struggling through page after page of its confusing symbols and endless pictures and words seemed overwhelming. She lacked resolve.

It always came back to this: where would she go? what would she do? who would she be? A free woman, she was no longer special, no longer remarkable. She was but one of a multitude, a common woman from Cornwall, empty of learning, and of prospects. Would she cook? Would she sew? Would she clean or serve? Would she marry? After years of confinement and hardship and showing her ability to survive and prevail against adversities that few would ever face let alone survive, was it the prospect of daily drudgery and an ordinary life that would defeat her? Perhaps at bottom she was unwilling to accept the ordinary. In any event, she felt utterly incapable of braving this new world.

Or the old. Fearing her father's ire and Fowey's shaming, finger-pointing gossip, neither would she risk a return to her roots. She wouldn't even risk a return to Newgate with its odorous corridors, rattling locks, and heavy, booming doors. She felt guilty about the continued imprisonment of James. Although they were convicted on the same day, she alone was free. Why? Was it Boswell? Was it because she was a woman?

Imprisoned by her fears, Mary B once again watched her life expire day by day. She tended towards brooding, and nurtured a nagging anger with herself - for her timidity, her ignorance, her inadequacy. When she tired of blaming herself, she blamed others. From the prison on Newgate to the prison on Little Titchfield, how typical! James had abandoned her with a baby, and Will with two. Mackesey had abandoned her in Newgate, and Tench had abandoned her on Little Titchfield like a sack of potatoes. And now Boswell! How was she supposed to find her way around this noisy, confusing warren? Men! They were all the same! Her litany of blame would grow until even she saw how foolish it was, and she'd make herself stop. But then she'd start again, blaming herself, blaming James, blaming Will. Racketing in frustration and loneliness, she frequently cried.

That morning she'd again followed her routine. After breakfast tea and toast were brought to her room, she brushed her hair, then gathered her bonnet and shawl, her prayer book and little purse, and took her seat by the window as if to collect herself before setting out. But today she would not spend hours looking idly out the window, lacking the courage to act. Today was different. Today Boswell was coming.

On Mary B's first day of freedom, she'd surprised and disappointed her champion. She'd shrunk from the teeming scene. To Boswell, London was grand theater, its flux and flow an ever-changing backdrop to a thousand entertainments. He relished the city's noise and vitality. After Tench dumped her off on Little Titchfield, Boswell scooped her up on a dazzling itinerary of shops and clubs and sights. He'd tippled wine and laughed and squeezed and recounted endless stories about these famous streets. But even his tipsy enthusiasm couldn't dislodge Mary B's thin-lipped reserve. Only later did Boswell realize how overwhelmed she'd been by the newness, and that she needed to immerse herself in this intimidating metropolis like a too-hot bath, bit by bit. Accordingly, he withdrew. "Don't worry about money," he said, "I'll see to your lodgings. You've suffered enough." He left her pocket money, and then left her alone, never thinking that in addition to coins in her purse she needed a hand to hold.

But Boswell was active on her behalf. Unbeknownst to Mary B, he began soliciting friends to build a trust fund for her independence. Feeling good when he'd raised almost twenty pounds, he stopped by to see how she was doing, and was taken aback by her frostiness.

His first impulse was to tax her for ingratitude, but his good sense prevailed. Perhaps he saw a frightened, petulant child. Instead of scolding, he ordered up tea, then coaxed and probed until Mary B at last began to talk about her dilemma. As Boswell came to understand her difficulty, Mary B saw the injustice of her secret blaming, and they parted with the crack in their friendship repaired. Ever the gallant, as Boswell rose to go he promised to take Mary B to someplace special, to St. Paul's, and it was for this event that she waited on this sunny June morning.

Mary B absently noted a post-delivery approach and a few moments later was surprised to hear a knock at the door. The serving girl curtsied and handed her an unsealed letter addressed in a roundish hand. Mary B looked blank. She unfolded the letter and stared at the meaningless symbols. With a hope she knew was foolish she asked the girl, "Did he say what this is about?" meaning the post-delivery. The serving girl shook her head. Proffering the paper, Mary B asked, "Can you read this, Beth?"

Beth half-curtsied and ducked her head, "Oh, no, mum, not me!" At Mary B's perplexed expression, she said, "Mistress reads! Ow, but she's gone out, mum."

Mary B studied the letter as if by sheer willpower she might divine its meaning. "Did he say who sent it?" The girl shook her head again. "Well, don't you know anybody who can read?" Mary B's tone indicated she held Beth to blame for this latest problem.

The girl reddened as she thought for a long moment, then shook her head, "No, mum, not cook, and none that I go to."

Mary B's face showed perplexity. With sudden inspiration she thought, Mackesey! At last Mackesey has written! The room seemed to brighten. "Of course," she said aloud, slapping her brow in a second insight, "How foolish of me! Mr. Boswell can read it. He'll be here any minute."

As an hour passed and Boswell failed to appear, Mary B entertained herself with the imagined good news. The letter announced Mackesey's return. She wondered if he'd written anything tender, anything about the future. She indulged herself with fantasies of their reunion. As a second hour dragged by, Mary B began to doubt. The letter was not from Mackesey. It was unsealed. And how could Mackesey know where she lived? This realization came as a blow. Then she wondered if the letter was from Mackesey by way of Boswell. Or perhaps from Boswell himself? Perhaps it announced a delay, or gave notice that he was unable to attend her. She began to fret. Was it Mackesey or Boswell? Or Tench? She paced, pausing at each turn to look out the window for signs of Boswell. Was that Boswell at the end of the street? Was that Boswell in the chair just turning the corner?

She hurried down to see if Madame Duval had come back, then returned irritated that the woman gadded about. Her agitation increasing, she waited until church-bells rang the noonday, then as if the bells were a call to action she flung on her shawl, picked up her purse, and set out with the letter, oblivious to everything save her purpose. She stopped the first gentleman she saw to ask if he'd read the letter for her.

It was from Boswell's young son, Jamie. Boswell had been attacked and robbed in the street the night before. Badly beaten, brought home by a passer-by, he was in bed under a doctor's care. Jamie conveyed his father's apologies for his broken engagement.

So different from what she'd imagined, the message stunned Mary B. Kind Mr. Boswell, beaten and robbed by footpads! How selfish she was! How childish! She was responsible for herself! She could not just sit by and wait, wait, wait and take, take, take. She had to do for herself! She had to give, too!

She blinked back tears and tried to speak. Her voice was unnatural, "Please, sir, does the letter say where it comes from?" The gentleman, who'd observed the play of emotion across her face with widening eyes, recollected himself and glanced down. "From Great Portland Street," he said, "Number 47," and returned the letter.

"Please, sir," Mary B asked, taking a big breath and gathering her shawl as if setting out on a journey, "Can you tell me which way, which way to Great Portland Street?"

The gentleman's eyes widened again, this time in mild disbelief. "Why, madam," he said, pointing up the street, "It's just there, just past the next block."

~~~~~~~~~~

Ned and Lame Johnny finished harnessing the last water butt and signaled the midshipman in the stern of the launch. "Ready below!" the midshipman called up. The second mate peered down. "Ready!" he acknowledged, "Stand clear below!" He turned to the winch-crew. "Stand by to heave, boys. Heave away. Heave!" Ned felt his heart thump at the significance of the moment.

For almost a week Ned and Lame Johnny had been hauling water to the *Sugar Cane*, each trip filling five of the huge water casks at Rio's jetty. Ned remembered the water-jetty and the pyramidal fountain, the very fountain where his father, Tench, and Hester Thistlethwaite had waited all night for him. Recalling the scene brought a lump to Ned's throat, and he resolved to send Tench a letter, perhaps from Cape Town.

The watering had gone slowly. With just one hose to serve the busy anchorage, Ned and Lame Johnny spent much of their time waiting their turn with crews from other ships. It was boring work, and on the second day their midshipman overseer deserted them to wander the plaza with juniors from other ships. Left to their own devices, Ned and Lame Johnny examined an endless succession of wares proffered by peddler-slaves and chatted and napped and visited with men from other crews, usually in pantomime and pidgin. Then a boatload of Americans appeared at the jetty, men they could talk with. From a bearded young man named Pease they learned the Americans' little brig was returning from a costly sealing expedition to the Falkland Islands. They'd lost a boat and two men in wild surf, and just a week earlier had lost four more to British impressment. Making liberal contributions of tobacco juice to the bay, Pease squirted for emphasis, a *ptooey*.

"Goddamn Brit frigate brought us under her guns like they own the Goddamn seas! Took four of our men just like that! Goddamn! What's that mean, eh? We're Americans! Goddamn Brits! How're those men going to get their share of our skins, eh? Goddamn!" *Ptooey!* "Down six men! We'll have the Devil's own time! Goddamn!" After reflection, he decided his story was ended. "Goddamn!" he said. *Ptooey!* Ned and Lame Johnny exchanged looks.

"Ready? Heave!" The men on the winch-line began their next haul. The winch boom groaned and the line creaked taut with the weight of the cask. "Heave!" the men cried in unison, and the sheave squealed to protest its burden. The cask lifted free and the launch rocked in relief. "Heave!" and the quarter-ton of water rose another two feet. At every "Heave!" a tail-man snugged the gain so the men on the line could take a fresh grip. "Heave!" "Heave!" "Heave!"

The rhythmic sounds of the work were like a manly poem, Mackesey thought. He grunted with his effort, the muscles of his arms and legs and back warming with the work. Even shackled, he pulled his weight on the winch-line, but now he pulled with extra energy, fueled by suppressed excitement. He wanted hard work. He wanted his body loose and ready. The time was near for his escape.

462

Controlling the swaying water butt with hand-lines, the two youths watched the creeping progress of the cask. Their last trip. Captain Musgrave had them hurrying to take advantage of an offshore breeze and the ebb tide. Ned turned at the approach of the pilot-boat. His heart beat faster. He wished dusk would fall more quickly. Soon they'd be aloft in the yards, Mackesey unshackled, and time and place be joined.

~~~~~~~~~~

A gangling young woman nearly six feet tall took a long moment to connect the distraught, somewhat breathless woman who stood before her with the Amazon her father so admired. She recognized the dress as one of her own before she connected the name. "Oh, yes," Euphemia said, but hesitated before admitting Mary B, as if the time wasn't good to visit.

But when Boswell learned Mary B was downstairs he insisted she be brought to his bedroom. As well as his son Jamie and a brother who lived in town, Boswell had servants and daughters aplenty to minister to his needs, but he was clearly touched by Mary B's concern, and he showed his pleasure by patting his bed for her to come closer. He held her fingers a few moments.

Mary B thought he looked awful. In his nightshirt, propped up by a half dozen pillows, a smeary palette of ugly purples and greens marred the swollen flesh of his right cheekbone. His right eye had swollen nearly shut. A balloon of bandage topped his head. The knuckles of his right hand were bruised and bloodied. A bandage encircled his right wrist.

In a small voice he asked how Mary B was and if she would take tea with his daughters. Then, as if the effort had taxed him completely, he closed his eyes.

Going to see Boswell freed Mary B. Although she was unable do anything meaningful for her benefactor, he was always glad to see her, and she found that subsequent visits gave her energy she hadn't known since the escape. Extending her outings to visit Boswell, she began exploring.

Her neighborhood was graced with several broad avenues, and Mary B found she liked the comfortable regularity of neat row houses stretching into the distance, with occasional greens offering carefully tended walkways. Life in this new part of London seemed so pleasant, so peaceful, so orderly, and such a contrast to the old city's twisty, cluttered streets, traffic snarls, and noisy vitality. St Paul's was her beacon for these excursions, a looming landmark some two miles from her room at Madame Duval's, but she held off the pleasure of her long-anticipated visit to the cathedral until Boswell might join her.

And the day finally came when she ventured back to Newgate. Her four friends inquired eagerly after her life, which she suddenly realized was good, and she tried to encourage them with the assurance they too would soon enjoy life outside.

Only the final moment proved awkward, when James asked, "Can you think of why you're free and I'm not?" Mary B could only shake her head, unable to meet his eyes. Through Boswell's efforts James had finally heard from his wife's family, but the news wasn't encouraging. Nonetheless, James desperately wanted to meet the son he'd never seen.

One day as she walked in the old city, Mary B chanced on a knot of people gathered around a street performer. The gawky figure swayed gracefully to an internal song, his arms moving above his head in a slow, languorous rhythm that reminded Mary B of one of Rami's dances. His eyes closed, oblivious to the tittering on-lookers, the man mouthed words that Mary B recognized as his name. She hurried on, hoping Pottle wouldn't open his eyes and see her.

Although publicly Boswell professed recovery, privately he continued to harbor hurts and debilities that worked at his spirit, as if his beating were an intimation of his mortality. Vexed at having been waylaid while staggering home drunk, he resolved to clean up his life. He spent more time with his children, drank only water or ale or an occasional glass of wine, or perhaps just two, or if the occasion were special, three, and on rare occasions four, but hardly ever more. He read the Bible, took his children to church, and worked at revising his *Life of Johnson*, adding new material and copious footnotes. But he still found time to look in on Mary B, who one day expressed dismay regarding the expense.

"I want to make my own way, Mr. Boswell, I want to find work." Boswell sought to deflect her. He said people of means wanted to see her do well and that a week's keep was a fillip for his friends. After all her suffering and loss, she should take her time. Besides, wasn't she enjoying her life? Mary B felt trapped by his question. Yes, she thought, I am enjoying life. She was exploring and learning, again looking at Mackesey's chapbook. But she also felt unconnected and lonely, as if she didn't belong. Other than Boswell and the men in Newgate, she knew no one, and had no one to talk to. She wanted to feel she belonged. Work, she wanted to say, would help me feel I belonged. But not wanting to seem dissatisfied or ungrateful, she found herself unable to say those words to Boswell. So she nodded and let the matter drop.

On a Sunday morning in mid-July a man named Elector Kestle knocked on Boswell's door. Boswell almost giggled at the frown-faced man worrying the brim of a black tri-corn. In his early thirties, with a slightly palsied head, Kestle wore an ill-fitting brown wig, a yellow coat with over-size red buttons, green sateen breeches, and white stockings with blue gaiters. His high-heeled, yellow shoes sported large brass buckles. He introduced himself as a glazier, and informed Boswell that he knew the Broad family in Fowey and they'd asked him to look Boswell up with a view to getting in touch with Mary B. Boswell cocked an eyebrow, ready to shut the door in the man's face. If ever he saw a sharper, one stood before him now. "I know her sister, Your Lordship. She's here in London. The family wants her to return to Fowey...Mary Broad, I mean, not her sister, who is most anxious to see her."

"You mean Mrs. Puckey?" Boswell referred to Mary B's older sister, who he knew was married to a pleaseman clerk named Puckey.

"No, your Lordship, that's who wrote. I mean her younger sister, Dolly. She's here in London, a cook, just over on Charlotte Street."

Kestle wore down the brim of his hat while Boswell digested this information. He swung the door wide. "Come in," he said finally, "And start at the beginning."

~~~~~~~~~~

Mackesey eased himself into his canvas hammock and put his hands behind his head. The sea-motion rocked his hammock like a cradle. Hammocks are a decided advantage, he thought, so restful. Maybe too restful for British ships. He eyed the bales of ripening sealskins packed all around, thinking how odd it was that his quarters on the prison ship were better than on this sealer. His mind drifted, and he found himself reliving the thrill of his free-fall, of that magnificent moment of commitment, of risking all for his freedom. It had been a perfect dive that plunged him near the struggling boys and sent him deep beneath the *Sugar Cane*. Did Johnny and Ned believe it had cost him his life?

The boys had performed their roles wonderfully. As the anchor was hove to in the dying light, they were shaking out the sails when Ned announced he felt dizzy. A couple faces turned to see him teeter, grab ineffectually for a line, then lose his balance and plunge forty feet with a piercing "Yi!" and hit the water with a horrible smack. "He'll drown!" Lame Johnny cried, and followed Ned into the water feet first. Mackesey waited interminable seconds. Some of the sailors laughed and scarcely glanced down, others looked concerned. As the boys continued their noisy struggle in the water, more of the ship's company crowded the rail to look down at their commotion. Mackesey hesitated. "If I stay, I live one life, if I jump, I live another." His hand tightened on a line. Memories swarmed. Mary B beckoned in brilliant light.

He launched himself silently, a blurred figure that knifed headfirst through the dusk to disappear near the boys.

His dive took him deeper into watery blackness than he'd ever gone before. As he oriented himself to the light above, his foot brushed something huge and moving, a slippery mystery. Unsure of his direction, he swam until his diaphragm pumped for air, then realized in a panic he was swimming in the wrong direction and reversed his course. He heard the struggling boys again and finally, giddy and weak-limbed, ready to give up on the idea of escape, he let himself rise and was surprised to scrape the barnacled hull. He surfaced with a gasp on the opposite side from the boys, controlling his desperate desire to pant for air. Under the wales, sheltered from the hubbub of Lame Johnny and Ned screaming at each other and those on deck, he gave himself time to recover his breath. He imagined the boys thrashing the water, pretending to fight over lines thrown to their aid. He heard voices. "He went too deep!" "Did you see him come up?" "Who was it?" "He broke his neck!" Mackesey took a huge breath, submerged, and began the first leg of his long swim to the *Nancy*, shotsure that an American master who'd lost four crewmen to British impressment would welcome a deserting Irish seaman.

"Two fine boys," Mackesey thought as he prepared to fall asleep, "They know what's right." He wondered if their paths would cross again, and hoped they would. He wanted them to know they'd succeeded, that their plan had worked perfectly, that he was alive and free. Mackesey felt like crying with relief.

A memory stirred. *Should auld acquaintance be forgot.* The night was perfect for music. He flexed the fingers of his right hand. He touched each of his fingertips to his thumb in turn, then tried them on his tongue. The feeling was almost all back. Just the tip of his little finger still had no feeling. In New York he'd get a guitar. A guitar and a ship. A guitar and a ship and Mary B. What more could a man want? Forget the ship! He'd happily be done with ships.

Mary B. Where was she now? She had to be freed from prison by now! Had she given up on him? Had she gone back to Fowey? Was there someone else in her life? Where should he begin to look, London or Fowey? He was probably safe going back to England, at least for a while. In Ireland they thought he was on his way to Botany Bay. On the *Sugar Cane* they thought he was drowned. He thought, in a way I don't exist. It's like I'm being born again to a new life.

~~~~~~~~~~

As matters turned out, Elector Kestle was telling the truth. Dolly *was* in London, working an assistant cook in a household on nearby Charlotte Street. A week after Kestle's visit to Boswell, the long-separated sisters were reunited in Mary B's room.

Boswell was rather taken with Dolly. Now a 20-year-old, slight, energetic, and gay, a pretty girl who affected a coquettish smile, Boswell found her fetching. The sisters clung to each other with such heartfelt happiness Boswell choked up.

But reunion with her sister was just one reason for Dolly's good spirits. The other was the news that their father stood to come into a great deal of money, although how was somewhat foggy. While Kestle favored the little assembly with his latest speculation concerning the family's sudden wealth, Dolly wouldn't let go of Mary B's hand. William Broad's fortune, Kestle pronounced, surely had to do with his maritime service. "Prizes, ladies and gentlemen, prizes! Prizes will make a mighty lord of a common man, er, begging Your Lordship's pardon!" Through her tears Dolly beamed at her intended and promised Boswell 1,000 pounds for his kindness to Mary B.

Mary B asked, "Have you heard from Little Bill? Anything?"

Dolly shook her head, "Only that he went to sea. A man told Mr. Kestle that he went to America and might be on a whaler or a sealer. But we don't know."

Mary B nodded sadly. She heard a repeated refrain from Cox's song.

Around, oh, around, oh
To see the world around

At her introspective silence the room quieted and it was several moments before Mary B recovered and noticed everyone looking at her with concern. She wiped her eyes and affected a smile as she began to recount an anecdote from Little Bill's boyhood. Dolly told another, and soon the sisters were telling family stories that had the whole room laughing. Then someone noticed Mary B was crying again. She suddenly threw herself into Dolly's arms. "Eight years!" she sobbed, "Eight years gone!" The room quieted again. Boswell discretely begged his leave and signaled Kestle they should leave together. He still felt mistrust for the palsied glazier.

The two strolled side by side, Kestle freely delivering opinions on his uncle's concoction for the treatment of kidney-stones, the treachery of the French, the price of malt, the virtues of tobacco, and several other coffeehouse topics. The only useful thing Boswell learned was the origin of Kestle's given name. It seems his mother and father disagreed on naming him George after the Hanoverian monarch, so they compromised on Elector.

That evening Boswell returned to Mary B's room, curious about how she was reacting to the news of her family's possible fortune. She snorted contempt. "My father! All my life he's spouted stories like that! If dear Dolly's head wasn't turned by that ridiculous man she's going to marry she wouldn't give the matter a minute's thought!"

Boswell seized Mary B's hand to smack it with a kiss. Puzzled, Mary B wondered if she'd passed some kind of test.

Boswell's visits continued, although for a time in mid-August he was distracted by publication of the second edition of his *Life*, which sold very well for several weeks and put him briefly at ease about money. Despite his many distractions, he continued to press Henry Dundas about the release of James Martin and the others. He called twice on Arthur Phillip, now ailing in lodgings near Covent Garden, on the verge of resigning his governorship of New South Wales. Whether it was ill humor arising from his poor health or genuine indifference, Arthur Phillip remained unmoved by Boswell's appeals. He said he vaguely remembered James Martin as a convict who might have worked on his house. His interest rose a little when Boswell told him Allie had served on the *Ramillies* about the same time as His Excellency. But Arthur Phillip soon sank into his shell again, and Boswell gave up. He left feeling Arthur Phillip wanted no more truck with convicts.

So passed the days and weeks.

Mary B saw Tench once more. He looked her up when he came to town to autograph his new book for subscribers. He read her a few passages, including a footnote about the escape. Noticing her well-thumbed copy of Mackesey's chapbook, he asked about her progress in learning to read and promised that if she wrote to him in her own hand by way of the Portsmouth Division he'd send her a copy of his book.

"Do you intend to stay in London?" he asked. Confusion crossed Mary B's face. She kneaded her handkerchief while her eyes searched the threadbare, stylized figure of a flower in the rug. Did she? The question seemed to unlock a trunkful of avoided issues. Why was she staying in London? What did she want here? What did she plan to do? Was she waiting for Mackesey? Mackesey, she had to acknowledge, had been absent for almost nine months! Even if alive he probably wasn't coming back to her!

Her gaze still fixed on the faded flower, she began slowly to shake her head. "No," she said so quietly that Tench barely heard, "No, I'll not be staying in London." Later Tench thought his offer to send a book to her was rather silly, and he wondered if he'd ever hear of her again.

~~~~~~~~~~

On an uncommonly warm and humid September afternoon, Boswell sat with Mary B drinking tea, both uncomfortable. Despite open windows the room was stuffy; glancing sunlight seemed to glare on every surface. Coatless, his neckpiece in his pocket, Boswell fanned himself with Mackesey's tattered chapbook. Mary B had undone the top two buttons of her dress and waved a handkerchief to stir the air near her neck. No, she insisted, it wasn't the disappointment of a phantom family fortune. No, it wasn't that her four friends still sat in prison. No, it wasn't that Boswell had failed her in some way. Oh, no! he'd been most wonderful and she'd be ever grateful!

"Well, what is it then?" Boswell pressed. He'd been trying to get her to talk about what bothered her, unsuccessfully. Her depression upset him. She avoided his eyes.

"Well, aren't you going to tell me?" he asked, aware of a trickle of sweat disappearing between her breasts. He fanned himself more vigorously. Inexplicably, for the first time since seeking her out in Newgate, the near presence of Mary B provoked in Boswell a penile twitch. He felt himself swell a little. He had a sudden insight. "Listen," he said, leaning closer, "Is it, ah, something else?"

Avoiding his eyes, Mary B shook her head. The silence grew as Boswell sat poised at the edge of his chair, surprised that his erection continued to develop. No wine, no busty serving girls, but something else! He was unsure whether to feel proud or embarrassed. His breathing became heavier. Perhaps Mary B sensed something. She looked up. Boswell wore an odd expression, as if on the verge of speaking, for his lips worked without words. She looked at him blankly, waiting. His eyes dropped briefly to her lap. Her own gaze dropped to his and she suddenly flushed deeply, "Oh, no, Mr. Boswell! It's not that!"

"Then what is it, woman!" he cried in exasperation, "I want you to tell me. You must tell me!"

"I can't!"

"I demand it!"

"I won't!"

"Damn you, woman! tell me!"

"Damn you, sir, go to Hell! It's Mackesey!"

Boswell sat stunned. "What?" he said at last, "That man? That...that...Irish republican!" Head bowed to hide her tears, Mary B nodded. "Mackesey?" Boswell was disbelieving. Mary B kept nodding. Boswell slumped, his erection a shrinking memory. He felt disappointed at her rejection and at the same time relieved their relationship would remain uncompromised. He was sure he could have her if he insisted, but he didn't want that. Still unclear why, he wanted Mary B to think well of him, as if his behavior were a test. So far he'd succeeded in keeping his motives concerning her altruistic, something he knew was too rare in his life when it came to women. *Memento mori.*

Mary B was surprised to hear Boswell laugh gently. She looked up to see him wipe his eyes and send her a smile of genuine friendship. Mary B wiped her own eyes and returned his smile. She dabbed at her nose and laughed her own little laugh. "Well, Mary B," Boswell said, "Wherever he is, he's a lucky man. Tell me about him again."

He was surprised when Mary B suddenly sprang forward to kneel and embrace his legs and bury her face in his lap and begin bawling. Boswell rubbed her moist, heaving back with a sad, avuncular satisfaction.

~~~~~~~~~~

In the setting sun, New York's East River was a rose-tinted jungle of weathered wood, canvas, and hemp. Above, bright white gulls rode a cool autumn breeze in perpetual patrol. A bucket of slops brought down a hundred, screeching and swooping in frenzied competition, a scene repeated a half dozen times an hour.

Oblivious to the scene, Mackesey sat on a box and fingered his little guitar, his third in just two years. His color was good. He felt healthy and whole. Satisfied that his fingers would do their work, he plucked the introductory notes of a song he'd carried for months. He strummed accompaniment, his soft baritone barely audible:

> *Oh, weep tonight, my love*
> *That I may come to you*
> *And kiss your tears away*
> *Though in your dreams*
>   *But if you chance to dream*
> *And I do not appear*
> *Then dream again my love*
> *And I'll be there*
>   *Oh, weep no more, my love*
> *I come, I come to you*
> *To hold you close again*
> *My Mary B*
> *My Mary B*

~~~~~~~~~~

In mid-October, Boswell came with his son to see Mary B off to Fowey. He found her subdued and weepy. She tried to pass off her mood by saying she was sad at the prospect of parting. While they waited for a carriage, Boswell asked if she'd to tell his son Jamie about her escape. With writing material from Madame Duval, Boswell wrote as she talked, covering two large sheets on both sides. "Here, Mr. Boswell," she said when he finished writing after almost an hour. She took his quill and awkwardly tried to form her initials at the end of the narrative. When her effort resulted in a chicken scratch that seemed to please her, Boswell smiled generously and kissed her hand. He folded and presented the pages to Jamie.

The carriage came at last. While Jamie carried Little Bill's battered box down the stairs, Mary B made her farewells to the household, and they were off. Boswell dropped Jamie off at Dilly's bookstore to await his return, then directed the carriage to Beale's Wharf in Southwerk, where Mary B would board the *Ann and Elizabeth*. There they put her box aboard and crossed the street to the Stinking Dog to sit in the kitchen; women were not permitted in the public room.

His abstemious period safely behind him, Boswell ordered a bowl of rum punch and settled back to wait while the ship finished loading. Nursing her cup of punch, Mary B started to cry again. Boswell drank steadily while Mary B said she missed her children and was sad to be returning home without them. "I left with nothing, and I return with nothing," she said, "Everything I did, my children, my friends, are all just memories. My life has come to nothing." Boswell drained his cup and dipped another, once again the pathos of her story moving him deeply. He squeezed her hand and, by way of distraction, invited the passing proprietor to join them. He explained who Mary B was, and the proprietor invited them into the public room and drank her health. After a time the master of the *Ann and Elizabeth* also joined them, and the three men drank her health several times.

When he was alone again with Mary B, somewhat besotted, feeling expansive, Boswell judged the moment right for his little surprise. He'd paid for Mary B's passage and for her new travelling outfit, and now in the Stinking Dog he laid a five-pound note and a guinea coin on the table. He wagged an instructive finger. The money represented the first installment of a ten-pound annual pension. "Many besides me are interested in your welfare," he told her.

It was a lie, of course. After the first month or so, contributions towards Mary B's keep and annuity dried up, and Boswell was footing the entire expense himself. Through her tears Mary B nodded.

Boswell's words washed over her. He'd send her the money twice a year in care of an elderly vicar who lived near Fowey, a Reverend Baron he met on his excursion to Cornwall the previous year when he visited his friend Temple. "The Reverend has promised he'll look in on you from time to time," Boswell said. He dipped himself another cup of punch and drank it off. "This money will make you independent," he counseled, "so if things don't work out with your family you'll always have something to fall back on." Mary B continued to cry and to hold his hand. She wanted to remind him that they'd never gone together to St. Paul's. And never would.

~~~~~~~~~~

How long do things matter? Dreams and passions are borne to the grave, houses decay and collapse, the bones of the just and the unjust look the same. But listen! Can you hear the distant clamor of the Paris mobs? Virulent argument in the young American Congress? Can you hear the sad, hopeless murmurs of the landless, the hungry, the oppressed? The disingenuous assurances of the privileged? No? But listen! Their inchoate voices rise from the yellowed pages of old books, from crazed paintings dulled by time, from dusty artifacts in museum storerooms.

This is fanciful? These are the modern times? But listen! Listen close! Amidst our agitated rush of getting and spending, of buying and selling, can you hear the murmuring poor, the hungry, the oppressed? The disingenuous assurances of the privileged? Their voices are everywhere, a Grecian chorus of the living and the dead.

Oh, the olden times! The past lies before us like a lovely landscape, its farther reaches blurred by time's relentless brush, simplifying, simplifying. What is it that wants one to bring a distant scene to better view, an antique age, an obscure woman, a Mary B? It is a strange alchemy. Was this the life she lived? Was this her? Was this really Mary B?

She rests in late afternoon sun on a wooden bench in front of a tiny cottage, her head fallen back against withered vines that climb the wall. She has fallen asleep. Crow's feet track the corners of her eyes, and gray sneaks out to tinge the temples of her rich, auburn hair. Her mouth is slightly open. In her lap a hand rests on Mackesey's tattered chapbook. She dreams of an orange halo around a rich, manly laugh, of oranges tumbling down a ladderway, bouncing and bounding, of laughing women. She holds oranges in each hand, big as melons, heavy with pulp and juice, citron-fresh, rare. She feeds pieces of orange to Charlotte and laughs at her baby's mouth so full that a tail of orange wriggles between her working lips. The juice runs down Charlotte's chin. Her little face wears a serious expression. The women laugh with girlish voices.

Mary B wakes with a start. The three young girls at the end of her path are giggling as they take turns vamping the walk of a grand lady, mock expressions of haughty solemnity on their smooth-skinned faces. They wear crude garlands of grass, their crowns. When they see Mary B they exchange looks, giggle again, and run forward. Mary B knows them. They're the daughters of the vicar in Fowey. The eldest approaches, curtsies, and proffers a thin parcel tied with cord.

"Good afternoon, Mistress," Dorothy says, "Father asked me to bring this. He said we should stay if you need help." She curtsies again. Mary B smiles and takes the parcel, a little befuddled by sleep.

"It was sent over by Reverend Baron at Lostwithiel," Meg volunteers, the middle child. "He says he's feeling indisposed."

"Oh, dear," Mary B says. "I hope it's nothing serious. But thank you for coming all the way up the hill. You must be very thirsty. Would you like something to drink?" The girls confer with their eyes and nod. Mary B likes their shy smiles.

In her cob house no bigger than the hut she and Will shared in Sydney, Mary B dips water in a pitcher, scrapes sugar from a cone, and stirs in lemon juice from a flask. She carries out the pitcher and a glazed mug. "We'll take turns," she smiles, and pours the lemonade. "Littlest first," she says as she hands the mug to Amy, perhaps seven.

"Thank you, Mistress," Amy says, dipping a curtsy. She takes a sip. "Oh, my, this is de...de...ah...good!" The older girls and Mary B laugh, and Amy joins with her own "Hee-hee." They share the cup around, and Mary B fills it again.

"Now," she says, "let's see what you brought." She picks at the knots until she can wind the loose cord around three fingers and lay the little garland on the bench. Her actions are leisurely, as if time has no value, but her heart pounds. She spreads the paper and withdraws a pamphlet. Tucked inside is a small, folded sheet with close writing. She recognizes her name. Heart racing, her eyes dart over the page. No sign of Mackesey. Just words. At the bottom she tries to puzzle out the name. She knew as soon as she saw the parcel that she hoped against hope. She keeps telling herself that Mackesey is dead, or will never return. But her hopes will not die.

She studies the letter for a full minute. Meaningless. She sighs. "Dorothy, I wonder, will you help? I don't know this kind of writing very well. Can you tell who it's from, who wrote?"

Dorothy takes the letter but it's Meg who looks over her shoulder and announces, "It's signed "James Martin.'"

"Meg!" Dorothy cries, "she asked me!"

Dorothy turns away to hoard the letter, then begins reading. "Exeter, December 18, 1793, Dear, Mary, B," she reads, hesitating badly. She labors over the next words, mouthing them several times: "I, hope, this, letter, finds, you, well, well." She stops and blows past a jutting underlip to cool her brow, where little beads of sweat have appeared. The paper trembles. "Here, Meg," she suddenly orders, "You do it! Father says you read best anyway."

"All right, Meg, you try," Mary B says. To Dorothy she says, "Reading is hard, isn't it? Here, let's all sit, and Meg will read."

Mary B fills the mug a third time and presents it to Dorothy. She draws Amy between her legs and rests her chin atop the little girl's head. She remembers her dream about the oranges and Charlotte and kisses Amy's hair.

Meg looks up with a serious mien. "This isn't written in a fair hand," she says, "But I can read it."

"Good!" Mary B says, "Tell us what it says!" Meg clears her throat and begins James Martin's letter again. With words crossed out, punctuation absent, the sheet spattered with ink-blotches, the letter is difficult. Meg reads slowly, sometimes pausing to puzzle over a word, sometimes repeating a sentence once she has the sense.

Dear Mary B,

I hope this letter finds you well. I got five pounds to write this book, which I wrote in Newgate, which I am pleased to send you. Nat remembers you and wishes you good health. Pottle remembers you and wishes you good health. He helped me. We were all put free on November 2 and went to see Mr. Boswell but he was gone. Allie remembers you and wishes you good health. He saw Mr. Boswell. I send you my book which I hope you will remember me. Mr. Jeremy Bentham gave me five guineas for my book. He was in France and showed me his new prison on a table which I crawled inside and looked out at the walls. Mr. Boswell gave me two pounds by Allie. He gave us all two pounds. God bless Mr. Boswell. I am sorry I never saw him again. Butcher joined the NSW corps and goes back to Sydney. I know nothing of Nat or Allie. I call my book memorandoms which is about our escape. I am happy with my wife and son now, a fine boy named James. I hope I am never a fool again. You are in my prayers which God give you health and peace which yours, etc, etc.

James Martin

Meg hands the letter to Mary B, who regards it for a long moment, then folds and carries it into the house. She comes out with three pennies and presents one to each of the girls. She doesn't speak, but strokes their hair in turn and smiles at some inner affirmation of what she sees in the young faces. "Thank you," she's finally able to say, "Thank you all very much."

The girls skip to the path that drops steeply to Fowey, excited by their new wealth. At the path-head they turn to wave. Mary B waves back and blows them a kiss. She sits on the bench and picks up the pamphlet. After a while she carries the pamphlet to the path-head and looks down the steep valley of the Fowey River, to the bay, to the Channel beyond. In the distance a small, westbound vessel is disappearing past the headland. Sensing a change in the light, Mary B turns to see a horizon-wide bank of blue-black clouds advancing from the north. The cloudbank looks cold and promises strong winds, but for now a winter sun sinks yellow-rose over the coastal hills and glows on the face of the eastern headland. Near the water tiny houses sit in shadow, with here and there a swirl of smoke promising warmth and security.

She thinks of her homecoming. Welcome home, indeed! Scarcely through the door before her father was after her about Boswell's money! Thank God he was at sea again now. She cares not if she ever sees him again, and hopes her mother will go live with the Puckeys. But she knows her mother lacks the will to leave their bad marriage. For her own part, Mary B would not lack resolve again. Although she waits for Mackesey, she's already decided that if he fails to return by spring she'll go to America. A fresh start. A new life.

Reflective, Mary B stands at the brow of the hill until a blast of cold wind pastes her dress and whips her shawl. The last sliver of sun disappears and dusk unfurls its gray cloak. She hurries back to the house, tosses a few sticks on the fire, and lights her candle. The yellow light gleams on Mackesey's chapbook and on the prayer book Boswell presented her on her first day of freedom. She told Boswell the prayer book would do no good, that she was unable to read and had doubts about.... But he wanted her to have it. "Just keep it," he said. "A keepsake." Since then she's sometimes looked at the words she knows are prayers, but the thought of prayers always brings sad memories of her children disappearing in pale bubbles. She holds the prayer book as a comfort sometimes, more as a talisman than anything, and thinks of Katie and Seedy and Liz and Hannah and all the others, but she gets more comfort from Mackesey's chapbook, especially when she hears his gentle, reassuring voice sounding out the letters and words.

She gently smoothes the paper cover of James's pamphlet and places it on a shelf with the others, then stands stroking its thin spine as she recalls shadowy memories of their stolen moments in Exeter and aboard the *Dunkirk*. Of her brief, giddy happiness. She'd been a foolish girl in a different world then, and now that world was gone. Never meant to be.

"No!" she suddenly says aloud, "I wasn't a fool! I was in love!" Maybe her love was foolish, just prison love, but she'd have no regrets. She'd got through it, got through it all, and James had been mostly a good part of getting through it. "I wish you well, James," she says to the pamphlet.

She caresses the prayer book. Kind Mr. Boswell! Why was he so good to her? What was it he wanted? Even if she never saw another farthing she'd be forever grateful to him. And what had she given in return? Sometimes when he looked at her with such woebegone sadness and loneliness she'd wanted to cradle his head to her bosom!

She sighs and carries her hairbrush and Mackesey's chapbook to her chair by the fire. As she absent-mindedly brushes her hair, she thinks, poor Mr. Boswell, so much success, so much wealth, such a fine family, and yet so unhappy. Did he think that helping me would bring him happiness? All I ever did was listen.

She remembers listening to Tench read from his second book. Looking up after a passage, he said it was really true, that he always thought of her with pity and astonishment, that their escape had been a heroic struggle for liberty.

But that was only a part of it, she thinks. What about the rest? I'm more than a convict! I'm a whole life!

Tench! He'd also read to her on the *Gorgon* from his first book and had pricked her anger, for he said on the voyage from

Rio to Cape Town that nothing happened except the drowning of a convict. "Do you remember who that was?" he asked her. "I should have written his name down, but I was sick and, well, maybe it doesn't matter. I think it was a man named Brown. Or was Brown the one I had flogged?" He shrugged and continued reading, oblivious to Mary B's reaction.

Nothing happened? My God, Charlotte was born! My first child! Nothing? That little girl who was so brave through all those hard times! Those wet, cold, miserable days! And you say nothing happened?

"I beg your pardon, Captain!" she says aloud to her empty dwelling. "Oh, no, excuse me, Major! But you're wrong, wrong, wrong! That was your story, not mine!"

She sets her hairbrush aside. And what of Quint's book? Did Governor Wanjon still have it? It seemed to her that by possessing Quint's book he possessed the very lives and souls of Quint and Seedy and Will. And Seedy would never know what became of Quint's book.

Mary B puts her things away. The fire is dying and the wind moans in the chimney. She stirs the ashes to find embers, places a few more sticks to catch, and toasts some bread. She eats it with butter from a crock while she stares at the dying fire. A sad wind in the chimney seems to be trying to find words. She banks the fire, takes her shawl from a peg, and steps outside. It's become nearly dark, and turned cold. The sky is clear again, however. She watches the cold wind swirl her chimney-smoke. When she returns, shivering with chill, she bolts the door and washes, smoothes a scented oil on her face, places the candle by her bed, quickly undresses to her shift, snuggles beneath the covers, and blows out the candle. Her eyes adjust and find the faint glow of the coals, the dim rosy light reflected on the hearthstones, on her chair. She likes the way the light in her house changes at night. Later a full moon will rise and bathe the room in silver through the tiny window.

What is her story? she wonders. Tench shared so much of their experience, yet missed the mark. So close for so long, and yet so far apart. He never said what his dreams were when he dreamed about her. If he'd written about his dreams he would have told a different story! Nothing happened? She'd given birth! She'd brought a new life into the world! But Tench would never write about his dreams. He'd be too afraid.

Did Boswell dream about her? Did James? Katie? Seedy? Rami? Was she in their dreams as they were in hers?

What is a dream? Is it a story about what really happens, or a story made up? Did she make up her dreams, or did something bring them to her sleep?

Dreams are so real! When she dreams about Charlotte and Emmanuel they're really there. Even though Charlotte wasn't even born, when Mary B feeds her an orange in Rio she's really there! Her children smile and laugh and she can feel their arms around her neck. But then they disappear. Where do they go?

Was she in Mackesey's dreams? Was Mackesey alive? Oh, Mackesey, please be alive! Please come back to me!

She thinks, if I die tonight everything will die with me - my stories about things I've done, things I've seen, things about me I never told anyone - they'll all be gone. People who knew me will forget, and I'll start to disappear. And someday I'll be gone forever.

Mary B roused herself with a realization. But I can live in dreams! I can come into dreams of people I know and even people I never knew, just as they come into mine. I can tell my stories in dreams. Aye, even in dreams of people I never knew.

~~~~~~~~~

Afternoon is dying on the shortest day of the year. A solitary figure trudges resolutely on the road from Bodmin, his back hunched against a cold, blustery wind from the north. The wind shakes the sere grasses and sends swirls ahead of road-dust from his footsteps, playful harbingers of his arrival. At the old bridge at the edge of Lostwithiel he stops to examine the signs. Fowey - 6 miles. He blows on his hands and flexes his fingers while he contemplates the sign. From Bristol to Fowey in four days. He bends each leg in turn to ease the knee, and adjusts the bundle on his back. If he hurries he can beat the darkness, but if not, a fortuitous full moon will rise.

He glances behind as he touches the hard shape of the small gold brooch hidden in the waist of his trousers, a precaution against robbery on this King's Highway. He remembers how his heart stopped when he first glimpsed the woman's profile on the brooch. The resemblance to Mary B was uncanny, and his thoughts tumbled as he tried to imagine how a brooch with her likeness came to lie in the glass-front case of a New York pawnshop. Only after moments of confusion did he realize the likeness could only be coincidence.

Still, his discovery of the brooch made him realize their connection was special and no matter where his search led he'd try to find her, for his heart knew that while destiny brought them together, and chance intervened to test his resolve, it was love, that eternal salvation, which would someday unite them again.

He begins walking again, faster. Bouncing on his back in rhythm with his steps, his little guitar bumps a monotonous, twangy tune.

The End

Afterword

Over the distance of more than two hundred years, much is lost or simply conjecture about the later fortunes of those in this story. But we know Boswell died unexpectedly in 1795 from a penile infection. He was 55, his *Life of Johnson* an enduring legacy, his kindness to Mary B a little-known footnote to his life. Although his journal tells us he wrote down Mary B's recollections of her escape on the day he took her to the *Ann and Elizabeth*, Boswell's notes of those recollections remain undiscovered.

We also know Arthur Phillip recovered both his health and his spirits, married a much younger second wife, and after many more years of service retired as an Admiral of the Blue. He died in 1814, some believe by suicide. That was the year the British seized Cape Town from the Dutch.

John Hunter, captain of the ill-fated *Sirius*, succeeded Arthur Phillip as Governor of New South Wales in an administration fraught with difficulties. Meanwhile, weary of waiting for government help, the Reverend Richard Johnson built Sydney's first church at his own expense.

John White took up with a convict woman and fathered a son. He brought the boy back to England, and resigned his position as a naval surgeon in 1795. Years later his son returned to Australia to find his mother.

Just before he left Sydney, Major Ross fought a duel, but not with Arthur Phillip or Tench. Ross subsequently served a stint under William Bligh on the *Director*, but disappears from history about 1808, still a major in the Marines.

Judge Advocate David Collins remained in Sydney until 1796, went back to England, then returned in 1802 to establish a new colony at what is now Hobart. There he died unexpectedly in 1810.

On their return to England most of the battalion officers received quick promotions, including plump Lieutenant Creswell, who died in Portsmouth in 1804, only in his forties. Even the Navy career of Captain Edward Edwards prospered.

William Bligh seemed to attract mutinies. He was seized in the Nore mutiny in 1797 when thousands of British seamen rose up to protest their harsh treatment. In 1808, when he too served a stint as governor of the Australia penal colony, he was overthrown by a mutiny of the New South Wales Corps. That was the year a ship accidentally discovered Pitcairn Island and the last two surviving mutineers from the *Bounty*, aged patriarchs of a small, devout community peopled by descendants of the original mutineers and their Tahitian wives. William Bligh died in 1817, a Vice-Admiral of the Blue, remembered best for his troubles on the *Bounty*.

Butcher returned to Sydney with the New South Wales Corps, completed his enlistment, and received a grant of 25 acres. We don't know if he brought his land to perfection. Katie Prior married John Arscott and also took to a life of farming, but of Allie and Nat Lilley, nothing more is known, nor of Dolly, Cox, Tartop, Nat Mitchell, John Coffin, Seedy Haydon, Liz Bason, Hannah Jackson, and many others who appear here.

The manuscript of James Martin's *Memorandoms* was discovered among Jeremy Bentham's papers almost a century and a half after the events of this story, but nothing more is known of James. The fate of Ned Thornapple is conjecture, as are the fortunes of Hester Thistlethwaite, Nell Thornapple, and Lame Johnny, whose last name remains conjecture. Thinking to get rich quick by selling his land allotment, Porter signed up as a settler, only to learn he had to work his land for five years before he could get title. After a year he gave up farming, enlisted in the New South Wales Corps, was sent to Norfolk Island, and disappears. There is intriguing evidence that Stork emigrated to America, for the passenger list of the American brig *Independence*, docking at Boston in 1795, lists a Peter Pottle of London. Of Mackesey's life nothing is known. Perhaps he too sought refuge in America.

Watkin Tench married the unsullied Anna Maria Savage and wrote a third book, this one about his experience as a prisoner in the war with France. He remained childless, but adopted the orphaned children of his wife's brother. He reached the pinnacle of his career as commandant of the Plymouth Division of the Royal Marines and retired a Lieutenant General, dying in Devonport in 1833. That was not long after the death of John White, both of them ripe old men. Mary Wollstonecraft also married, but died in 1797 of complications arising from childbirth.

According to Bligh's account of his second voyage to get breadfruit plants in the *Providence*, he set a clerk to work copying Will Bryant's *Remarks* in Kupang, but the work was never finished and has disappeared. In recent years a search for Bryant's original manuscript revealed that Quint's book and most of the Dutch archives were probably used by the British to make paper cartridges when they were besieged in Kupang in the early 19th century. Thus, like so many people who appear on these pages, Quint and Will Bryant's only tangible legacies are not their books, but their graves, also long lost.

Boswell's Samuel Johnson once observed that a writer will turn over half a library to make a book. Granted, this story draws on material from official records, journals, letters, and narratives, and granted, the chronology of historical events is as correct as might be surmised, and granted, the ships and historical figures came and went as described, and all the convicts and officers and officials and many others depicted here actually lived and did what they did as described - nonetheless, this book is not a history, but a story, and the characters are purely a work of the imagination. Any similarities between what we might regard as historical truth and the people and events of this story are happy coincidence.

Mary B probably left Fowey in 1795. That was also the year of a pleasing discovery in New South Wales. The cattle that disappeared in 1788 on the night of the first celebration in Sydney of the Prince of Wales' birthday were discovered peacefully grazing some fifty miles away, their numbers increased tenfold. "*Omnia tempus revelat*," one might say. In any event, the last record of Mary B uncovered to date concerns Boswell's pension payment to her in November, 1794. On a receipt for the payment she made her mark, and disappears from history, where or with whom, we know not.

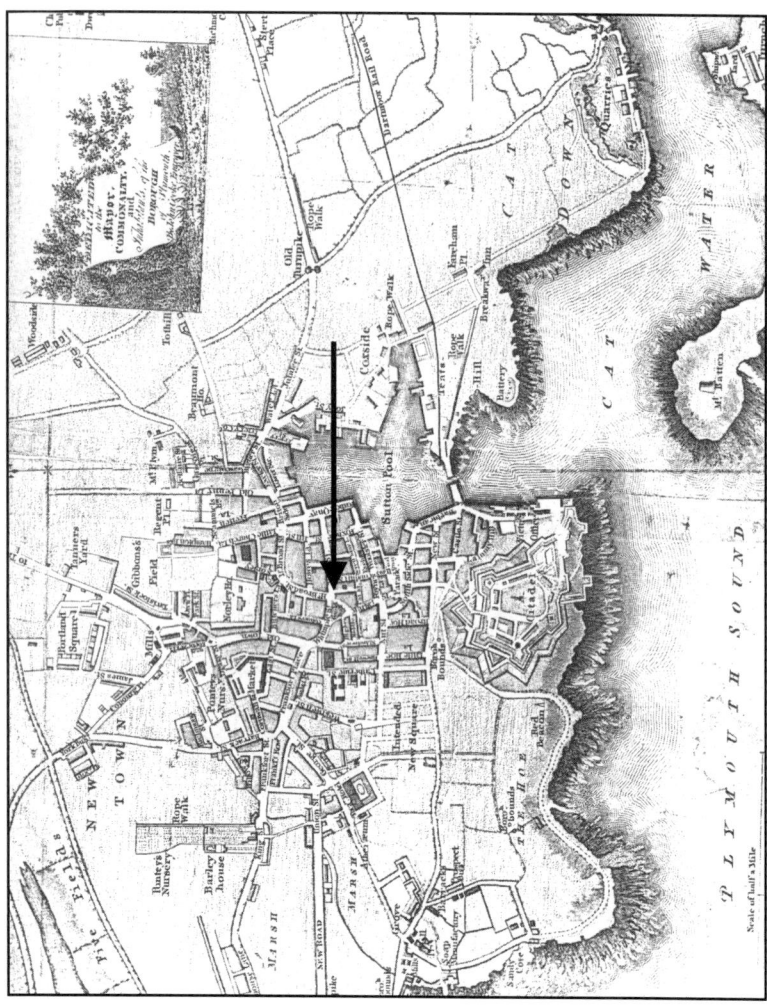

A view of Plymouth about the time of Mary B's birth in 1765. The arrow marks the location of the Guildhall, which contained cellar cells for felons and debtors rooms above. Dock (today's Devonport and not shown) was situated to the left, as were the Marine barracks and a large military hospital and the hulk *Dunkirk*, which was anchored in the Hamoaze, a confluence that enters Plymouth Sound. Mary B was imprisoned in the Guild Hall for several months before standing trial in Exeter. Ropewalks near the city produced the endless lines and cables needed in a world of sailing ships.

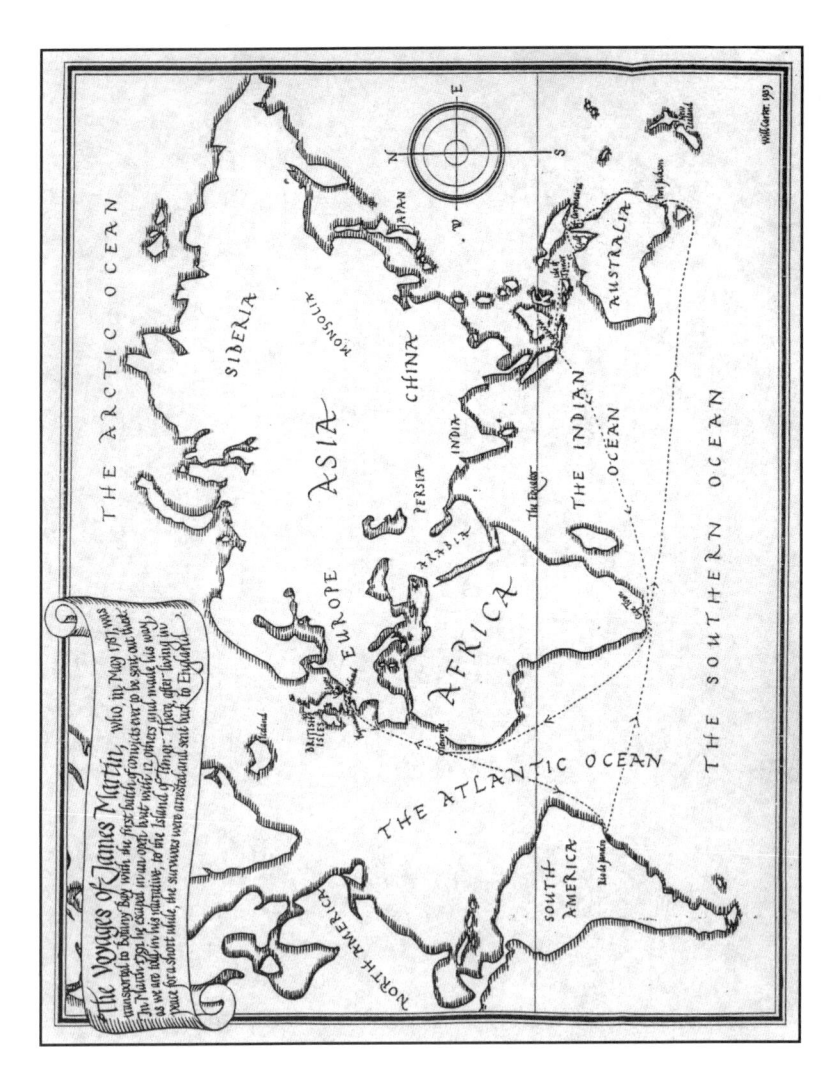

Published in 1937 as a monograph by Rampant Lions Press of Cambridge, James Martin's *Memorandoms* includes this map showing the ocean voyages of James and Mary B. *Memorandoms* remains the most complete account extant of the convicts' adventuresome flight to freedom. *Memorandoms* also appears in *True Patriots All* by Geoffrey Engleton (Angus and Robertson, 1952).

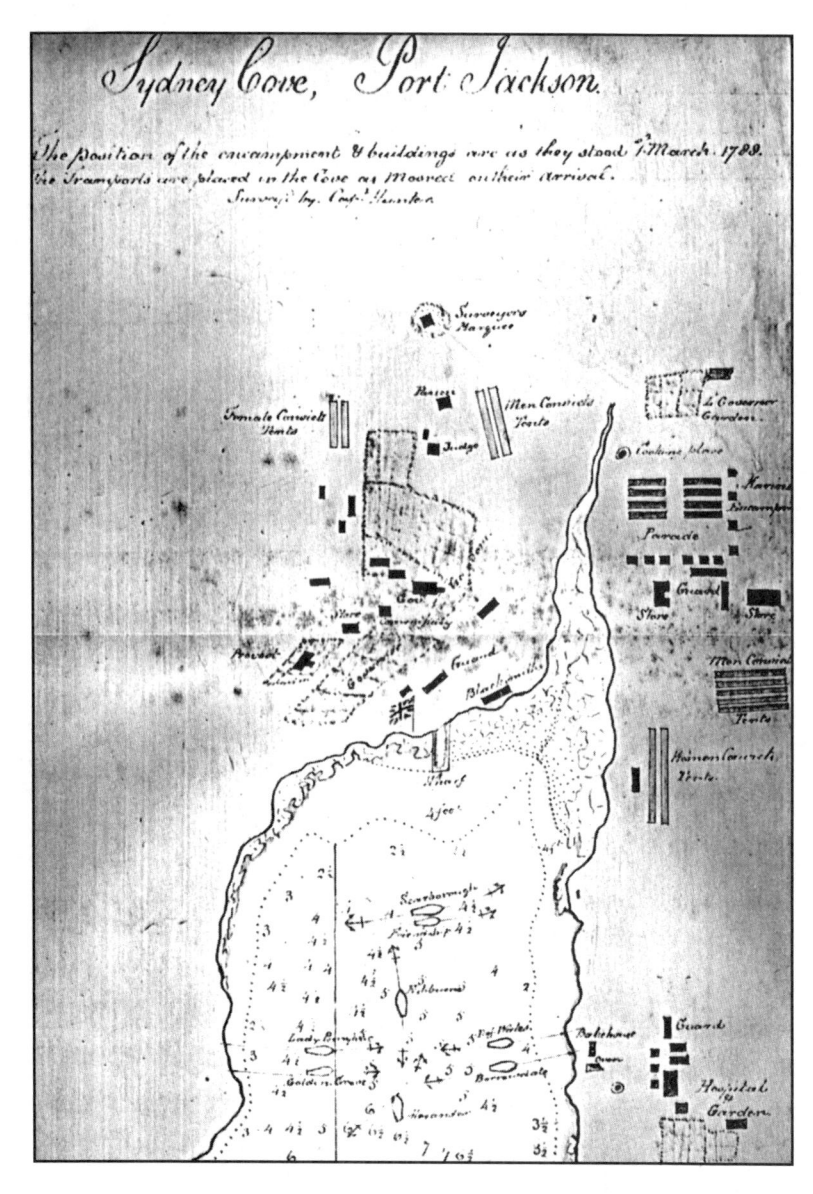

Captain John Hunter of the *Sirius* was one of a half dozen in the First Fleet who wrote about his experiences. This representation of "the encampment & buildings" of the settlement in 1789 appeared in Hunter's *Historical Journal of Transactions at Port Jackson and Norfolk Island*, 1793. Note his soundings of the cove in fathoms.

The first page of the Reverend Richard Johnson's marriage register for the convict colony shows the entry for Mary B and Will Bryant at the bottom. Illiterate, many marriage celebrants "signed" their names by making X's, including Mary B. Will Bryant's signature is his own. The complete entry for Mary B reads: *The solemnization of matrimony between William Briant and Mary Brand -- Married this 10th day of Feb'y 1788 -- by Mr. Rich. Johnson, Chaplain This marriage was solemnized between us {William Bryant/Mary Brand Witness our hands in the presence of -- A Baird/Wm Freeman, Sam'l Barnes.* Misspelled names (*Briant* for Bryant) and *Brand* for Broad) were commonplace. The number 23 near Mary B's name represents her age at the time of her marriage and was probably written on this record later.

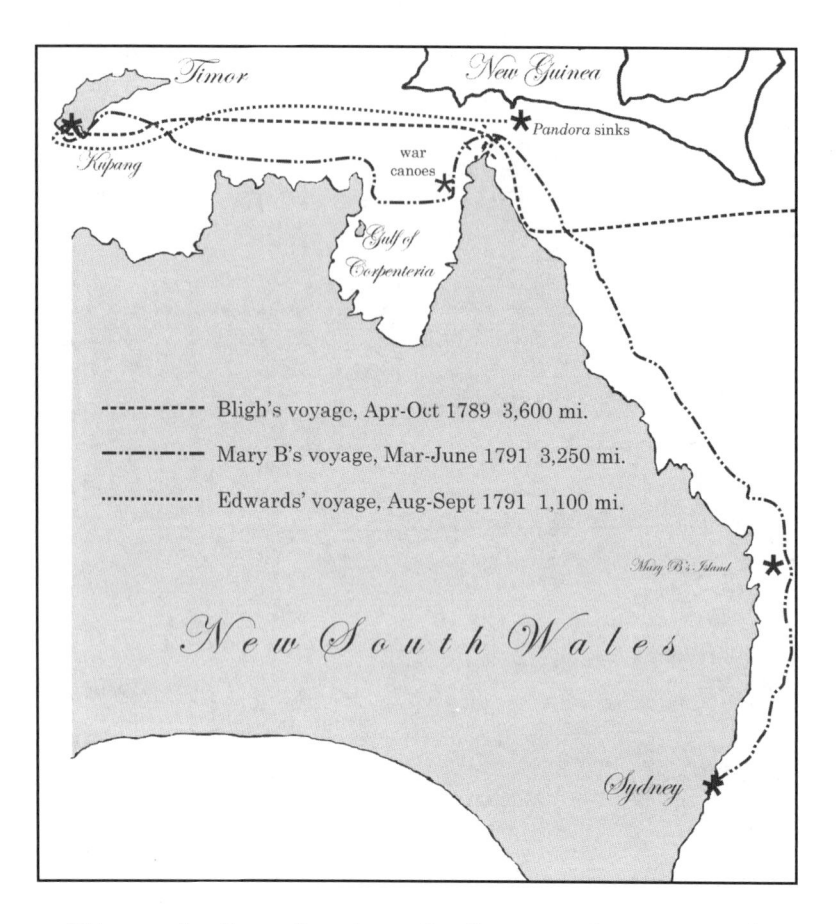

This map by the author shows the three open boat voyages that made Kupang their objective. Until William Bligh reached Kupang after being deposed by mutineers from the *Bounty*, such long ocean voyages in undecked boats were thought impossible. News of Bligh's success arriving at the penal colony with the Second Fleet probably prompted Will Bryant to plan his escape to Kupang by open boat.

The locations of Mary B's Island and the confrontation with the war canoes are deduced from the sequence of events recorded in James Martin's account of their voyage. Discovered among Jeremy Bentham's papers almost a century and a half after the events described, his account was perhaps not published in Bentham's time, but the penologist apparently had a professional transcribe Martin's work into a coherent whole in preparation for publication. The discovered papers included the copyist's transcription and the convict's original narrative written on 23 small sheets in various styles of handwriting. At the top of the first sheet of Martin's account was written, "Memorandoms by James Martin."

A. *The Keepers House*
B. *Lodges for the Turnkeys*
C. *Tap Rooms*
D. *The Arcade under the Chapel*
E. *Closets*
F. *Stair Cases*
G. *Cells for the Refractory*
H. *Passage to the Condemned Cells*
I. *Passage to the Sessions House*
K. *Wards*
L. *Bed-Rooms for Turnkeys*
M. *Cellar-Stairs*
N. *Passages, a few on the Cellar Floor.*

Plan of Newgate Prison, where Mary B, *et al* were jailed after their return to England. From her quarters in the women's quadrangle Mary B would have had a good view of nearby St. Paul's Cathedral, the most prominent structure in London. Considerable information about England's jails and hulks in Mary B's time are available in a survey of the kingdom's penal institutions made by John Howard, an advocate for penal reform who published *The State of Prisons in England and Wales* in 1784. Howard continued his work, revisiting the Hamoaze hulks in 1787 and again in 1788.

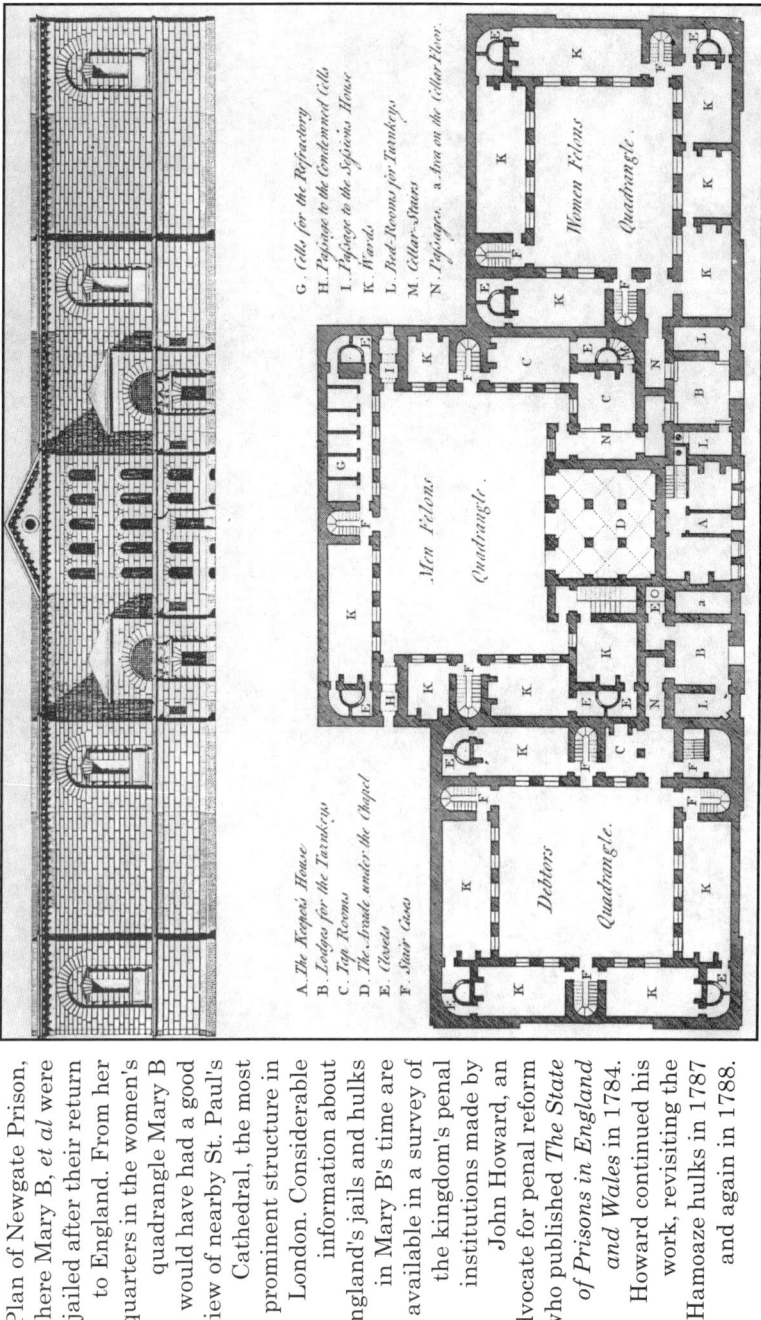

489

Pictured is James Boswell's home in Scotland, Auchinleck Castle (pronounced *aff-leck*). As eldest son, in 1782 Boswell inherited the manor, the lands, and the family title, Laird of Auchinleck. The estate later passed to descendants named Malahide, whose Malahide Castle in Ireland proved a trove for Boswell papers when they were brought to light in the 1920's. Auchinleck Castle is now owned by The Landmark Trust and is available for vacation rental.

The last known record showing Mary B's mark (the big + amidst the words, "The mark of Mary Broad") is this receipt for Boswell's final annuity payment of £5 before his sudden death. Dated Lostwithiel, November 1, 1794, the receipt reads: "I return grateful thanks to Mr. Boswell for five pounds paid this day by the Rev. Jn— Baron to me." Boswell had made the acquaintance of the Reverend Jonathan Baron when visiting an old school friend, William Temple, while Mary B was in Newgate. Lostwithiel is near Fowey. The image is courtesy of the Beinecke Rare Book and Manuscript Library, Yale University.

Chronology of Events

May 1, **1765**	Mary B baptized in Fowey, Cornwall
Feb, **1782**	English Parliament ends war with United States
Mar 20, **1784**	Will Bryant sentenced to 7 years, Launceston assizes

1785

July 27	Mary B, Katie, and Seedy jailed in Plymouth
September	James Boswell publishes *Journal of a Tour...*

1786

March 20	Mary B, *et al* sentenced to 7 years, Exeter assizes
May 20	Mary B *et al* transferred to the hulk *Dunkirk*
December	Mary B becomes pregnant

1787

March 12	*Charlotte* sails from Plymouth to join First Fleet
May 12	First Fleet sails from Portsmouth
June 10	First Fleet sails from Tenerife, Canary Islands
September 4	First Fleet sails from Rio de Janeiro
September 8	Mary B bears a baby girl, later baptized Charlotte
November 13	First Fleet sails from Cape Town
November 14	Katie bears a baby boy, baptized John Mathew
December 12	*HMS Bounty* sails for breadfruit

1788

Jan 18	First Fleet anchors in Botany Bay
Jan 26	First Fleet anchors in Port Jackson (Sydney)
Feb 10	Mary B and Will Bryant married
Feb 27	Thomas Barrett hanged for theft

1789

February 14	Convicted of selling fish, Will Bryant is flogged
March	• Tench's *Narrative of the Expedition...* published
	• Six Marines hanged for theft
April 28	Mutiny deposes Bligh from the *Bounty*
July 14	Storming of Paris Bastille
October 16	Bligh lands at Kupang with men from the *Bounty*

December 24	*HMS Guardian* strikes iceberg
	1790
March 14	Bligh reaches England with men from the *Bounty*
March 28	Mary B bears a baby boy, baptized Emmanuel
April 14	*HMS Sirius* wrecks off Norfolk Island
June 20	*Lady Juliana*, van of Second Fleet, relieves famine
November 7	*HMS Pandora* sails to find *Bounty* mutineers
December 17	*Waaksamheyd* arrives with cargo of rice
	1791
March 5	*HMS Supply* sails for Norfolk Island
March 28	• *Waaksamheyd* sails with crew of *Sirius*
	• Will Bryant escapes with Mary B, children, and seven convicts
May 17	Boswell publishes *Life of Samuel Johnson*
June 5	Escapees reach Kupang, say they are shipwrecks
August 28	*HMS Pandora* wrecks on reefs off Cape York
Sept 15	*HMS Pandora* survivors reach Kupang
Sept 21	*HMS Gorgon* arrives at Sydney with Third Fleet
October 6	*Rembang* sails with Bryant *et al*, *Bounty* mutineers
December 1	Emmanuel Bryant dies in Djakarta
December 18	Tench and Marines leave Sydney on *Gorgon*
December 22	Will Bryant dies in Djakarta
	1792
Mar 23	Mary B *et al* transferred to *Gorgon* in Cape Town
April 6	*Gorgon* sails with Mary B *et al*, *Bounty* men
May 6	Mary B's Charlotte dies off coast of West Africa
July 7	Mary B *et al* imprisoned at Newgate prison, London
August 10	James Boswell sees Mary B *et al*, works for release
December 11	Arthur Phillip leaves Sydney on *HMS Atlantic*
	1793
May 2	Mary B pardoned, released from Newgate
August 18	Boswell brings Kestle to see Mary B
August 25	Mary B, sister Dolly, Boswell, and Kestle meet
October 12	Boswell puts Mary B on *Ann & Elizabeth* for Fowey
November 2	James Martin *et al* pardoned, released
Nov 1, **1794**	Mary B receives last pension payment from Boswell

The Women of the Charlotte*
(those embarked from the Dunkirk)

Ann Carey, Taunton, twice convicted of stealing linen, sentence unknown, died October 1788.

Ann Combes, Taunton, stealing petticoats, sentence unknown, married John Bryant, a convict, February 1788.

Ann Lynch, Bristol, receiving stolen goods, 14 years, no references extant.

Ann Smith, Southhampton, married, burglary of a bank note and watch, sentence unknown, married Edward Elliot, a convict, September 1791.

Catherine (Katie) Prior, Exeter, highway robbery, 7 years, married John Arscott, a convict, December 1792.

Elizabeth (Liz) Bason, Bristol, married, stealing cloth, 7 years, married James Hatherley, a ship's carpenter, November 1792.

Elizabeth (Sick Liz) Cole, Exeter, housebreaking, sentence unknown, married Joseph Marshall, a convict, February 1788.

Fanny Anderson, Southhampton, dealer in stolen goods, 7 years, married Simon Burn, a convict, February 1788.

Hannah Jackson, Bristol, stealing three yards of cloth, sentence unknown, no references extant.

Hannah Smith, Winchester, stealing clothes, sentence unknown, married Edward Pugh, a convict, February 1788.

Jane Fitzgerald, Bristol, crime unknown, sentence unknown, married William Mitchell, a Marine, October 1788.

Jane Meech, Exeter, married, stealing four iron chains, sentence unknown, no references extant.

Jane Poole, Somerset, burglary of a silver watch, sentence unknown, no references extant.

Margaret Jones, New Sarum, receiving stolen goods, sentence unknown, no references extant.

---

* From John Cobley, *The Crimes of the First Fleet Convicts*, 1970

Margaret Stewart, Exeter, shoplifting, sentence unknown, no references extant.

Mary (Mary B) Broad, Exeter, highway robbery, 7 years, married William Bryant, a convict, February 1788, escaped in 1791 but was recaptured and returned to England.

Mary Cleaver, Bristol, burglary, sentence unknown, married John Banghan, a convict, February 1788.

Mary (Cedelia, Seedy) Haydon, Exeter, highway robbery, 7 years, no references extant.

Mary Phillips, Taunton, stealing two aprons, sentence unknown, married John Pye, a convict, December 1791.

Mary Wickham, New Sarum, receiving stolen goods, 14 years, married John Silverthorn, a convict, February 1788.

Boswell's Farewell to Mary B

In addition to being a great biographer, James Boswell was also one of England's great 18th century diarists. The following entry records his last hours with Mary B.[*]

Saturday, October 12, 1793

...I had fixed that Mary Broad should sail for Fowey in the *Ann and Elizabeth*, Job Moyse, Master, and it was necessary she should be on board this night, as the vessel was to be afloat early next morning. Having all along taken a very attentive charge of her, I had engaged to see her on board, and in order to do it, I this day refused invitations to dinner....

I went to her in the forenoon and wrote two sheets of paper of her curious account of the escape from Botany Bay. I dined at home, and then went in a hackney-coach to her room in Little Titchfield Street and took her and her box. My son James accompanied me and was to wait at Mr. Dilly's till I returned from Beal's Wharf, Southwark, where she was to embark.

I sat with her almost two hours, first in the kitchen and then in the bar of the public house at the wharf, and had a bowl of punch, the landlord and the captain of the vessel having taken a glass with us at last.

She said her spirits were low; she was sorry to leave me; she was sure her relations would not treat her well. I consoled her by observing that it was her duty to go and see her aged father and other relations; and it *might* be her interest, in case it should be true that money to a considerable extent had been left to her father; that she might make her mind easy, for I assured her of ten pounds yearly as long as she behaved well, being resolved to make it up to her myself in so far as subscriptions should fail; and that being therefore independent, she might quit her relations whenever she pleased.

Unluckily, she could not write. I made her leave me a signature, "M.B.", similar to one which she carried with her, and this was to be a test of the authenticity of her letters to me which she was to employ other hands to write. I saw her fairly into the cabin and bid adieu to her with sincere good will.

---

[*] *Boswell: The Great Biographer, 1789-1795*, ed. Marlies K. Danziger, McGraw-Hill Publishing Company

James had tired at Dilly's, waiting so long, and was gone home. I followed him.

I paid her passage and entertainment on the voyage, and gave her an allowance till 1 November and £5 as the first half year's allowance per advance, the days of payment to be 1 November and 1 May.

Notes

p. 9 "...feloniously assaulting..." Quoted from official records by Geoffrey Ruxton, *The Strange Case of Mary Bryant*, 1938, 283

p. 16 "Long is the way..." *Paradise Lost*, Book 1, line 432

p. 19 "One Friday morn..." "The Mermaid," English sea-song

p. 29 "By arriving..." *Historical Records of New South Wales (HRNSW)*, Vol I, 50-51

p. 29 "*Omnia tempus revelat*" Time reveals all.

p. 32 "When lovely woman..." "Song," Oliver Goldsmith, 1766

p. 56 Adapted from "How Can I Keep My Maidenhead," 18th century ballad

p. 68 "*De la mano*...." From the hand of my love. "*De la main*...." From the hand of my friend.

p. 101 "I hope to sail..." *HRNSW*, I, 113

p. 102 *Ah, Vossa Excelencia.... Vice-King*: Ah, Your Excellency, my country is always made richer by your presence. *Arthur Phillip*: And I, Your Excellency, am always enriched by coming here.

p. 112 "At breakfast...." Entry for September 19, in James Boswell, *Journal of a Tour to the Hebrides*, 1785

p. 138 "*Pardonnez la liberté*...." *Tench*: Pardon the liberty I take, Madame Labonne, but my name is Captain Tench, and I came from England to make your acquaintance. *Labonne*: Good day and welcome. Sit, sit here. I think you speak French better than I, no?...Oh, Captain, you are very charming, very charitable, a real gentleman!...Ah, sir, I daydream, pardon me...That's nice, children. Grandchildren are the crown of old men, no?...That's nice. Be good to them, because the glory of children are their fathers [*Proverbs*, 17,6] *Tench*: And their mothers? *Labonne*: Their mothers! Ah, a foolish man despises his mother, and women! [*Proverbs*, 15,20] Ah, good, once again I stand, thank God. Excuse me, Captain, but I'm very tired. Goodbye, and God go with you and your wife.

p. 143 "Sirius, at the Cape...." *HRNSW*, I, 118

p. 156 "Oh, we've come...." Mary B's song is to the tune of "Malbrouck"

p. 182 "To our trusty...." *HRNSW*, I, 62

p. 204 "Dearly beloved...." From "The Form of Solemnization of Matrimony," *The Book of Common Prayer*

p. 233 "What doth it profit...." *James* 2, 14-16

p. 238 "For it must be...." Watkin Tench, *Sydney's First Four Years*, 1961, a reprint of his *Narrative of the Expedition to Botany Bay* and his *Complete Account of the Settlement at Port Jackson*, 32

p. 240 "Goddamn, you scoundrels...." Sir John Barrow, *The Mutiny of the Bounty*, 1978, Gavin Kennedy, ed, 64

p. 253 "It was by far...." William Bradley, *A Voyage to New South Wales*, a facsimile reproduction, 1969, 183

p. 253 *Memento mori* Remember death

p. 258 "Pease porridge...." English nursery rhyme

p. 264 "How sweet...." Verse appears in an anonymous letter dated, "Sydney Cove"

p. 272 "He strack the tap-mast...." Anonymous ballad, "The Demon Lover" *Six Centuries of Great Poetry*, 62

p. 273 "Come live with me...." "The Passionate Shepherd to His Love," Christopher Marlowe, 1600

p. 279 "Sir, prepare yourself...." Tench, 164

p. 283 "Dismal accounts...." *HRNSW*, I, 330. "We are now at less...." *HRNSW*, II, 704. "...a country and place...." *HRNSW*, I, 333

p. 290 "July 2, 1790 Buried this day...." Names appear in John Cobley, *Sydney Cove, 1789-1790*, 238-244 "When the ships...." *HRNSW*, I, 362

p. 312 "...seven years will...." *Genesis*, 29, 20

p. 315 "Sydney, New South Wales...." *HRNSW*, I, 472

p. 331 "Drink to me only...." Poem by Ben Jonson set to music sometime in the 18th century

p. 353 "In Plymouth town...." English sea-song, also known as "A-roving"

p. 356 "*es wrak*" is touched, or mad

p. 378 "Oh, the roast beef...." Henry Fielding, *Grub Street Opera*, act III, sc. 2

p. 385 "My wife is my plague...." Plutarch, *Morals, on the Tranquillity of the Mind*

p. 395 "My father sailed...." 18th century sea-song

p. 396 *J'espère seulement....*" I just hope they're happy. "*Oui, oui....*" Yes, yes, because he's a good boy, I know it. "*Ecoutez l'instruction....*" Listen to the instruction of a father [*Proverbs*, 4,1] Get wisdom. Yes, and to the best of your power get understanding [*Proverbs*, 4,7]

p. 400 "What though the field...." *Paradise Lost*, Book 1, lines 105-108. "Vain wisdom all...." *Paradise Lost*, Book 1, line 565

p. 409 "In sure and certain hope...." From, "The Order for the Burial of a Child," *The Book of Common Prayer*

p. 413 "On Saturday, James Martin...." Reprint of article from the *London Chronicle* in G. C. Ingleton, *True Patriot's All*, 1952, 11

p. 418 "...nothing harsh shall be done...." Quoted by C. H. Currey, in *The Transportation, Escape, and Pardoning of Mary Bryant*, 1963, 37 "*...Privilegium non...*" Edmund Burke, *Reflections on the Revolution in France*, 1790

p. 434 "I want to see...." William Godwin, *Memoirs of Mary Wollstonecraft*, 1969, 223

p. 435 "My Mary's asleep...." "Sweet Afton," *The Poetical Works of Burns*, 1974

p. 436 "For auld lang syne...." Auld Lang Syne," *Ibid*.

p. 438 "May it please your Honor...." *HRNSW*, II, 4

p. 449 "Judge not, that...." *Matthew*, 7, 1-2

p. 471 "Oh, weep tonight...." Adapted from 18th century ballad

499

Did you enjoy *The Odyssey of Mary B*?

Then you will also enjoy John Durand's gripping historical novel of New Mexico, *The Taos Massacres*.

What others say about *The Taos Massacres*...

- "...rich in historical and physical detail and complex characterizations...I'm impressed!" (Dave Wood, *Wood on Books*)
- "...the tension of a good mystery...a good historical presentation...the research...is detailed and exemplary.... (*Southwest BookViews*, Summer, 2004)
- "...I find myself saying, just one more page... just one more page and suddenly it's 1:00 am. Heck, I've got a business to run! Did you have to make the novel so intriguing? So fun to read?" (David Wright, ColoradoHistory.com)
- "...a masterful combination of historical fact and irresistible suspense...." (Phil Ginsburg, author, NY Times bestseller, *Poisoned Blood*)

On a cold January night in 1847, Col. Sterling Price was awakened at his Santa Fe headquarters with news from an exhausted rider. A mob in Taos had killed the new American governor and was on the rampage. Price marched north. Next day his make-do army met a rebel force on the move that outnumbered him 5 to 1.

Here is a complete telling of that brief time after America's occupation of New Mexico, when neighbor turned against neighbor, patriot fought patriot, and the passion for revenge overwhelmed reason and restraint.

Faithful to the documented history. Peopled with real-life characters caught up in this bloody conflict. Includes updated U.S. military battle maps and a first-ever chronology of this history-changing event.

281 pages 5½ x 8½ (paperback only)

Available on-line from Amazon, Barnes & Noble,

ColoradoHistory.com and Puzzleboxpress.com

Order this great read today!

Do you want someone else to enjoy
The Odyssey of Mary B and *The Taos Massacres?*
Order copies signed by the author

Yes, I want signed copies! Please send me...

_____ copies of *The Odyssey of Mary B* @ $16.95 each

_____ copies of *Taos Massacres* @ $15.00 each

I am including $3.^{00} shipping and handling for one book, and $1.^{00} for each additional book.

Wisconsin residents, please add tax as follows:

The Odyssey of Mary B $1 each

The Taos Massacres $.75 each

(Canadian orders must include payment in U.S. funds, with 7% GST added)

Payment must accompany order. Please allow two weeks for delivery.

My check or money order for $_____ is enclosed.

Send to:

Name _____

Address _____

_____

City _____

State & ZIP _____

Make check or money order payable to

Puzzlebox Press

Send to:
**Puzzlebox Press
PO Box 765
Elkhorn, WI 53121**

Thank you!

www.puzzleboxpress.com

Reader Notes